Tom ALLEN
25719

ELECTRIC UTILITY SYSTEMS AND PRACTICES

ELECTRIC UTILITY SYSTEMS AND PRACTICES

Fourth Edition

Prepared by

**Electric Utility Systems Engineering Department
General Electric Company**

HOMER M. RUSTEBAKKE, *editor*

A Wiley-Interscience Publication
JOHN WILEY & SONS
New York • Chichester • Brisbane • Toronto • Singapore

Copyright © 1983 by John Wiley & Sons, Inc.

All rights reserved. Published simultaneously in Canada.

Reproduction or translation of any part of this work beyond that permitted by Section 107 or 108 of the 1976 United States Copyright Act without the permission of the copyright owner is unlawful. Requests for permission or further information should be addressed to the Permissions Department, John Wiley & Sons, Inc.

Library of Congress Cataloging in Publication Data

Main entry under title:

Electric utility systems and practices.

   "Wiley-Interscience publication."
   Includes index.
   1. Electric power systems. I. Rustebakke, Homer M. II. General Electric Company. Electric Utility Systems Engineering Dept.
TK1005.E457   1983   621.31   83-3640
ISBN 0-471-04890-9

Printed in the United States of America

10 9 8 7 6

CONTRIBUTORS

HOMER M. RUSTEBAKKE, *Editor*

J. C. Appiarius
J. Berdy
G. J. Bonk
J. A. Brander
G. D. Breuer
P. G. Brown
M. L. Crenshaw
R. C. Degeneff
F. J. Ellert
D. N. Ewart
H. J. Fiedler
S. R. Folger
R. L. Hauth

T. E. Jauch
L. K. Kirchmayer
D. W. Krauter
J. J. LaForest
R. G. Livingston
W. D. Marsh
S. A. Miske, Jr.
H. R. Propst
M. M. Schorr
R. M. Sigley, Jr.
J. B. Tice
C. J. Truax
D. J. Ward
T. D. Younkins

FOREWORD

This book is intended for use as a college-level course or an in-house seminar for power system engineers. Its objective is to provide an overview of the technical challenges and opportunities that exist in power system engineering. The book achieves its objectives by presenting an excellent description of each power system component and how they integrate into the system. The reader will be introduced to such topics as how power system components are designed and operated, and the complexity of a power system and its flexibility and limitations. It is written in clear and concise terms and concentrates on concepts; formulas and mathematics are kept to a minimum to improve readability. I sincerely recommend this book and hope that the student finds it as much a pleasure to read as I did. The time spent reading this book will provide a firm understanding of electric utility systems and practices.

ARMANDO J. PEREZ
SOUTHERN CALIFORNIA EDISON COMPANY

Rosemead, California
July 1983

PREFACE

This book provides an in-depth survey of the electric utility system. Previous editions have been successfully used by electric utilities for educating young engineers in the design and nature of the complicated electric power system. This updated version will also prove useful to engineers beginning careers in the electric utility industry and for experienced engineers to improve their technical understanding of the entire system. Some electrical engineering background is a prerequisite for thoroughly comprehending all this book has to offer. However, the nonengineer with an interest in this subject will also be able to gain new insight into the nature and complexity of the electric power system.

Chapter 1 describes the history and development of the industry and some of its unique features. Chapter 2 briefly examines the various components in the power system. Included is a discussion of generating stations, transmission and subtransmission systems that convey the energy from the primary energy source to the load areas, and distribution systems that deliver energy to the customers. Chapter 3 comments on sources of available energy and the methods for converting energy into electricity. Various types of generating plants are described, including fossil (fuel), combustion turbine, combined cycle, hydroelectric, and nuclear. The energy conversion processes using fossil fuel, nuclear fuel, and water are described in more detail in Chapters 4, 5, and 6. Chapter 4 also reviews various aspects of an electric generator including design, performance, excitation, and control.

After the primary energy has been converted into electricity, it must be transferred into the load area. This is accomplished through the transmission system (Chapter 7). Since transmission lines operate at voltages higher than generator-rated voltage, transformers (Chapter 8) are necessary. At this point, the power system is completed in regard to power transfer capabilities. It is still lacking system protection and the ability to switch the power circuits. Chapter 9 surveys the various devices and components (switchgear) that control and protect the electrical parts of the system. The high-voltage switches, called circuit breakers, are included in this discussion.

The power system load is not a single concentrated load, but is made up of houses, shopping areas, commercial buildings, factories, and industries which usually cannot be connected to high-voltage transmission lines. The power voltage is reduced by transformers in substations. The transmission voltage may be reduced to primary distribution level, or to subtransmission voltage with another substation for reduction to primary distribution voltage. Substations are discussed in Chapter 10 and distribution in Chapter 11.

A vital element of the electric power system is protective relaying. The relay detects a problem and will initiate the tripping of circuit breakers. Due to the large number of fault types and situations, there are many types of relays. Typical relays used on a power system and their application are discussed in Chapter 12.

The power system must be analyzed for stability, that is, all generators must operate in synchronism. This is discussed in Chapter 13 along with procedures for improving and maintaining stability.

Some of the complex controls required to operate the power system are introduced in Chapter 14. The present controls are the results of many years experience, technological innovation, and computer application. Chapter 15 gives insight into the various planning and design techniques used to expand the power system.

Earlier editions of this book were written by

members of General Electric's Electric Utility Systems Engineering Department and were edited by Robert Treat. The authors of the present material may be found under the listing of Contributors, with Dr. Homer M. Rustebakke as Contributor and Editor. This work was completed with major contributions from Celeste Colangelo for secretarial assistance; Jean Fellos, Janice Nolan, and Holly Powers for typing the manuscript; and Lucienne Walker for editorial, proofreading, and production assistance.

HOMER M. RUSTEBAKKE

Schenectady, New York
July 1983

CONTENTS

1. THE ELECTRIC UTILITY INDUSTRY   1

Electric Power Systems, 1
The Genesis of the Electric Utility Industry, 1
The Growth of the Electric Utility Industry, 2
Industry Growth Rate, 3
Electric Power Companies, 3
Unique Features of the Electric Utility Industry, 5
Summary, 9
Appendix I.A, 10
Appendix I.B, 10
Appendix I.C, 12

2. THE POWER SYSTEM   13

Introduction, 13
The Modern Power System, 13
Investment, 22
System Evolution, 23
Load, 23
Load Curves, 24
Inspection and Maintenance, 25
System Design, 25
Summary, 26

3. POWER GENERATION   27

Sources of Energy, 27
Thermodynamics of Steam-Electric Generating Plants, 31
Combustion Turbine Power Plants, 37
Combined Cycle Power Plants, 39
The Costs of Power Generation, 39
Power Generation Reliability, 40
Power Plant Siting, 40
Summary, 42

4. POWER FROM STEAM AND COMBUSTION TURBINES   44

Thermal Electric Generating Stations, 44
Increase in Size of Units, 45
Production of Steam, 45
Turbine, 51
Condenser, 52
Heat Rejection, 53
Auxiliaries, 53
Feedwater System, 56
Forced- and Induced-Draft Fans, 57
Power for the Auxiliaries, 58
Nuclear Plant Auxiliaries, 60
Combustion Turbines, 60
Generator Design, 63
Generator Configuration, 75
Generator Performance, 80
Excitation System, 80
Controls and Instrumentation, 90
Summary, 97

5. POWER FROM NUCLEAR FUEL   98

Introduction, 98
Physics of Nuclear Processes, 98
Reactor Core Behavior, 103
The Nuclear Fuel Cycle, 110
Summary, 110

6. POWER FROM WATER   111

Introduction, 111
Plant Size and Load Factor, 111
Variation in Cost per Kilowatt of Hydroelectric Developments, 112
Pumped Storage Hydro, 113
Control of Stream Flow, 113
Power and Energy Equations, 114

Essential Elements of a Hydroelectric Development, 114
Types of Hydraulic Turbines, 114
Effect of Head on Type of Development, 116
Specific Speed, 116
Speed Regulation, 117
Runaway, 118
Generator, 118
Amortisseur Windings, 119
Generators for Pumped-Storage Hydro, 119
Auxiliary Power, 120
Automatic Hydro Stations, 120
Summary, 121

7. POWER TRANSMISSION 122

Introduction, 122
Defining Transmission, 122
Components of a Transmission Line, 124
All-Weather Transmission, 127
Lightning, 128
Lightning-Proofing a Transmission Line, 129
Mechanical Design, 130
Electrical Design, 133
Corona, 134
Bundle Conductors, 134
Economics, 135
Direct-Current Transmission, 136
Summary, 144

8. TRANSFORMERS 146

Description and Use, 146
Ratio, 146
Taps, 146
Single-Phase versus Three-Phase Units, 147
Autotransformers, 147
Phase-Angle Transformers, 147
Cores, 147
Windings, 148
Losses, 148
Cooling, 148
Oil Preservation, 150
Reduction of Fire Hazard, 152
Effect of Temperature on Organic Insulation, 153
Impulse Voltage, 154

Dielectric Tests, 155
Summary, 156

9. SWITCHGEAR 158

Definition, 158
Function, 158
Components, 158
Short Circuits, 158
Calculation of Fault Current, 159
Interruption of Fault Current, 160
Interruption of Leading Current, 162
Methods of Arc Interruption, 163
Arc Interruption in Fuses, 170
Current-Limiting Devices, 170
High-Speed Operation, 171
EHV Circuit Breakers, 171
Additional Requirements of Breakers and Switches, 172
Outdoor Gear, 174
Control Point, 175
Summary, 176

10. SUBSTATIONS 177

Definition, 177
Bus Arrangements, 177
Circuit Breakers, 180
Transformers, 180
Reactive Power Compensation, 181
Protective Relaying, 185
Overvoltage Protection, 185
Additional Substation Design Considerations, 187
Other Types of Substations, 189
Unattended Substation Control, 191
Summary, 197

11. DISTRIBUTION 199

Introduction, 199
Subtransmission System, 199
Distribution System Scope, 199
Circuits by Type of Load, 200
Service to Industrial Loads, 200
Service to Commercial Loads, 200
Service to Residential, Suburban, and Rural Loads, 204
Summary, 214

12. PROTECTIVE RELAYING — 215

Introduction, 215
Definitions and Required Characteristics, 215
Basic Relay Types, 218
Fundamental Application Principles, 226
Application Practices, 229
Summary, 240

13. STABILITY — 241

Introduction, 241
Steady-State Stability, 243
Transient Stability, 250
Higher-Transmission Voltage, 260
Series Capacitors, 261
Summary, 261

14. SYSTEM OPERATION — 263

Introduction, 263
System Operation Tasks, 263
Role of Automation, 264
Control Functions, 264
System Security, 268
Hierarchy of Action Centers, 270
Functional Service Requirements, 273
Communication Alternatives, 274
Data Display, 275
Summary, 275

15. SYSTEM DESIGN — 276

The Early Systems, 276
System Design and Changing Conditions, 276
System Planning, 278
Summary, 287

Index — 289

ELECTRIC UTILITY SYSTEMS AND PRACTICES

1
THE ELECTRIC UTILITY INDUSTRY

ELECTRIC POWER SYSTEMS

Electric power systems (also called electric energy systems) are an integral part of the way of life in all developed countries. The electricity generated by these power systems has proved to be a most convenient, flexible, clean, safe, and useful form of energy.

Our electric supply will continue to be reliable unless electric utilities are prevented from installing new equipment to maintain the reserve capacity necessary to meet their traditionally high reliability. At the present time, the lack of electricity (a blackout) is newsworthy. Electricity is so vital to the lives of people that its unavailability causes inconvenience, loss of production, and danger to many individuals, including those who are hospitalized. A prolonged blackout, which is inevitable when a growing load is combined with no generation additions, could lead to social disorder, and even national tragedy.

Companies that provide essential electric service build energy conversion stations, known as generating stations or plants, or simply power plants. The voltage is usually transformed to a high level at the power plant. The electrical energy is then transmitted to the general vicinity of ultimate use via high-voltage transmission lines. At this point, the voltage is reduced and transmitted to a location near ultimate use via subtransmission lines. Here another voltage reduction is experienced and the energy is delivered to the user via a distribution system. Building on this simplified system, there may be many more voltage transformations, thus the connection between the power stations and loads is not a simple path but a complex network. The entire system of generating stations, transmission lines, and so on, is called a power system. The power company will plan, design, build, and operate this power system. Companies will interconnect their systems for mutual operating and economic advantages. The result is one of man's most complex developments.

THE GENESIS OF THE ELECTRIC UTILITY INDUSTRY

Although arc lighting companies had existed before, the electric utility industry is generally considered to have begun when Thomas A. Edison's Pearl Street Station started operation in New York City, on September 4, 1882.* This station supplied power for lighting to 59 customers within an area of approximately 1 square mile. As additional dynamos were started, the number of customers rapidly increased to 300, and then to 439 by the end of the first year of operation. For a few months no bills were sent out so that the first year's operation showed a loss of about $6000. In the 3 months after billing was instituted, there was a profit of $491, and at the end of the second year the profit was $28,000.

Satisfied with the results of the New York City operation, Edison issued licenses to local business men in various communities. These licenses

---
* The state of the industry at this time is described in a letter written by Samuel Insull, who was then Edison's private secretary. This letter appears in Appendix I.A to this chapter.

granted them the privilege of organizing and operating electric lighting companies similar in all respects to his own New York City system. By 1884 about 20 local electric lighting companies had been started, mostly in scattered communities in Massachusetts, Pennsylvania, and Ohio. In 1885 there were 31 of these local companies; in 1886, 48; and in 1887, 62. Thereafter, the number multiplied steadily year after year. All of these companies furnished energy for lighting incandescent lamps, and all operated under Edison patents.

The year 1882 brought two other achievements in the power art; a water wheel-driven generator was installed in Appleton, Wisconsin, and the first transmission line was built in Germany to operate at 2400 volts direct current over a distance of 37 miles (59 km).

All of the early Edison systems were two-wire, constant-potential, dc systems. Their load consisted almost entirely of lamps, despite the expectations concerning the use of electric energy for operating motors. These expectations were met in 1884, when Sprague started to produce a motor designed to operate on Edison systems.

The introduction of a practical motor and the constantly increasing use of incandescent lamps made necessary the further expansion of the Edison systems. During this period the three-wire system was developed. This allowed the loads to increase somewhat before getting into voltage troubles again. But the three-wire system was not enough. By 1886, the Edison Companies were experiencing limitations because they could deliver energy only a short distance from their stations. These limitations were overcome by Stanley's development in 1885 of a commercially practical transformer and the ac system.

THE GROWTH OF THE ELECTRIC UTILITY INDUSTRY

Transmission

In 1889, the first ac transmission in the United States took place between Oregon City and Portland, 13 miles (21 km) away.* Two 300-hp water-

---

* The June 4, 1889 issue of the Oregonian carried the following news item:

The Williamette Falls Electric Company started up one of their brush arc dynamos last evening, and the electric-

wheels drove a single-phase generator, and with a capacity of 720 kW, the power was transmitted at 4000 volts to Portland.

The first three-phase line went into operation in 1891 in Germany. Power was transmitted 112 miles (179 km) at 12,000 volts. The first three-phase line in the United States (2300 volts and 7.5 miles [12 km] long) was installed in 1893 in California by a predecessor of the Southern California Edison Company.

In 1897, a 44,000-volt transmission line was built in Utah. It extended from Provo Canyon 32 miles (51 km) to Mercur, a booming gold mining camp. Mercur thus became the first completely electrically equipped mine and mill in the history of that industry. In 1903, a 60,000-volt transmission line was energized in Mexico. In 1913, transmission voltages rose to 150 kV; in 1922 to 165 kV; in 1923 to 220 kV; in 1935 to 287 kV; in 1953 to 330 kV; and in 1965 to 500 kV. Quebec energized its first 735 kV line in 1966 and 765 kV was first put into operation in the United States in 1969. Transmission voltages of over one million volts are being studied.

Frequency

In this early ac period of the electric utility industry, frequency had not been standardized; the most common frequency was 133 Hz. In 1891, the desirability of a standard frequency was recognized and 60 Hz was proposed. In 1893, 25 Hz came into use with the synchronous converter. For many years 25, 50, and 60 Hz were standard frequencies in the United States. Much of the 25 Hz was railway electrification and this was gradually retired over the years. The city of Los Angeles Department of Water and Power and the Southern California Edison Company both operated with 50 Hz. The city converted to 60 Hz at the time that Hoover Dam power became available. The Edison Company completed its conversion to 60 Hz in 1949. The Salt River Project was originally a 25-Hz system; most of it was

---

ity was sent from Oregon City for lighting one of their ten o'clock circuits in this city. It worked magnificently and conclusively demonstrated the fact that our city can be lighted successfully from the falls. The result was a pleasing surprise to the company, the percentage of loss of electricity by transmission being less than their most sanguine expectations. The work of moving the machinery from the station here to the falls will be carried on as expeditiously as possible. Another large dynamo will be moved up today.

converted to 60 Hz by the end of 1954 and the balance by the end of 1973.

The first direct-current transmission system in the United States (30-kV system, 20 miles [32 km] long) was installed in 1936 to transmit power from Mechanicville to Schenectady, New York. The next high-voltage direct-current (HVDC) system was the Pacific Intertie Line (±400 kV and 850 miles [1360 km]) installed between Oregon and California in 1970. There have been three other HVDC systems up to 400 kV in the United States, and a total of 24 HVDC systems in operation worldwide with voltages up to ±533 kV. Nine HVDC systems are currently under construction.

Generation

In 1895, water-wheel generators of 5000 hp were installed in Niagara Falls, New York. Beginning in 1941, a series of 18–108,000-kVA hydro generators were installed in Grand Coulee Dam, Washington. In 1966, two 158,000-kVA units were put in service at Smith Mountain, Virginia. A 231,600-kVA unit was added at Dworshak Dam, Idaho in 1973. The third powerhouse at Grand Coulee, containing three 615,000-kVA units, was installed in 1975 and 1976. The first of three additional units with ratings of 718,000 kVA went into service there in early 1978. The output rating of hydro units is not the only factor affecting their physical size. Other factors are normal speed, runaway speed, generator reactances, and inertia.

Two 4400-kVA pumped-storage hydro units were placed in service in 1928. In 1956, the size in service was up to 70,000 kVA and to 204,000 kVA in 1963. Six 388,000-kVA units went in service in 1973 at the Luddington station, Michigan. In 1978–79, four 425,000-kVA units went in service at Raccoon Mountain station, Tennessee.

Fossil-fueled steam turbine generator units grew in size from the first 500-kW unit in 1903 to a 160,000-kW single-shaft 1800 r/min unit in 1932 with a 200,000-kVA generator. A cross-compound multishaft 1800 r/min unit at 208,000 kW installed in 1929 was the record size unit until 1955 when modern cross-compound 3600/1800 r/min units reached 217,260 kW. A 500,000-kW unit of this type went in service in 1960, 650,000-kW units in 1963, and a 1,130,000-kW unit in 1970.

The size history of 3600 r/min fossil-fueled steam turbine generators started with a 4000-kW unit in 1914, reached 15,000 kW in 1933, and 40,000 kW in 1937 with a 50,000-kVA hydrogen-cooled generator. Single-shaft 3600 r/min units of 173,500 kW (216,000 kVA generator) were in service by 1953, 350,000 kW (506,000 kVA) in 1963, 737,600 kW (907,000 kVA) in 1969, and 884,040 kW in 1974 with a 1,120,000-kVA generator. Cross-compound 3600/3600 r/min two-shaft units grew from 100,000 kW in 1940 to 450,000 kW in 1960, and to 1,300,000 kW in 1973.

Light-water nuclear units operate as single-shaft units at 1800 r/min and their growth is shown by units of 90,000-kW capacity in 1957, 210,000 kW in 1960, 590,000 kW in 1967, and a 1,177,600-kW steam turbine in service in 1976 with a 1,280,000-kVA generator. Nuclear units with steam turbine ratings up to 1,288,000 kW are on order.

INDUSTRY GROWTH RATE

The growth of the electric utility industry in the United States of America over several years is shown in Figure 1.1. It is from this growth pattern that the rule-of-thumb "doubling in a decade" was derived. This figure indicates that very roughly the capability of the nation's power system doubled in about 10 years over a period of 70 years. Thus, in each succeeding 10 years another electric utility, equal in capacity to the existing system, had to be built. This was accomplished with maximum economy, and with no service discontinuity except for the depression years in the 1930s. In addition, systems were added that met public acceptance, i.e., they had reasonable appearance and low air and water pollution. This task requires the best of enginering talent as well as other essential talents, such as financing, public relations, etc.

Since the energy crisis of 1973–74, the growth rate has slowed appreciably. Some areas of the country had little, if any, growth in electricity demand for several years. A recent forecast projected a 3% annual rate of growth in peak load through 1991.

ELECTRIC POWER COMPANIES

Companies have been formed to provide electric service. Over the years, they have grown from small companies to industrial giants by load growth and mergers. Like other businesses, the power companies obtain money on the open market, em-

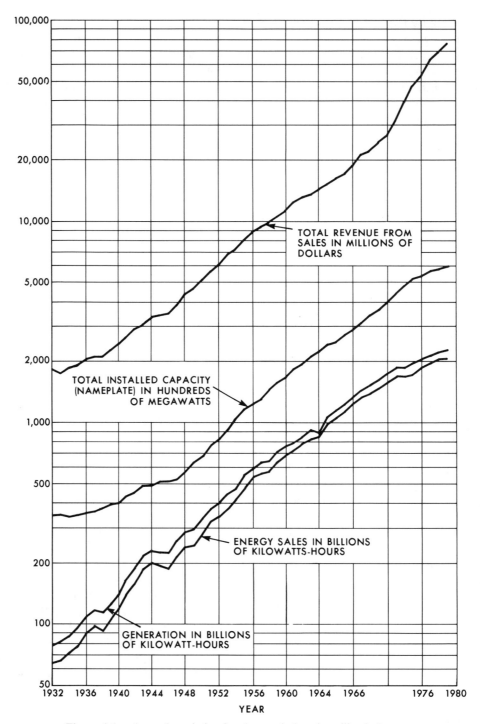

Figure 1.1. Annual statistics for the total electric utility industry.

ploy people to perform the many necessary tasks, sell service to the public, and compete for existing markets; they also must make a satisfactory return over operating expenses to attract new capital and stay in business.

Since the electric utility business is a very high investment business, duplicate facilities to serve a geographic area would be inefficient. For these reasons, the utility is inherently a monopoly, and franchises are granted to serve given areas. Govern-

TABLE 1.1. Capacity and Annual Consumption of Electricity Per Capita

Country	Annual Use in kWh Per Capita	Installed Capacity in kW Per Capita
Norway	21,704	4.48
Canada	14,803	3.29
Sweden	11,360	3.25
United States	10.592	2.81
West Germany	6,115	1.33
United Kingdom	5,366	1.33
Japan	5,030	1.22
USSR	4,695	.97
France	4,524	1.06
Italy	3,178	.77
Spain	2,849	.78
Brazil	1,000	.24
China	293	.05
India	167	.04

SOURCE. *1979 Yearbook of World Energy Statistics,* United Nations.

ment regulation replaces direct competition. However, there is still competition for new markets that have the freedom to locate in any of several areas or generate their own electricity, and there is competition with alternate energy sources such as oil, gas, and coal.

The reference to "high investment business" can be amplified by citing some numbers applicable to the year 1979. The electric utility industry had approximately 89.5 million customers who paid over $77.7 billion for about 2079 billion kWh of energy. The total generating capacity of the electric utility industry was over 598 million kW. The assets of investment in electric utility plants for the investor-owned companies was over $196 billion, and these investor-owned companies employed about 463,000 people.

Table 1.1 shows the capacity and consumption per capita of 14 countries in the world with the highest installed capacities.

UNIQUE FEATURES OF THE ELECTRIC UTILITY INDUSTRY

The electric utility industry has unique technical characteristics which give it certain unique business characteristics.

Technical Characteristics

Intangibility. The customer cannot directly detect a kilowatt-hour with any of his physical senses. The result of the kilowatt-hour of service (light, heat, work output from a motor) can be detected and for this he is willing to pay. Devices called watt-hour meters are used to measure the intangible electrical energy delivered to a customer.

Quality. Although electricity is intangible, it can be measured in amount and quality. The quality of service can be measured by service continuity or reliability, uniformity of voltage at the proper level, and proper and uniform frequency of the alternating voltage. Also, the shape of the voltage–time curve and its smoothness or wave shape is a measure of quality, but it is seldom a problem since irregularities can invariably be traced to the customers' load characteristics. System frequency is held very accurately except during infrequent times when abnormal situations exist. Both constant voltage at the customers' load and reliability of service can be improved by installing additional equipment which increases the cost of service. The principal job of the utility management and engineering is to provide good service at minimum cost—conflicting requirements that must be watched with diligence.

Product Storage. Unlike most businesses, a utility must create its product simultaneously with its use because there is no storage of electricity. The storage battery has chemical, not electrical, storage, but the production of large blocks of power by chemical action so far is both technically and economically unsatisfactory. This simultaneous generation and transmission with use requires a system capacity to meet the load demand at whatever moment it occurs. Thus, a maximum or peak demand of only a moment's duration requires that the necessary equipment be in place throughout the entire year.

Responsibility for Delivery. Because the utility delivers its product to the customer's premises, it must assume responsibility for the safe and reliable delivery of its product. The necessary transmission and distribution lines must be available and kept in service (or promptly returned to service) despite wind, lightning, ice, kites, and so on.

Franchise. The circuits over which electric energy reaches the customer must use paths across intervening land, and rights-of-way or easements across fields or forest, or franchises along public streets and highways must be secured for this purpose. The utility does have the right to condemn privately owned property to traverse it, but this action is typically used as a last resort. The arrangement may be in perpetuity for a single payment or in some cases, annual "rent" is paid.

Public Safety. Electricity is a very useful servant when it is kept under control; however, it can be very destructive if control is lost because it can shock, burn, and kill. The utility must provide reasonably adequate protection for the public and its own skilled workers. Property should not be exposed to conditions such as fire hazard as a result of inadequate equipment.

Business Characteristics

Slow Capital Turnover. In many kinds of businesses, the gross annual income is several times the amount invested in the business. The reverse is true for the utility. A utility may have nearly $4.00 invested for each $1.00 of annual revenue, the railroads about $2.00, and total manufacturing about $.40 cents for each $1.00 of sales. This shows a high fixed or sunk (investment) cost in a utility, which gives rise to a high annual revenue requirement in addition to the revenue required for operating expenses.

Pricing System—Rates. Since the initial operation of the Pearl Street Station until today, many methods of charging for electric service have been proposed. The rules for charging are known as rate structures and are extremely important—it is through rates that the utility receives the money necessary to operate.

The history of electric utility rate structures is one of gradual development to recognize the many elements of cost entering into the supply of electric service. The first rate charged a fixed amount for each lamp. Since electric motors had not yet been developed, only lamps were used and the company was able to operate under this charging system. The next step in the evolution of rates was to use an ampere-hour meter and charge for current, not energy. The development of the induction-disk, watt-hour meter enabled a charge to be made on the basis of energy. The concept of a demand charge was a later development. Demand is the rate at which electric energy is delivered to a customer, expressed in kilowatts, kilovolt-amperes, or another suitable unit at a given instant, or averaged over any designated period of time.

The cost of electric service can be described as:

1. *Customer cost.* This varies with the number of customers and includes investment charges and expenses of a portion of the general distribution system, service drop, metering equipment, meter reading, billing, and accounting. It represents the costs directly associated with the individual customer.

2. *Load cost.* This varies with the magnitude of the customer's load and location. It includes a portion of the investment charges and the expense of generating plants, transmission lines, substations, and the distribution system not included under customer cost. The peak-demand and power factor of the load will determine this cost, not the average energy required. The service voltage will affect the cost, as will the continuity of service required.

3. *Energy cost.* This is principally a fuel expense and varies with the energy supplied.

The following is a list of the names of rates that have been used, with a brief description of each.

1. *Flat demand rate.* Billing made on the basis of connected load. No metering is used.

2. *Straight-line meter rate.* This is a constant cost per watt-hour. Customer and demand costs are not recognized.

3. *Step meter rate.* This is a forerunner of the block meter rate and is not used today. In this rate scheme, a certain unit charge is made based on the total consumption. In going from one consumption rate to the next, the total bill will decrease in a step, then monotonically increase to the next rate change.

4. *Block meter rate.* Charges are made on the basis of so much per kilowatt-hour for various kilowatt-hour blocks. The price per kilowatt-hour decreases for succeeding blocks. In its pure form the demand element is not recognized.

A minimum charge is usually associated with this rate, which approximates a customer charge. The customer feels he gets something for a minimum but tends to regard a customer service charge as a penalty, not understanding it is a utility cost. This reveals one of the essential elements of a rate structure; it must be acceptable to the customer. This is more important than to be technically correct. A technically correct rate would be too complicated for most customers, hence not understandable or acceptable. The block rate is widely used today, particularly for residential rates.

5. *Hopkinson demand rate.* This is a two-part rate consisting of separate charges for demand and for energy. This rate or a modification of it is widely used.

6. *Wright demand rate or hour's use rate.* This rate also considers both demand and energy. The energy is priced in blocks with decreasing prices for succeeding blocks, and the size of the energy blocks increases with the size of the load.

If a customer charge is included with any form of demand-energy rate, it is called a three-part rate, that is, separate charges for customer, demand, and energy.

7. *Time-of-day rates.* The cost of generating electricity varies over a 24-hour period. It is low during the off-peak hours when the demand is low, and it increases to a higher value during the on-peak hours when the demand for electricity is high. Time-of-day rates are designed to reflect this variable cost of electricity. This rate structure is also called indirect load management. Dates may also vary by season (seasonal dates) to reflect the seasonal variability of the cost of producing electricity.

The rate designer must be a diplomat as well as a technician. He must recover the companies' costs at a profit, with competitive rates, and not be unduly discriminatory.

Regulated Monopoly

The characteristics of a utility that make it inherently a monopoly have already been discussed, but are repeated here since this is such an important business characteristic. Government regulation is the substitute for direct competition.

Regulation of Rates

In a "free" economy, competition will tend to regulate prices and protect the customer from overcharges. A monopoly has no direct competition and government regulation has evolved as a substitute for direct competition to protect the customer. State utilities commissions operate in nearly all states. When a power company operates in more than one state or has been determined to deal in interstate commerce, it is also under the jurisidiction of the Federal Power Commission. The principal objective of state regulation is to limit earnings of utilities to a fair rate of return on the fair value of the property. The regulatory authorities require that a high-quality service be rendered. They also prevent direct competition, as mentioned earlier, by issuing certificates of public convenience giving companies the right to do business in designated areas. The regulatory authorities take interest in accounting practices, reasonableness of operating expenses, depreciation policies, and so on. The commission must keep in mind that earnings must be sufficient to attract new money, which is essential for system expansion to meet load growth.

Commitment to Serve

The electric utility must be prepared to serve anyone requesting service in its allotted area. The utility may require protection of its investment incidental to providing service to a customer, but it cannot deny service. Rather then denying service, the utility historically has promoted the use of electricity by encouraging greater use by existing customers and by attracting new customers. Promotional rates to attract new customers were justified since growth was also an advantage to existing customers. At the present time, the cost of new facilities to meet load growth is greater than the historical cost of equipment and service.

Because the design and construction of electric power generating and transmission facilities requires several months to a few years, the prediction of load growth and system planning are very important to the utility. Since the Oil Embargo of 1973–74, new factors have affected the utilities' ability to predict load growth and to plan for expansion. Energy conservation has become a key concern due to rapid oil price inflation and the United States' price deficit balance of payments in world trade. Since utilities are now investigating alternate energy sources, they can no longer easily predict the magnitude and type of energy sources that will be developed or available to serve the load growth. Projections based on past trends are less reliable than has been the case historically because of new factors, such as the substitution of other forms of energy or fuel for the critical oil import, the new balances that are being sought between alternative fuel costs, and the effort to utilize renewable energy sources such as solar and wind.

Direct Competition

The privately owned electric utility is a regulated monopoly, which is granted a franchise to service a given area, thus freeing the utility from direct competition. There is competition from new industry, new or old industry generating its own electricity, alternate energy sources, and so on. Direct competition, however, can develop from utilities that are owned by cooperatives or governments.

Government systems and cooperatives are exempt from paying taxes and can obtain money at a lower rate. Some payments have been made to local communities in lieu of taxes. The taxes paid by investor-owned companies as a percentage of the total revenue for the electric department is shown in Table 1.2.

TABLE 1.2. Taxes in Percent of Revenue, National Average for Investor-Owned Utilities

Year	Federal Taxes	State and Local Taxes	Total Taxes
1979	5.7	7.7	13.4
1978	6.5	8.2	14.7
1977	6.9	8.4	15.3
1976	6.5	8.8	15.3
1975	6.0	8.8	14.8
1974	4.8	9.2	14.0
1973	6.2	10.3	16.5
1972	6.1	11.1	17.2
1971	6.2	11.2	17.4
1970	7.0	11.3	18.3
1969	9.8	11.1	20.9
1968	11.1	10.9	22.0
1967	10.8	10.8	21.6
1966	11.6	10.5	22.1
1965	11.7	10.6	22.3
1964	12.5	10.5	23.0
1963	13.0	10.3	23.3
1962	13.2	10.3	23.5
1961	13.4	10.4	23.8
1960	13.8	10.2	24.0
1950	12.4	8.6	21.0
1945	13.7	8.0	21.7
1940	8.3	9.4	17.7
1937	5.0	10.2	15.2
1930			10.0
1927			9.4

Source. Edison Electric Institute Statistical Year Books of the Electric Utility Industry published 1946 through 1979.

Municipal Plants

When cities began to light their streets electrically, some of them elected to put in their own generating plants instead of buying energy from the local utility. Some cities took the next step and started to sell service to their citizens. This put them, in some cases, in competition with the privately owned utilities operating in the same area. This competition still persists in varying forms and degrees.

For example, in Los Angeles, Seattle, and Tacoma, electric energy is now supplied entirely by municipally owned utilities. Cleveland and Columbus are served by both privately and municipally owned utilities. In Detroit, the municipal plant has confined itself to street lighting and other municipal uses; it does not compete with Detroit Edison for nonmunicipal business.

Reclamation Projects

For many years the Reclamation Bureau of the Department of the Interior has encouraged and fostered the reclamation of arid lands in the West by irrigation. When water for irrigation requires damming a stream, there is always the possibility of byproduct hydroelectric power. Any such power becomes the property of the "water users association" or "irrigation district," or whatever it may be called. This power, even though sold only to the members of the cooperative that developed it, is competition for the utility that might otherwise have served those areas.

Recapture Clause in Federal Licenses

Federal licenses for the erection of hydroelectric plants on navigable streams are for a period of not over 50 years; they also contain the proviso that when the license expires, the federal government, under certain conditions, may take title to the plant. This proviso contains a very real threat of future competition from the federal government.

Hoover Dam

The Reclamation Bureau built Hoover Dam (Arizona–Nevada boundary) on the grounds that it was too big an undertaking for any privately owned utility. The Bureau retains title to the dam, powerhouse, and contents, including the generators. The utility customers, some privately owned and some municipally and publicly owned, operate the generating station and buy the power on a "come and get it" basis. That is, they built the transmission lines to the powerhouse, and they own and operate them.

Tennessee Valley Authority

The idea of TVA was sold as a multipurpose project. Among the stated purposes were:

1. Flood control
2. Navigational improvement
3. Power development
4. Development of recreational areas
5. Reforestation
6. Establishment of power rates, which could be used as a "yardstick" with which to compare the rates charged elsewhere by privately owned utilities, to determine if the latter were excessive.
7. Development of the Tennessee River and its watershed through an integrated plan which would assure the accomplishment of items 1–5.

In its original concept, power was to be only secondary to the other purposes. It has now, however, become the prime purpose; and the TVA system, which is one of the largest single power systems in the world, contains a high percentage of thermal generation. History soon showed that privately owned utilities were unable to exist in an area where they were forced to compete with the very low prices per kilowatt-hour charged by the government. In a few years, the Tennessee Electric Power Company was forced out of business. Now TVA has no competition in the area it serves.

Rural Electrification Administration

REA encouraged the formation of "electric membership cooperatives," lent them money to build distribution systems, and rendered engineering assistance to its borrowers. It also sought to reduce the costs of these rural systems through standardization of equipment and through mass buying and manufacturing.

Many of these cooperatives buy their energy wholesale wherever they can—usually from the privately owned utilities serving the area. In a number of cases, several cooperatives have united to form a super cooperative to build and operate generating and transmission facilities to serve them all.

The "public utility district" is another form of cooperative with purposes and function very similar to the electric membership cooperative. Its funds are derived from private rather than public sources, and it is not sponsored nor directed by REA.

Bonneville Power Administration

The BPA is an agency of the federal government under the Department of Interior. It is a transmission agency located in the Pacific Northwest that markets the power produced at the federal hydro projects in the Columbia river basin.

SUMMARY

In 100 years, electric power has become a vital factor in modern living, in agriculture, and in industry. As an example, the electric utilities were a $100 billion business in 1979. Historically, this business grew at such a rate that it tended to double in less than 10 years. Since the energy crisis of 1973–74, the rate of growth is less, and a recent industry forecast projected a 3% annual rate of growth in peak load through 1991.

The industry differs technically from other forms of business in a number of respects. The product is intangible. Since it cannot be stored, it must be manufactured and shipped at the same instant it is used, and it must be handled with respect. These and other unique characteristics set the electric utility company apart from other forms of enterprise.

The business is inherently monopolistic, but it is subject to regulation as to rates and practices by federal, state, and municipal governments.

REFERENCES

1. Annual statistical issue of *Electrical World*.
2. Annual statistical issue of *Electric Light and Power*.
3. Edison Electric Institute, *Statistical Yearbook of the Electric Utility Industry 1979*.
4. Electrical World Editorial Staff, *The Electric Power Industry, Past, Present and Future*, McGraw-Hill, New York, 1949.
5. Marsh, W. D., *Economics of Electric Utility Power Generation*, Oxford, New York, 1980.

6. North American Electric Reliability Council, *Electric Power Supply and Demand, 1982–1991.*
7. Vennard, E., *The Electric Power Business,* McGraw-Hill, New York, 1962.

APPENDIX I.A. SAMUEL INSULL'S LETTER

The following is an excerpt from a letter written by Samuel Insull, Edison's private secretary, to friends in England. The letter was evidently written just after the start of Edison's first electric light plant.

You ask me about electric light. Well, I have seen 700 lights burning, the current generated from the same dynamo-electric machine for the whole lot, all of them getting their current from the same main (i.e., street cables) of no less than eight miles in length. Edison gets eight lights or thereabouts of 16 candles each per indicated horsepower, which allows of his competing with gas. Into the details of the cost I cannot go, as it is not told to anybody. Suffice to say that here in New York he can produce light and get a handsome profit on it at a charge to the consumer which would ruin the gas companies. There is not, however, that vast difference between the cost of the two lights which will allow him to be utterly oblivious of his friends, the gas producers; but his estimates show that he can compete with them and do it at a handsome profit. Besides he can furnish power by means of electric motors, which will give him an enormous pull over the gas companies as he will not have the greater part of his plant lying idle during 365 working days of the year, as the gas companies with but very slight exception must, as the business is at night; but he can sell electricity for power purposes by day, which means that his plant is never idle, his capital is never running to waste, but is always earning money by night and by day alike. Edison will work just as the gas companies do. He will have central stations where the current will be generated (probably one station of about 15,000 lights to each square mile). This current will be conveyed along the streets underground by means of copper wire embedded in two-inch iron pipes insulated with a special form of insulation of his own invention. Branch pipes will be led into each house, and the electricity, whether for light or power (to use it is all the same), will be sold by means of a registration on an electric meter, which is the most ingenious and yet the simplest thing imaginable. The district which he will light up first in New York has about 15,650 lights in the various buildings in the district and a great deal of power in varying amounts. He is getting contracts just as fast as his canvassers apply for them, and we have large gangs of men wiring the houses in anticipation of the time when we can lay our mains, erect our dynamo machinery and light up. I suppose this district will be all lighted up in from three to four months, and then you will see what you will see.

Menlo Park is practically abandoned. All experiments are finished; all speculation on the probable results are dismissed; and Edison thinks, and so does everyone else who has looked in to the matter, that success is assured. Of course, time alone can prove this. As for myself, I am not competent to judge but I can use my eyes, can see the success with which the houses, the fields, roads and depot have been illuminated here, and I can see nothing to disprove the assertions. His lamps last about 400 hours; at all events that is the estimate by a time test, i.e., by running them at about four times their ordinary candle power until the carbons break; but this estimate is every day falsified, and experience points to the conclusion that the life of his lamps will be much longer than the estimate. As for rivals, Edison has but little to fear, in fact, none, from them.

To carry out the gigantic undertaking of fighting the gas companies we have much to do. A great difficulty is to get our machinery manufactured. This Mr. Edison will attend to himself. He personally has taken a very large works for this purpose, where he will probably within the next six months have 1500 men at work. The various parts of the machines will be contracted out, one firm making one part in large quantities, another firm another part, and so on. At Mr. Edison's works ("Edison Machine Works"), all these parts will be assembled and put together. Then there is the lamp factory, in which Mr. Edison owns almost all the interest, for manufacturing lamps and which is now turning out one thousand lamps a day, the Electric Tube Company (of which I am secretary and Mr. E. President) for manufacturing our street mains. So you can imagine what Mr. Edison has to do, as he is the mainspring and ruling spirit of everything. And you can imagine also what I have to do as his private secretary.

APPENDIX I.B. STANDARDS IN THE ELECTRIC UTILITY BUSINESS

Interchangeable equipment is demanded by economics. This requires standard sized mounting holes, shafts, and so on. Also technical definitions must be comprehended by both manufacturer and user. The performance characteristics required by the user must be designed into the manufactured equipment. To facilitate this need, standards have been produced by those concerned with the particular technology. These standards are the result of years of effort in discussing, testing, and compromise.

For most economical manufacture, General

APPENDIX I.B. STANDARDS IN THE ELECTRIC UTILITY BUSINESS

Electric management long ago found it desirable to draw up certain standard specifications for the purchase of materials and to adopt standard processes and practices in manufacture; as volume grew, it was found advantageous also to have standards for the products sold. A standard is no more than a description or specification of the best and most appropriate thing or practice that is known when adopted.

In a highly competitive business, it is not sufficient to have only company standards; both sellers and buyers find the need for industry standards. Before the turn of the century, people participated in developing industry standards through some of the engineering societies, like the American Institute of Electrical Engineers and the American Society of Mechanical Engineers. For years, standard tests and testing procedures for the materials used in manufacture have been developed by the American Society for Testing Materials.

Years ago, the electrical manufacturers established product standards through the Electric Power Club and the Associated Manufacturers of Electrical Supplies. In the early 1920s, these two organizations were merged into the National Electrical Manufacturers Association, which is the trade organization having the most to do with making product standards for the electrical manufacturing industry.

To present the viewpoint of the users or purchasers in the electric utility business, engineers of the utility companies developed specifications in committees of the National Electric Light Association, which later became the Edison Electric Institute (EEI).

Another utility organization is the Association of Edison Illuminating Companies (AEIC), which started as an organization of companies licensed to produce electricity under the Edison System patents. That organization has been particularly interested in steam turbine progress, lamp testing, and promoting the quality of appliances by setting up standards for testing them. At their meetings, they have followed the practice of reporting good and poor quality in apparatus as a result of operating experience. Many of the electric power companies belong to both the EEI and AEIC, and together they represent the thoughts of the privately owned electric utility companies. In many standards activities involving apparatus for the generation and distribution of electricity, the EEI and NEMA representatives work together through joint committees.

In addition to the electrical industry, many other industries began to feel the need for standards. In all there are now more than 500 trade associations, engineering societies, institutes, and similar organizations, each representing some industry or group interested in standardization. It was recognized that not one of these organizations was in a position to set up national standards or speak for the country as a whole and present a national viewpoint. During World War I, the desirability of some national agency became clearly evident, and in 1918, five of the engineering societies formed the American Engineering Standards Committee (AESC). The purpose of this committee was to provide a means for facilitating the development of standards in engineering fields and to eliminate duplication and overlapping among the activities of various existing standards bodies in the country. After this committee got started, many other standards organizations made application to join. The membership grew fast, and in 1928 the activity became so large that the AESC was reorganized into the American Standards Association (ASA).

In 1963, the U.S. Department of Commerce established the Panel on Engineering and Commodity Standards to review the broad requirements for industrial and commodity standards in the United States and to make recommendations as to activities important to meeting national requirements for standards, with particular emphasis on the role of the Federal Government and the Department of Commerce. The La Que Panel, as it came to be known due to its Chairman, F. L. La Que, reported in early 1965. Among its findings the panel noted "due to lack of support by industry and government, ASA has not always been able to handle efficiently and effectively its responsibilities and assignments in the coordination of standardization activities, elimination of redundant efforts, promulgation of standards, and representation of the USA in international standardization." In view of this and other findings, the panel strongly recommended the early establishment, by legislative action, of a national coordinating institution for voluntary standardization in the United States, with international recognition equivalent to that of national standards bodies of other countries having officially recognized national standards organizations. The United States coordinating institution should have a federal charter and the standards promulgated by the institution should be designated as USA standards. Further, it was suggested that preference be given to reconstituting the existing national standards organization, ASA, rather than creating an entirely new body.

In the fall of 1966, ASA was replaced with the United States of America Standards Institute (USASI). The Institute adopted the 2800 American Standards approved by ASA. These standards and those subsequently adopted by the institute are now USA standards. In 1969 the USASI changed its name to the American National Standards Institute (ANSI).

The technical work of the institute is under the direction of the member body council with liaison and support from the company member council and the consumer council. The company member council advises the board of directors on institute policy, procedure, and planning. The consumer council represents the consumer, protects his interests in the national standardization program, informs him of the function of standards and standardization.

Through these councils, all segments of the nation's complex and expanding economy can voice their needs for up-to-date standards. Through them, all interested groups, including departments and agencies of the federal government, will have an opportunity to cooperate in a standardization program that will provide the United States with a set of dynamic USA standards that do not conflict with or duplicate each other and that have been accepted by all national groups who have a substantial interest in their application.

The International Electrotechnical Commission (IEC) is a world-wide organization for developing international electrical standards. It is the electrical division of the International Organization for Standardization (ISO) and has consultative status with the Economic and Social Council of the United Nations. The member bodies of the IEC are national committees organized in various ways but usually representing the important segments of the electrical industry of the particular country involved. At present there are 40 national committees participating in IEC.

The U.S. National Committee (USNC) is affiliated with the American National Standards Institute, but operates with a separate set of by-laws. There are about 30 member organizations of the USNC including NEMA, EIA, IEEE, ASTM, Underwriters Laboratories, and several branches of the Federal Government.

The USNC manages its technical work by appointing a technical advisor and an advisory group for each IEC technical committee in which U.S. industry has agreed to participate. At present the USNC participates in this way in 56 of the 61 IEC technical committees now in operation.

APPENDIX I.C. ENERGY VERSUS POWER

The words energy and power are often used in this text as though they were synonyms, which they are not. Energy is a basic concept that is used in other definitions. We might employ, however, an elementary definition of energy such as the "capacity for doing work." Power, on the other hand, is the rate of energy change or flow or dissipation. Then kilowatts are power and kilowatt-hours are energy. The electric utility can deliver kilowatts (power) as a function of its system size, that is, rating of generators and network capacity. It delivers energy by employing its capacity for a period of time. The electric utility could be called an energy company but is customarily called a power company in the United States. The word "power" has gained such widespread use that this text will not attempt to do otherwise. Power system, power stations, etc., will be used in the conventional manner. Every month we pay our "power bill" even though it is computed on the basis of kilowatt-hours.

2
THE POWER SYSTEM

INTRODUCTION

The present form of power systems would be impossible without alternating current. The development of a practical transformer freed the utilities from the limitations imposed by the low voltage inherent in direct current. These limitations were beginning to be felt by the time the Edison dc systems were only 3 years old. After the advent of transformers, much of the load was served by alternating instead of direct current. Then it became possible to generate power and transmit it to the load area at voltages appropriate for those functions, while continuing to serve the utilization devices at the low voltage considered safe for use in homes, offices, and factories. This arrangement freed generation from the necessity of being within 110 volts (volts, not kilovolts) transmission distance of the load. Power stations no longer had to be in the congested load areas; they could be built in locations that were better from many standpoints, for example, accessibility of water and fuel. Also, water power could be used to far better advantage. It was no longer necessary to locate the mill or factory right at the water wheel. Now the wheel could be made to drive an electric generator with the power sent electrically to wherever it could be used to the best advantage.

The modern power system must recognize the public's dependence on electric service. Service reliability must be very high. Many years ago, the operator of the village electric plant would shut down the village generator at nine or ten o'clock on a moonlit night. Today, the electric clocks are expected to keep correct time throughout the year.

With today's emphasis on environmental consideration, many constraints are imposed on the utility systems. Power systems are being designed to lessen the environmental impact with regard to aerial pollution, thermal discharge, and the esthetic aspects of siting. Many power companies are retrofitting existing facilities to meet the many stringent new federal, state, and local antipollution regulations with regard to aerial pollution and thermal discharge. The relative cost of various fuels burned by power plants is seriously affected by differences in the pollution controls that must be implemented.

THE MODERN POWER SYSTEM

The basic elements of a modern power system are shown in Figure 2.1. Obviously not every possible element of a power system is shown, for example, combustion turbines, circuit breakers, and fuses are not included. In this figure, for convenience, geographical separation is used to distinguish between bulk power supply and distribution. The generation plants, transmission lines, and primary substations are shown above the dashed line; the load (except the large industrial customer) and distribution below the line. In most actual systems, all the elements are more or less intermingled geographically.

Generation

Five generating stations are shown. A small utility may have only five or even fewer; a large one may have a hundred or more. Again, the size of each plant may be anything from a hundred to many millions of kilowatts, or even more. Companies situated in an area where abundant hydro power was available developed power systems supplied only by hydro. As loads grew and exceeded the capacity of available hydro sites, thermal generation was

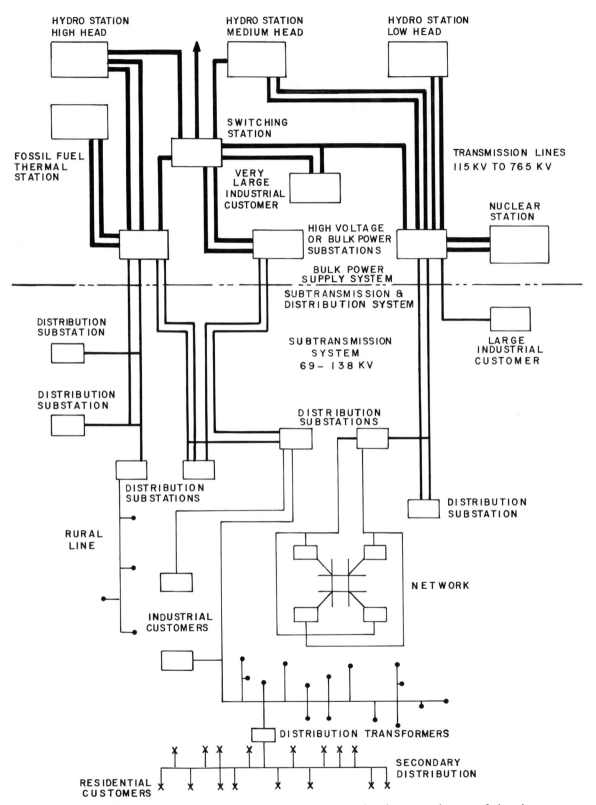

Figure 2.1. Basic elements of a modern power system showing several types of electric generation.

THE MODERN POWER SYSTEM

Figure 2.2. Typical medium-head hydroelectric station. (Courtesy of Southern California Edison Company.)

Figure 2.3. A low-head hydroelectric station. (Courtesy of U.S. Army Corps of Engineers.)

necessary to deliver firm power. Firm power is the power or power-producing capacity intended to be available at all times during the period covered by a commitment, even under adverse conditions. Figures 2.2 and 2.3 give an idea of the appearance of medium- and low-head hydroelectric stations. A fossil-fired steam station is shown in Figure 2.4, and a nuclear station in Figure 2.5.

Some utilities have no generation facilities; others lack sufficient generation to supply their customers. Therefore, they must purchase their requirements from other utilities. For the purpose of this discussion, the assumption is that the utility generates its own power.

Transmission

Transmission is an indefinite term that usually designates the highest voltage or voltages used on a given system. The voltage may be anything from 115 kV up. Transmission voltages above 230 kV are usually referred to as "extra-high voltage" (EHV) and those above 800 kV are referred to as "ultra-high voltage" (UHV). Power from the largest and from the most distant generating stations usually reaches the load area over the highest voltage circuits, and it is considered to be "transmitted." Interconnections from neighboring utilities, for the exchange of economy or emergency power, are

Figure 2.4. A fossil-fired steam station. (Courtesy of United Illuminating Company.)

made over "transmission" lines. Figure 2.6 shows a 500-kV transmission tower, and Figure 2.7 shows a 765-kV transmission tower.

In the United States, 345 kV, 500 kV, and 765 kV are generally accepted as EHV levels. The next step in voltage level (UHV) is expected to be between 1100 kV and 1500 kV. Extensive research is

Figure 2.5. A nuclear steam plant. (Courtesy of Tennessee Valley Authority.)

THE MODERN POWER SYSTEM

Figure 2.6. A 500-kV transmission tower, under construction. (Courtesy of Southern California Edison Company.)

underway at organizations such as the High Voltage Transmission Research Facility (HVTRF), Lenox, Mass. (Figure 2.8) to determine the appropriate voltage level that can be economically justified within the framework of modern technology. UHV transmission can have important advantages such as a very high power capability for each right-of-way and a reduction in the number of transmission lines required. The engineering concern of selecting the proper design constants for a UHV ac line must address such characteristics as radio and audible noise, corona-loss economics, conductor vibration, insulator and air-gap flashover from switching surges, 60-Hz breakdown of insulators due to surface contamination, and charging currents to vehicles and other objects on the ground at midspan.

Overhead rights-of-way in metropolitan areas are very expensive or are prohibited entirely. This has prompted system designers to consider underground ac and dc transmission by cable. The choice to date has been primarily ac cable, since the required lengths have not been sufficient to make dc cable economically feasible. Continued growth of metropolitan areas may find dc a viable alternative.

In cross-country overhead lines, savings in the cost of HVDC lines can offset the higher cost of HVDC terminals when compared with ac transmission. Today the breakeven point between alternating current and direct current is approximately 400 miles (640 km) for overhead lines and 20–25 (32–40 km) miles for underground cable. Combinations of both overhead and underground HVDC line segments frequently occur.

HVDC affects system performance in several ways. For example, limited dc ties can be put in without affecting system stability. Because of the asynchronous nature of HVDC, it allows power to be transferred into an area without increasing the short-circuit duties on the receiving ac bus. Also, system stability of ac systems can be improved by HVDC systems, since the terminals can be used to modulate dc power transfer during system disturbances, and thus help damp system oscillations.

Switching Stations

Switching stations are for the purpose of sectionalizing the system. Whenever a fault (short-circuit) occurs on a transmission line (or some other equipment like a transformer or a bus), the faulted equipment should be disconnected from the system. This protects the faulted equipment from further damage. A transmission line may be undamaged from a

Figure 2.7. A 765-kV transmission tower. (Courtesy of American Electric Power Service Corporation.)

Figure 2.8. View of High Voltage Transmission Research Facility (HVTRF), sponsored by EPRI/DOE.

short circuit and can be placed in service as soon as the arc products have moved away from the line.

Another reason for disconnecting the short circuit is to protect system service. There is usually an alternate circuit for power flow so system operation can be preserved by quickly removing the short circuit from the system. This quick removal from the system is accomplished by circuit breakers. Circuit breakers are high-voltage (also high-current) switches automatically operated. They can also be manually operated to isolate a line or other equipment for maintenance or new construction. When it is necessary to de-energize a circuit, it is desirable to keep in operation as much of the system as possible; for this reason, circuit breakers are used to cut off only the selected circuit. As a result, switching

Figure 2.9. A bulk substation. (Courtesy of United Illuminating Company.)

THE MODERN POWER SYSTEM

stations are substations whose function is switching circuits in and out of service. Figure 2.9 shows a bulk substation. Without the transformer, this would look like a pure switching station.

Primary Substations

High-voltage transmission lines are usually terminated some distance from the load they will serve because they are often not permitted in a populated area, and the loads are usually dispersed. The lines are terminated in substations, which are called "high-voltage substations," "transmission substations," "bulk-power substations," "major substations," "receiving substations," or "primary substations." At the primary substations, the voltage is stepped down to a value more suitable for the next part of the journey toward the load.

The equipment found in a primary substation would include power and instrument transformers, lightning arresters, circuit breakers, disconnect switches, capacitor banks, bus work, and a control house in which are located the station control equipment, protective relays, and so on. Some primary substations might include more equipment and some less, depending upon the functions they are intended to perform. Some stations are manually operated and others are completely automated.

Subtransmission

Subtransmission usually designates that part of the system between the transmission and the distribution systems. If a generating plant is located in or close to the load, there may be no transmission lines from that plant. It will then be connected to the subtransmission or distribution system.

The voltage of the subtransmission system is intermediate between the transmission and the distribution voltages. Some systems have only one subtransmission voltage. Frequently there are more than one. If there are, the reason may be historical. Several subtransmission voltages in one system may be an inheritance from the past, each voltage, in turn, having formerly been a transmission voltage until it became inadequate for that function. Then it was relegated to the role of subtransmission as higher-voltage transmission circuits became necessary.

On the other hand, economics may justify the use of two subtransmission voltages in series, even without any influence of the past. If the length and loading of the circuits point to a higher voltage than can safely be taken all the way to the distribution system, it may be cheaper to use two voltage levels, even though that necessitates the investment and

Figure 2.10. Two-circuit subtransmission line. (Courtesy of Duke Power Company.)

losses of an extra transformation in the intermediate substations. Circuits which take off across country and serve a number of small villages and crossroads communities might well need a voltage different from any used to serve the higher-load density areas.

Figures 2.10 and 2.11 show types of lines used for subtransmission.

Figure 2.11. Single-circuit subtransmission line. (Courtesy of Niagara Mohawk Power Corporation.)

Distribution Substations

The distribution system is energized through distribution substations. Prior to the development of a successful automatic reclosing relay, these substations had to be attended. They were of large capacity and supplied power to a large area through many feeders. Modern distribution substations are unattended; many are small and close to the load, controlling only a few feeders—sometimes only one (for example, a single-circuit unit substation). A distribution substation is shown in Figure 2.12.

Primary Distribution

Primary distribution takes the power from the distribution substations to the final stepdown operations, which, in residential areas, may be a distribution transformer as shown in Figure 2.13. Primary voltages are usually between 4 kV and 34.5 kV.

Secondary Distribution

This is the part of the system through which the power finally reaches a large proportion of the customers, excluding the industrials and other large-use consumers.

Secondary Network

The downtown commercial district of a city or a large shopping center is an area of high-load density. Because of the nature of the loads, it is very desirable to avoid interruptions to service. This part of the city is usually served by a secondary network which serves all the customers in its area. Usually the network, the network unit which energizes it, and all connections to and from it are underground.

Industrial Customers

Small industrial customers are served directly by the primary feeders, or possibly from the subtransmission system. Large industrial customers are served directly from the subtransmission system, and very large industrial customers may be served from the transmission system. Figure 2.14 shows POWER/VAC® switchgear which may be used in an industrial substation.

THE MODERN POWER SYSTEM

Figure 2.12. A distribution substation. (Courtesy of Seattle City Light.)

Rural Lines

Farms are served by long, lightly loaded rural lines. In the usual case, the line starts out as a 12.45-kV, three-phase, four-wire circuit. Branches are 7200 volts, single phase. Toward the far end, the main circuit itself may become single phase, one phase wire and neutral.

The Bulk Power Supply System

Some utility people use this term to designate the wholesale part of the business. It includes generation, transmission, and primary substations, and sometimes also subtransmission and distribution substations. This part of the system corresponds in function to that of manufacturer and wholesaler.

The bulk power supply system is under centralized control, including operation, routine maintenance, and construction of new facilities. It contains all the big pieces of apparatus on the system: the large steam-turbine generators, the high-voltage transformers, the water wheel generators, and the

Figure 2.13. Pole-mounted distribution transformer.

Figure 2.14. Typical metalclad switchgear used at industrial installation.

power circuit breakers. It also provides solutions for some design and operation problems, such as protective relaying, stability, and control of system voltage, load, and frequency.

The Distribution System

This is the retail part of the system. It serves the residential and commercial customers and some of the smaller industrials. It has the final responsibility for seeing that service is maintained to the customers, and at the correct voltage. On most systems, distribution represents from 35 to 45% of the total system investment and half the total system losses.

INVESTMENT

How much of the system investment is represented by these various components? The answer to this question varies considerably from one company to the next. Because some generation developments are more costly per kilowatt than others, a company whose generation is predominantly oil or gas has a smaller proportion of its capital in generation than a company having predominantly coal, nuclear, or hydro. A company that has several large industrial customers has less money invested per kilowatt in distribution than one that must deliver practically all its energy to many small customers. A company whose generation is close to its load has a relatively small investment in transmission. The following table shows the investment in each category in percent of total electric plant for investor-owned electric utilities for 1979. This is a national average, and the numbers for any given utility will differ from those stated.

TABLE 2.1. System Investment for Investor-Owned Electric Utilities

Type	Percentage of Total
Generation	49
Transmission	17
Distribution	31
General and intangible	3
Total system investment (gross) per kW of installed capacity	$391

SOURCE: Statistics of Privately Owned Electric Utilities in the United States, Department of Energy.

System investment costs for new generating plants and other utility plants are increasing rapidly due to inflation and the added costs for protection of the environment. Therefore, the total system investment per kilowatt of installed capacity is expected to increase significantly in the next few years.

SYSTEM EVOLUTION

What factors determined the evolution of an existing system? The fact is that it was not designed—at least not in the sense that someone, starting with no system, with no power supply facilities at all in the area, conceived and planned the present system. The present system may comprise what originally were separate operating companies which have been consolidated into the present system. Throughout this process, the load in the area of each of the components was increasing, and adequate facilities for supplying it had to be available all the time. At the time it was taken, each step in the evolution of the present system was deemed to be the most logical in view of:

1. Existing loads and current estimates of future loads; presently available apparatus, equipment, materials, knowledge, and techniques; economic conditions, current and predicted.
2. Limitations on freedom of choice imposed by policies adopted and actions taken previously (these limitations have sometimes constituted severe handicaps).

The present assembly of generating stations, substations, and transmission and distribution lines may be different from what would now be selected to serve the present load on the system.

Tremendous gains in the field of automated system planning have been made possible by the development of computing devices. Calculating techniques are in use that provide optimum economic design solutions, taking into account system expansion from the existing pattern, outages for whatever reason, interconnections with other utilities, and load characteristics. Included in these studies are considerations of reliability, adequacy of system under unusual circumstances, stability, short circuits, surge voltages, radio interference, protective relaying, and all other important aspects of system planning.

LOAD

The system load at any instant is the sum of the loads drawn by all the devices that happen to be in operation, plus the system losses. Some of the power-consuming devices are manually controlled (household radios and electric ranges), some are controlled automatically (refrigerators), and some are both (flatirons and ovens). The amount of load at any instant is determined primarily by the customers and is sometimes referred to as demand. Demand may be measured in either kilowatts or kilovolt-amperes; it may apply to a single device or piece of apparatus, to a feeder, to substation or generating stations, or to the entire system. The utility can exert some influence on the demand, but not much. If the load is predominantly resistive (e.g., lighting and heating), a reduction in voltage reduces the demand, and vice versa. Motor load is affected only slightly by changes in voltage, but it may be sensitive to changes in frequency. How sensitive it is to frequency depends on the composite of the speed-load characteristics of all the driven devices in operation at the moment.

Emergencies can be created when insufficient generation is available to meet the load demand. This could be caused by unscheduled outages of any type generator or even other equipment. When companies operated an isolated system, some load reduction could be obtained by reducing voltage and/or frequency. With interconnected systems, however, frequency reduction on one system is not possible, and local voltage reduction in excess of 5–8% can create undesirable var conditions. The most effective and direct way of reducing load in an emergency is "load shedding," wherein selected loads are automatically disconnected according to a planned schedule.

When any energy-consuming device is selected on or off, it affects the system frequency and voltage. Even lighting a 25-watt lamp tends to reduce the voltage at the lamp socket. The effect of a 25-watt lamp is infinitesimal, of course, and is completely masked by the simultaneous starting and stopping of other devices all over the system. Large rolling mills, on the other hand, cause large, sudden, and frequent variations in load. In some cases they cause undesirable changes in tie-line load between systems. Welders, arc furnaces, sawmills, mine hoists, and shovels are other loads having extremely variable demands. Sometimes they are merely a nuisance to other customers on the same feeder because of the voltage fluctuations and light flicker they cause; sometimes they present the utility with major problems of load, voltage, and frequency control.

Despite these occasional difficulties, the load on most systems is relatively free from objectionable fluctuations throughout most of the 24 hours. Major changes in load occur only with changes in the tempo of industrial activity, that is, at the beginning and end of the morning and afternoon shifts in the factories, and with the approach of daylight, of darkness, and of bedtime.

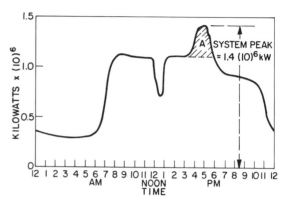

Figure 2.16. Typical load curve for a weekday during the time of system peak load.

LOAD CURVES

A plot of the kilowatt demand, or load, over a given period of time is known as a load curve. The period most often used is a day, from midnight to midnight. Under normal conditions the system load curves for Monday through Friday are about the same. The Saturday and Sunday load curves of companies that have a considerable commercial or industrial load, or both, show the effects of the weekend shutdown of shops and factories. The system load curves will generally vary with the season, depending upon such considerations as: darkness occurring earlier in winter, air conditioning load on a hot summer day, irrigation pumping, and so on. One daily load curve may be represented by the power-versus-time curve of Figure 2.15. Another, showing a higher system peak, is shown in Figure 2.16.

These curves bring out the difference between power and energy (see Appendix I.C of Chapter 1). The area under the curve represents energy, the kilowatt-hours which must be generated during

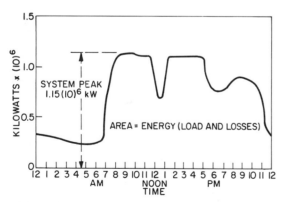

Figure 2.15. Typical load curve for a weekday.

each 24-hour period. The ordinates of the curve show the rate at which the energy must be generated. The highest ordinate shows the maximum rate, that is, the maximum power for the day. It is stated in kilowatts and is known as the peak load.

The ratio between the actual area under the load curve expressed in kilowatt-hours and the kilowatt-hours that would be generated if the load were constant at peak value during the entire 24-hour period is called the load factor.

For the system whose loads conform to Figures 2.15 and 2.16, the system peak is about 1.15 million kW in one and 1.4 million kW in the other. The utility must have firm generating capacity available (which may include purchased power, if it is firm) to carry these peaks. Although the higher peak requires 250,000 kW more generation than the lower peak, this 250,000 kW is needed for less than 3 hours a day—a very low load factor. The energy is represented by area A of Figure 2.16. Many systems have old steam units which have become too inefficient to be operated continuously. These units are still valuable for low-energy loads like A, where high economy is not very important, and for emergencies. Utilities are increasing the purchase of combustion turbine units intended specifically for peaking. Such units are characterized by their ability to be started and put into service quickly. Considerably lower investment costs and their easy adaptability to automatic control more than offset their higher fuel cost. For those utilities with low-cost sites available, pumped hydro installations are becoming popular for peaking generation. Also, conventional hydro units may be used for peaking, depending on the characteristic of the river flow and the storage available. It may also be economical to install additional generation at an existing dam site to obtain peaking capacity.

INSPECTION AND MAINTENANCE

Utility equipment such as switchgear, transmission lines, boilers, turbines, and so forth, require inspection and maintenance. A circuit breaker can be taken out of service, inspected, and even maintained without any service interruption to customers. When a boiler is taken out of service for scheduled maintenance, however, the associated generator is not available. Thus, maintenance should be done on a unit when it is not needed to carry the load.

The load curves shown in Figures 2.15 and 2.16 show the difference between an off-peak load and a peak load. Maintenance would be scheduled for the off-peak season. During the time of system peak, all units should be available. This was easy to schedule when the system peak occurred during the winter, probably in December, and the load dropped markedly after the peak. Air conditioning load during the summer, however, has now reversed the season for peak loads on many utilities. Many times the "system peak" might occur every 6 months. There are systems where the load does not drop substantially between system peaks, or one might say the system peak is difficult to determine. With such a load characteristic, maintenance cannot be scheduled during the "off-peak" season. Thus, the system design must be such that a reserve (extra capacity) is available so a necessary maintenance schedule can be kept.

SYSTEM DESIGN

As long as loads continue to increase, systems must also continue to grow. This requires what is known as "system planning." System planning is management's defense against being caught with not enough generation to carry the load, or with a system that cannot maintain proper continuity or voltage in some areas. Loads historically have doubled about every 10 or 11 years. That is a national average rate of growth: it is subject to increase or decrease by business conditions, world outlook, and, of course, local utility conditions. Although it is necessary to maintain the ability of the system to carry the load that may be demanded, it is expensive to have more facilities on hand than are needed. It is easy to see that management must be alert and exercise good judgement, especially when orders for new base load generation must be placed 8–12 years before it will be needed.

The emphasis in system design has been changing gradually over the years. This change is caused primarily by three factors:

1. Experience and development have greatly increased the reliability of each component part of the system.
2. As systems get bigger and bigger, any one piece of equipment becomes less and less important to the successful operation of the whole system.
3. The increasing use of interconnections between neighboring systems has had the effect of making each system larger, and, hence, less dependent on any one component.

Utility planners have learned that it is of supreme importance to build into the system as much flexibility as possible. "Flexibility" means several things. It means freedom to grow as necessary to serve ever-increasing loads. It means the ability to serve new loads, whatever their nature and whenever they may appear. It means the ability to take advantage of new tools and devices, new methods and techniques, and new operating, construction, and design practices, as they may be developed, invented, or discovered.

The development of a successful automatic reclosing relay changed completely the philosophy of distribution system design. As already noted, this relay made an unattended substation practical. When the substation no longer required operators, it could economically be made smaller and located nearer the load. Primary distribution feeders could be shortened, and the power could be brought nearer its destination over the more economical subtransmission system. The large distribution substation feeding a large load area over many primary feeders became a relic of the past. Many are still in operation, of course, but most of them have been made automatic (and unattended). This is only a small part of the whole story of the effect of automation on system design and operating, resulting in great savings in capital investment and in operating costs.

SUMMARY

A modern power system consists of generating stations, transmission and subtransmission systems to convey the energy to the load areas, and distribution systems to deliver it to the customers. A utility must keep its system in condition to serve its present load adequately. It must, at all times, have enough generation available for service, and enough of the transmission, subtransmission, and distribution systems in operation to assure continuity and good voltage to every customer despite planned outages for inspection and maintenance. It must meet these technique requirements, while at the same time operating the system so as to realize the maximum economies. Finally, it must plan and provide additions to the system to provide capacity for load growth.

REFERENCES

1. Kirchmayer, L. K., *Economic Control of Interconnected Systems,* Wiley, New York, 1959.
2. Kirchmayer, L. K., *Economic Operation of Power Systems,* Wiley, New York, 1967.
3. Miller, R. H., *Power System Operation,* McGraw-Hill, New York, 1970.
4. Sporn, P., *The Integrated Power System,* McGraw-Hill, New York, 1950.
5. Vennard, E., *The Electric Power Business,* McGraw-Hill, New York, 1962.

3
POWER GENERATION

SOURCES OF ENERGY

Power generation, in the electric utility industry, means conversion of energy from a primary form to the electrical form. The current sources of nearly all the electrical energy distributed by utilities come from: the conversion of chemical energy of fossil fuels, nuclear fission energy, and the kinetic energy of water which is allowed to fall through a difference of elevation. The installed nameplate capacity of the major generation for the United States is shown in Table 3.1.

Heat energy, obtained from the combustion of fossil fuels or the fission of nuclear fuels is converted into mechanical energy through the rotating shafts of steam turbines, combustion turbines, or internal combustion engines. These shafts in turn drive electric generators.

Figure 3.1 shows the relative utilization of the primary sources of energy to produce electrical energy in the United States. The curves show each specified primary energy source in percent of total. If the vertical axis of Figure 3.1 were in Btu or some common energy unit, the total would increase greatly from 1850 on. If the plot were energy per capita versus time or year, the curve would be greater in the late years by about a factor of 4 when compared with 1850. Examination of Figure 3.1 reveals not only the important primary fuels, but the fact that it takes many years, after the introduction of a new fuel, before that fuel becomes dominant. For example, petroleum and natural gas liquids increased during the 1970s. This trend should reverse as the plot is extended into the 1980s. Note that geothermal is not shown since it is still a small percentage of the total. In the near future, the use of geothermal energy, wherein the earth itself supplies steam, may be expanded somewhat.

Solar energy in the form of heat, direct conversion to electricity, or wind may also play an expanded role in producing electric energy. However, for solar energy to become a major primary source of energy for the generation of electricity, an economical-technical development that supersedes today's understanding is required.

Magnetohydrodynamics (MHD) and fuel cells also have the potential for becoming useful energy converters, i.e., to be able to convert the energy of a fuel to electrical energy. The state-of-the-art is such that it is unlikely that they will be useful to a utility for several years.

Primary energy sources each have their own characteristics. For example, solar energy appears to be abundant and free. However, it arrives very dilute and is of low quality since its energy density and temperature are both low. Nuclear energy is concentrated, but the fissionable material, uranium-235, is well under 1% of the natural or mined uranium. Uranium-238, which is the abundant isotope (99$^+$%), is used in the breeder reactor. The latter represents a very significant source of energy, although considerable effort would be required to develop it in economic form.

A nearly limitless primary energy source is the fusion of deuterium, or heavy hydrogen (one proton and one neutron), which is found in the waters of the earth. For this reason, the oceans contain a tremendous quantity of energy—enough to satisfy mankind's energy needs for hundreds of thousands of years. This energy source can be exploited as soon as an economical fusion reactor is developed. However, the technical problems are extremely

27

TABLE 3.1. Installed Nameplate Capacity for the United States

Type	Capacity (MW)	Percentage
Hydro	75,326	12.6
Fossil steam	412,324	68.9
Nuclear	54,594	9.1
Combustion turbine	50,562	8.5
Internal combustion	5,492	0.9
Total industry	598,298	100.0

SOURCE. *Energy Data Reports*, DOE/EIZ-0049(79), June 1980.

great and successful fusion reactors are still many years in the future.

During the next few decades, petroleum, coal, nuclear, and hydro energy will supply the majority of the energy necessary to produce electricity. Declining long-run domestic supplies of oil and gas, combined with high prices and uncertain availability of foreign oil supplies, make it difficult to plan for the continued use of oil and gas as major sources of electrical energy. Under these circumstances, development of new energy sources is essential.

Coal

The U.S. coal reserves can serve utility coal energy requirements well past the 21st century. For this to be accomplished, reserves must be developed by opening new mines and constructing coal handling facilities. The U.S. Geological Survey estimate of identified U.S. coal reserves is 1.7 trillion tons. Classified as hypothetical or undiscovered are another 2.2 trillion tons. The heating value of coal used by most utilities lies in a range from 9,000 to 13,000 Btu/lb.

Since coal contains sulfur and particulates, its burning presents environmental discharge problems. Sulfur content can range from 0.2 to 7.0% by weight, while ash content can range from 5 to 20% by weight. Because of these contaminants, virtually all utility boilers use pollution control equipment. Control technologies, such as precipitators and scrubbers, can be thought of as postcombustion attempts at reducing emissions. On the other hand, development is underway on technologies that will provide commercially feasible coal gasification and liquification equipment. Those pose the opportunity for cleaning the coal before it enters the combustion process.

Petroleum

While quantities of coal are found in continuous beds covering large areas and in close proximity to the earth's surface, oil and gas deposits tend to be much more distributed. In addition, the prediction today is that the world production of crude oil from conventional sources will reach its peak by the end of the century and decline thereafter.

The future use of oil (from conventional sources) to produce electricity is expected to decrease because of the projected price and availability concerns. The percentage of the world's energy supplies, produced from oil and gas, will probably decline as is illustrated in Table 3.2.

The average heating value for oil is found to range from 17,450 Btu/lb to 18,300 Btu/lb, as purchased. The term "as purchased" is usually based on the higher heating value (HHV), while fuel consumption requirements are sometimes given in terms of lower heating value (LHV). The difference between higher and lower heating value is the heat required to form the water vapor part of the combustion products associated with the burning reaction. The higher the hydrogen to carbon ratio of the fuel, the more water vapor is produced per pound of fuel. The HHV/LHV ratio is 1.11 for natural gas, and 1.03 for coal. A gallon of oil at 60°F, when burned, will yield from 8.2 to 8.9 pounds (3.7 to 4.0

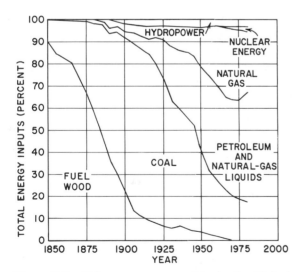

Figure 3.1. Primary energy sources in the United States.

SOURCES OF ENERGY

TABLE 3.2. Potential World Primary Energy Production

	As Percentage of Total			
	1972	1985	2000	2020
Hydrocarbons				
Oil	42.8	44.2	28.3	10.6
Gas	17.1	15.8	20.7	12.5
"Unconventional"	.0	.0	.6	4.0
	59.9	60.0	49.6	27.1
Coal	24.5	23.6	24.6	25.9
Nuclear	.7	4.7	12.8	31.4
"Renewables"				
Hydraulic	5.2	4.9	4.9	5.6
Other	9.7	6.8	8.1	10.0
	14.9	11.7	13.0	15.6
Total	100.0	100.0	100.0	100.0

SOURCE. World Energy Conference.

kg) of water. A review of the specific equipment associated with burning fossil fuel to produce electricity is found in Chapter 4.

Natural Gas

The natural gas supply may be depleted by the end of the first quarter of the 21st century. The estimated U.S. reserve of natural gas in proven and undiscovered recoverable reserves in 1980 was 785 trillion ft^3. Production of natural gas in 1980 was 19.3 trillion ft^3. In 1980, the natural gas purchased by the electric utilities for power generation had an average heating value of 1035 Btu/ft^3 HHV varying from 927 to 1133 Btu/ft^3.

Hydroelectric Power

Hydroelectric generation, as a percentage of energy production, is shown in Table 3.1. This energy source is described in detail in Chapter 6.

The potential energy possessed by water is the product of its weight and a usable difference in elevation, called head. Hence, the head is measured from the higher, or forebay elevation, to the lower, which we may call the plant elevation. For a reaction wheel, plant elevation is the same as tail-race elevation. For an impulse wheel, it is the elevation of the jet. Thus,

$$\text{potential energy} = WH \text{ foot-pounds}$$

where W is weight of a given quantity of water, in pounds, at a difference in elevation of H feet.

The hydroelectric plant capacity is a function of the water flow, head, hydraulic turbine and generator efficiency, and fluid flow losses. That is,

Hydroelectric plant capacity, kilowatts

$$= \frac{q(H - h_f)\eta}{11.8}$$

where q is water flow, cubic feet per second; H is head, feet; h_f is head loss because of fluid flow, feet; η is turbine-generator efficiency; and 11.8 is conversion factor, ft^4/kW-second.

The head value may differ considerably from the time the stream flow is in flood to the time of minimum stream flow. The head loss and turbine-generator efficiency vary with the load on the unit. The minimum net head limits the firm capacity of the plant.

Nuclear

One of the greatest potential energy sources in the United States is nuclear energy, using uranium as a fuel. The use of uranium as a fuel can take either of two directions. The light-water reactors use enriched uranium. Uranium, as mined, contains about 0.7% of U^{235}, the fissionable isotope. The enrichment process builds up the U^{235} content to about 3%. The principal isotope in natural uranium is U^{238}, with a content of approximately 99.3%. The U^{238} can be made to capture a neutron and form plutonium-239 after two steps of disintegration. Since Pu^{239} is a fissionable material, a "breeder reactor" can be made, this being the second use of uranium as a fuel. The available energy appears to be increased by the ratio 99.3/0.7 or about 140 times when comparing breeder reactors to light-water or heavy-water reactors. A more realistic figure may be one-half of this amount or 70.

Current domestic reserves of uranium ore are more than sufficient to supply the requirements of all the currently operating or planned nuclear reactors. A description of the currently utilized reactor requirements is provided in the following Chapter 4.

A discussion on the nuclear core physics and the nuclear fuel cycle is provided in Chapter 5.

Geothermal

Geothermal power is a primary source for the production of only a small portion of electrical power. The Pacific Gas and Electric Company established the first U.S. geothermal plant in 1960 at the Geysers, a steam field about 90 miles (145 km) north of San Francisco. The 1972 plant capacity was 322 MW. The potential of the Geysers field is estimated to be close to 2000 MW.

The Geysers field is a "dry" steam field. The moisture content of the steam is such that the wellhead steam can be used, after particulate filtration, directly by a steam turbine. The steam turbines are designed to handle the large volume flow required due to low pressure and temperature. The steam conditions at the Geysers No. 5 unit are 115 psia (0.8 MPa) and 355°F (179°C) which is slightly superheated [17°F (8°C) of superheat]. The condenser is of the direct-contact type, since the steam does not have to be condensed separately from cooling water and returned to a boiler. The condensate flows to a forced draft cooling tower where about 80% is evaporated, depending on weather conditions. Because the liquid condensate contains boron and ammonia, the condensate is injected back into the steam field rather than to the surface environment.

Geothermal power is of interest because the costs appear to be competitive with other energy sources. However, these costs are not well defined due to such considerations as exploration, drilling, and the troublesome corrosive nature of hydrogen sulfide present in the steam. (The steam, as it comes from the well, must often be cleaned before it enters the steam turbine.)

A potential exists for geothermal energy development due to the existence of "wet" steam fields which have favorable steam conditions of about 80–90% moisture. The Imperial Valley in Southern California is currently the most promising area with an estimated potential of 2000–4000 MW. Because of very high salinity and the possibility of earth slipping, all of the water will be returned underground.

The estimated worldwide potential for geothermal is insignificant in relation to total energy requirements. However, this energy source can take on local significance in places such as Iceland where geothermal is a major energy source.

Solar Energy

Solar energy can produce electricity by direct conversion to electricity, energy concentrations as in a solar tower that produces high-temperature steam, or indirect conversion in the form of wind energy. The cost of direct conversion cells has decreased for several years. Extrapolating, it appears that solar cells may become economical in a decade or two. A major problem with solar energy is its low-energy density which amounts to about a maximum of 1 kW/m^2. Using this figure, the area requirement for 1 MW would be 1000 m^2, with 100% efficiency and the sun overhead. The range of efficiency is actually 10% to 15%. Using 10% efficiency, 10,000 m^2 would be required to produce 1 MW. At locations where the sun's rays are not perpendicular to the earth's surface, it is necessary to divide by the cosine of the latitude to get the area of earth needed. This is strictly true if the sun were always over the equator. In addition, the cells (or reflectors) would normally be moved throughout the day to intercept the most energy, requiring spacing between movable units. By assuming a factor of 1.6 to account for spacing, and a latitude of 40°, 10,000 m^2 × 1.6/cos 40° is nearly 21,000 m^2. Converting this to megawatts per square mile gives 124 MW/mi^2. The 124 MW would be peak power that would be available in a clear atmosphere for several hours per day. No firm power would be available at night unless storage were provided. Solar power will be useful as an energy source or fuel displacement during the sunny hours under favorable economic conditions. If energy storage is required, the cost appears to be prohibitive.

Wind Power

Wind has been used as a source of energy throughout history. Prior to the Rural Electrification Act, wind power was extensively used for pumping water and generating electricity in the United States. Due to the scarcity of fuel during World War II, interest in wind-turbine generators once again increased. However with the decrease in fuel prices after World War II, wind turbines lost their appeal as a cost-effective means of generating energy. Today, with the current high cost of fuel, electric utilities are expressing increased interest in wind as an alternate source of energy.

Modern wind turbines are considerably more effi-

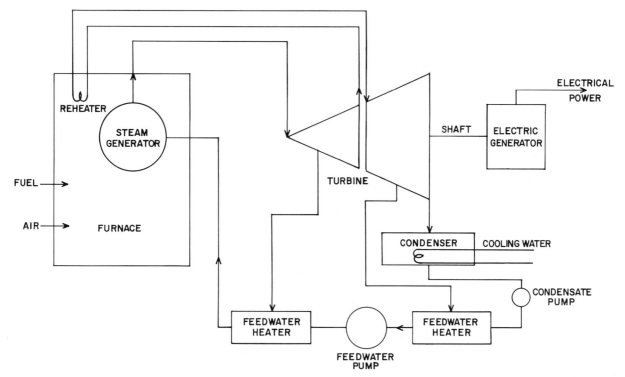

Figure 3.2. Sketch showing the principal components in a fossil fueled power plant.

cient than their predecessors. Since kinetic energy in the wind cannot be completely converted to mechanical energy, only about 59% of the total wind energy is available for use as an energy source. Under the appropriate conditions, today's wind turbines can extract about 43% of the wind's total power. The older wind turbines still in operation are capable of 10–15% extraction.

If wind turbines are to contribute to the world's energy needs, it is important that they be designed so as to reliably produce electrical power at low cost year after year. This requires the accurate determination of the loads acting on all the system's structural elements during start-up, shut-down, normal operations, and emergency periods.

Unlike gas turbines, which can depend on a regulated supply of fuel, wind turbines must contend with varying wind speeds, shifting wind directions and sharp positive and negative gust effects. In addition, the rotor is exposed to wind shear, the change in wind velocity with height above ground, and with tower shadow. These factors combine to create an extremely difficult environment, both for the structure and the controls.

Despite these obstacles, advancements continue to be made in wind turbine technology. Under continuing development programs, wind turbine costs are expected to become commercially competitive within the next few years in areas with good winds and high costs of competing energy sources (Macklis, 1982).

THERMODYNAMICS OF STEAM-ELECTRIC GENERATING PLANTS

Thermodynamics is the science that is concerned with the transformation of heat into other forms of energy (and vice versa). In the simplified sketch of the fossil-fueled power plant shown in Figure 3.2, there are many places where such energy transfers occur. For example, energy stored in the fuel is released as heat during the process of combustion. This heat is then transferred to the working fluid (water in this case) where it is stored in the form of internal energy of steam. The steam flows into the turbine where this internal energy is released and does work on the turbine-generator, where it is transformed into electrical energy. It is through the use of thermodynamics that we can analyze all of these energy exchanges to optimize plant output and efficiency.

The Carnot Cycle

In 1824, Sadi Carnot proposed a thermodynamic cycle which has the highest attainable efficiency of any cycle operating between the same sets of conditions and as such, it is a standard for comparison with other power cycles. The Carnot cycle is a reversible cycle composed of a series of reversible processes. A process is reversible if, after it is completed, it can be made to retrace, in reverse order, the various states of the original process; and if all energy quantities to and from the surroundings can be returned to their original states (work returned as work, heat returned as heat, etc.). No process can actually be reversible because of friction, turbulence, nonzero temperature differences required for heat transfer, and so on. Therefore reversible cycles cannot be built.

Before any further discussion of the Carnot cycle, or any other cycle, one must have a minimal understanding of the concept of entropy. Entropy is a thermodynamic property which is a measure of the amount of disorder in the system. A thermodynamic property is simply a characteristic of a system, similar in this respect to pressure or temperature. Any two independent properties are sufficient to describe the state of the system, just as any two coordinates are sufficient to locate position on a plane.

The change in entropy between any two states is defined by the following equation:

$$\Delta S_{1-2} = \int_1^2 \frac{dQ_{rev}}{T}$$

Q_{rev} is the amount of heat added or removed reversibly during the process from 1 to 2. Note that for heat transfer to occur reversibly, there must be no temperature difference between the object supplying and the object receiving the heat. T is the absolute temperature at which the heat transfer occurs. Absolute temperature is measured relative to absolute zero. Temperatures in Fahrenheit are converted to absolute temperatures on the Rankine scale by adding 459.67°. (Centigrade is converted to Kelvin by adding 273.15°.)

The usual units of entropy are Btu/lb-°R (kJ/kG-°K), when expressed on a specific (per unit mass) basis, or Btu/°R (kJ/°K), when considering the total entropy of the system.

The Carnot cycle is shown as the solid lines in the temperature-entropy (T-S) diagram in Figure

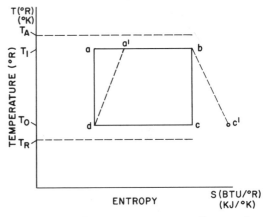

Figure 3.3. Temperature-entropy diagram for the Carnot cycle.

3.3. It consists of the following four reversible processes:

1. Isothermal (constant temperature) heat addition at T_1, represented by line a–b.
2. Isentropic (constant entropy) expansion, represented by line b–c. Since the process is reversible and occurs at constant entropy, we can determine that there is no heat transfer (adiabatic) from the definition of entropy change.
3. Isothermal heat rejection at T_0 (line c–d).
4. Isentropic compression, represented by d–a.

The only heat addition occurs during a–b, and the amount equals $Q_1 = T_1(S_b - S_a)$. Similarly, the only heat rejection is during c–d, with the amount $Q_0 = T_0(S_d - S_c)$ transferred (Q_0 is a negative number). The efficiency of a cycle is the net work divided by the heat input. By the first law of thermodynamics, the net work of a cycle equals the net heat transfer, so that the efficiency is:

$$\eta = \frac{\text{Net Work}}{\text{Heat Input}} = \frac{\text{Net Heat}}{\text{Heat Input}} = \frac{Q_1 + Q_0}{Q_1}$$

$$= \frac{T_1(S_b - S_a) + T_0(S_d - S_c)}{T_1(S_b - S_c)}$$

$$= 1 - \frac{T_0}{T_1} \quad \text{since } S_b - S_a = S_c - S_d$$

Therefore, the efficiency of the Carnot cycle depends solely on the absolute temperatures of the external heat source and sink, T_1 and T_0. Carnot's principal states that no cycle operating between

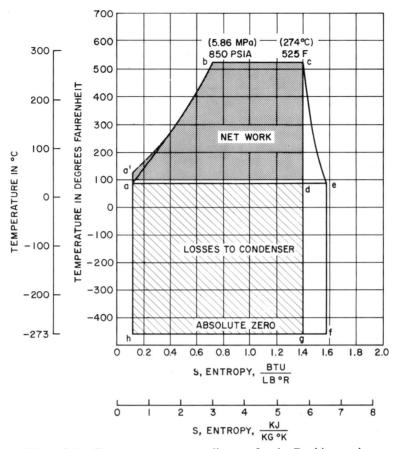

Figure 3.4. Temperature-entropy diagram for the Rankine cycle.

these same two temperatures will have an efficiency greater than that of the reversible cycle. As a result, the efficiency of this cycle represents the maximum achievable for a given source and sink.

Actual cycles have irreversibilities that reduce their efficiencies to values significantly lower than that of the Carnot cycle. For example, heat transfer occurs only with nonzero temperature differences ($T_A - T_1$ and $T_0 - T_R$ in Figure 3.3) and this introduces external irreversibilities to the cycle. Internal irreversibilities, such as those due to friction and turbulence, cause the paths followed during compression and expansion (d–a′ and b–c′ in Figure 3.3) to deviate from the ideal. The net result is a reduction in efficiency below that of the Carnot cycle.

In electric utility applications, the thermal performance of a power plant is normally expressed in terms of the plant heat rate, which is defined as the heat input to the cycle divided by its electrical output. Heat rate is inversely proportional to efficiency and is normally given in Btu/kW-hr (kJ/kW-hr). The conversion between heat rate and efficiency is as follows:

$$\eta, \% = \frac{100 \times 3412.14 \text{ Btu/kW-hr}}{\text{Heat Rate}}$$

$$\left(= \frac{100 \times 3600 \text{ kJ/kW-hr}}{\text{Heat Rate}} \right)$$

The Rankine Cycle

The simple Rankine cycle is the basis for the cycle used in steam generating power plants. Although water is the most commonly used fluid in the Rankine cycle, it can be run on many others such as: mercury, butane, freon, or ammonia.

Figure 3.4 shows the ideal Rankine cycle operating with a throttle pressure of 850 psia (5.86 MPa) and an exhaust pressure of 1.5 inches of mercury absolute (1.5 in. Hga or 38.1 mm Hga) on a T-S diagram. Starting at point a, feedwater from the condenser is compressed in a 100% efficient feed pump to state a', which is at throttle pressure and 130°F (54°C). The water then enters the boiler where it is heated to the saturated liquid state at point b and then vaporized to the saturated vapor state at point c. The steam then expands through the turbine where the energy available in the steam is converted into the output of the machine. If the turbine is 100% efficient, the expansion is represented by the vertical line c–d. Steam exhausts from the turbine at point d enters the condenser where heat is rejected; the stream is condensed back to state a, and the cycle repeats itself.

In the temperature-entropy diagram of Figure 3.4, the area enclosed by the path followed by a series of reversible processes represents energy. The total amount of energy added to the cycle is represented by area aa'bcdgha; the heat rejection by area adgha; and the net work by the difference between these two, area aa'bcda. Since the efficiency equals the net work divided by heat added, we have:

$$\text{Efficiency} = \frac{\text{Area aa'bcda}}{\text{Area aa'bcdgha}}$$

Actual Rankine cycles suffer efficiency losses that can significantly reduce the overall cycle performance. For example, turbines are never 100% efficient because of steam leakages around nozzles and buckets, nozzle and bucket inefficiencies, bearing losses, residual velocity losses from the last stage, and so on. The net result is that the expansion in the turbine follows the line c–e rather than c–d. Steam at state e has a higher energy level than that at d, so more heat must be rejected in the condenser in order to return to state a. By increasing the heat rejection, both net work and cycle efficiency are reduced.

Rankine Cycle With Superheat and Reheat

Carnot's analysis showed that the efficiency of a cycle improves as the temperature at which heat is added to the cycle from an external source is increased. The same is true for the Rankine cycle, even though the temperature at which heat addition occurs is a variable quantity. In this case, the objective is to increase the average value of this temperature. Superheat and reheat are two methods commonly used to achieve this goal.

Superheat in a cycle is achieved by adding heat to saturated steam, thereby increasing its temperature above the saturation temperature correspond-

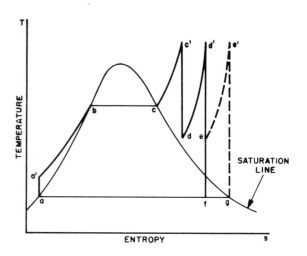

Figure 3.5. Temperature-entropy diagram for the Rankine cycle with superheat and reheat.

ing to its pressure. This is shown in the T-S diagram in Figure 3.5. In the simple Rankine cycle, saturated steam at point c enters the turbine. In the cycle with superheat, on the other hand, the steam is superheated to state c' before flowing into the turbine.

Reheat is also shown in Figure 3.5. At some intermediate point in the turbine expansion (point d), the steam is removed from the turbine and sent back to the boiler for reheating to state d'. This steam then re-enters the turbine and expands to the exhaust condition at point f. This cycle has a single reheat, although additional reheats can be used. For example, a second reheat is illustrated by the dashed set of lines ee'g.

In actual cycles, superheat and reheat have several additional effects on the overall cycle efficiency. One of the most important of these is the reduction in moisture losses as a result of the decreased moisture content of the exhaust steam. The reduction in moisture resulting from a second reheat is readily seen in Figure 3.5. Note how point g is closer to the saturation line than point f.

An analysis of ideal Rankine cycles would predict improved cycle efficiency with each additional reheat, although the gain becomes smaller with each successive reheat. In practice, losses in the actual cycle (such as reheater pressure drop) limit the number of reheats used to two. Heat rate improvements depend on the steam conditions used, but approximate values are 5% for the first reheat with an additional 2% for the second reheat.

THERMODYNAMICS OF STEAM-ELECTRIC GENERATING PLANTS

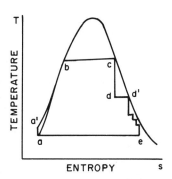

Figure 3.6. Temperature-entropy diagram for a non-reheat nuclear Rankine cycle.

The majority of fossil-fueled steam plants being installed in the United States use throttle temperatures of 1000°F (538°C), which is a superheat of 338°F (188°C) for the normal throttle pressure of 2400 psig (16.54 MPa). The first reheat is usually to 1000°F (538°C). A number of double reheat units have been built with second reheat temperatures in the range of 1000–1050°F (538–566°C).

Nuclear units normally operate with no superheat (slightly wet throttle), with all boiling water reactors (BWR), and most pressurized water reactors (PWR), to as much as 60°F (33°C) in some PWR's. The cycles can be either non-reheat or reheat. The T-S diagram for an ideal non-reheat cycle is shown in Figure 3.6. Steam at point c enters the high pressure turbine and expands to state d. The wet steam then enters an external moisture separator (MS) vessel where it is dried to the saturated vapor state at point d'. The dry steam then expands through the low pressure turbine, exhausting at e. The three horizontal jogs in the LP expansion represent internal stage moisture removal. Both internal and external moisture removal have the same purpose—to improve turbine efficiency by reducing the moisture content of the steam.

The reheat nuclear cycle is similar, except that the dried steam at d' is superheated before entering the LP turbine. Reheating to a given temperature is done in either one or two stages. In the single stage design, reheat is accomplished by heat transfer from steam extracted from the main steam line. With the two stage design, the first portion of the heating is done with extraction steam from the HP turbine, and the second part done with an extraction from the main steam line. In most nuclear designs, the reheat and moisture separator functions are performed in a combined moisture separator-reheater (MSR).

Regenerative Feedwater Heating

Carnot's analysis showed that one must increase the average temperature at which heat is added in order to improve cycle efficiency. So far, we have attempted to increase the average temperature by increasing the temperatures on the high side of the cycle, for example, through superheat and reheat. Another area where gains can be made is in the heating of feedwater from state a' (the feed pump outlet) to state b (saturated liquid) in Figures 3.4–3.6. If this could be done by an internal heat transfer, there would be a significant increase in the average temperature of heat addition with a corresponding improvement in efficiency. One way in which this might be achieved is shown in the upper part of Figure 3.7. Feedwater passes through the hollow turbine casing and picks up heat from the steam on the opposite side of the conducting partition. If one neglects the effect of the feed pump, one may show that the cycle abcd in the T-S diagram in Figure 3.7 has the same efficiency as the Carnot cycle.

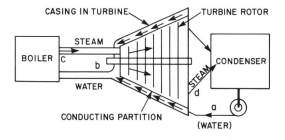

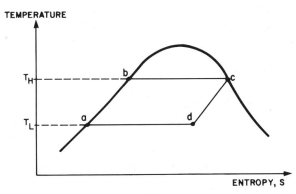

Figure 3.7. Idealized feedwater heating scheme and temperature-entropy diagram (from Faires).

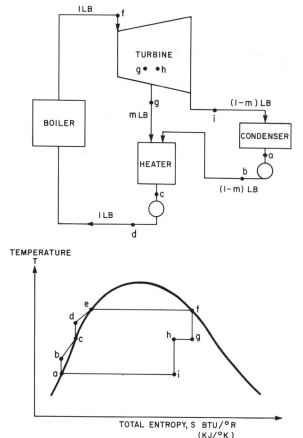

Figure 3.8. Feedwater heating and associated temperature-entropy diagram.

Obviously, one cannot build a turbine in this fashion with continuous feedwater heating. However, feedwater heating can be done in a number of discrete steps, and this will begin to approach the continuous case as the number of steps increases. An example of feedwater heating with a single heater, and the associated T-S diagram, is shown in Figure 3.8. The diagram is drawn on the basis of one pound of steam at the throttle. Steam expands through the turbine to point g, where m pounds of steam are extracted for heating. The remaining steam then flows through the turbine, condenses and then mixes with the extraction steam supplied to the feedwater heater. The pound of water that leaves the heater is then pumped back to the boiler and the cycle repeats itself. One may see that this T-S diagram begins to approximate that in Figure 3.7. As more heaters are added to the two diagrams, the similarity increases, as shown in Figure 3.9.

The effects of feedwater heating may also be explained in the following manner: for every pound (0.45 kG) of steam that enters the condenser, approximately 1000 Btu (1055 kJ) of heat is rejected from the cycle. If feedwater heating is used, the flow into the condenser drops to (1 − m) pounds, heat rejection is reduced by 1000m Btu (1055m kJ), and this amount of heat is transferred to the feedwater. In the boiler, heat input is reduced because the feedwater enters at a higher temperature and the net result is an improvement in cycle efficiency.

The amount of heat rate improvement resulting from feedwater heating will depend on the number of heaters, but it is normally in the range of 14–15%.

Stage Design

Large steam turbines are made up of a series of stages, which can number up to 20 or more depending on the steam conditions and type of stage design. Stages are composed of two rows of blades, one stationary and the other moving. The stationary blades are referred to as nozzles and moving blades are called buckets. Stage designs are broken into two categories, impulse and reaction. In the impulse design all of the stage pressure drop is taken across the nozzle, while in the reaction stage the pressure drop is evenly split between the two rows of blades. No actual stage is either pure impulse or pure reaction and most turbines have stages of both designs. In general, manufacturers will favor one design or the other and that type of stage will predominate in their design.

Stages operate as follows: Steam enters the nozzle passages and is accelerated to a high velocity; then it flows into the bucket passage where it is

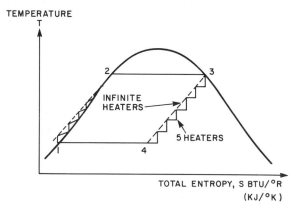

Figure 3.9. Comparison of T-S diagrams for cycles with five and infinite number of heaters.

COMBUSTION TURBINE POWER PLANTS

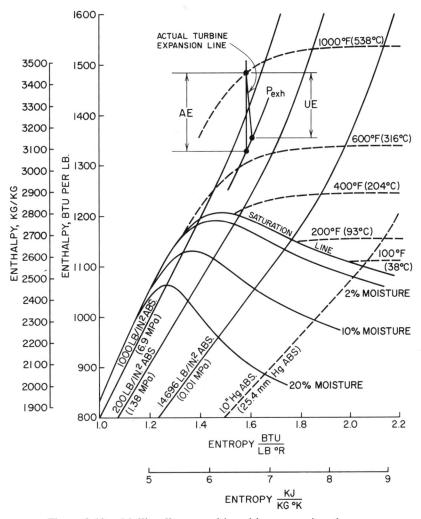

Figure 3.10. Mollier diagram with turbine expansion shown.

slowed down and gives up its energy to the bucket; finally, the energy transferred to the bucket is converted into electrical energy in the generator.

Turbine expansion lines are normally drawn on Mollier diagrams, which are plots of enthalpy versus entropy, as shown in Figure 3.10. Two quantities are shown in the figure. The first, available energy (or AE) represents the total amount of energy available in the steam from the given initial conditions down to the exhaust pressure, P_{exh}. This is the amount of energy that would be extracted by the turbine if it were 100% efficient (an isentropic expansion). Actual turbines suffer losses from friction and leakages that reduce their efficiency, so the amount of energy extracted (the used energy or UE) is always less than the AE. The ratio of the UE to the AE equals the section efficiency. In the turbine expansion shown in the figure, the efficiency is 85%.

COMBUSTION TURBINE POWER PLANTS

Combustion turbine power plants operate on the Brayton cycle in which: the working fluid is compressed; heat is added either through the transfer from an external source or through the direct combustion of fuel in the fluid; and the resulting high temperature gases expanded through a turbine. The exhaust from the turbine is either cooled in a heat exchanger, in the closed type of cycle, or vented to the atmosphere in the open cycle. Closed cycles operate on gases such as helium, while the open cycle operates on air.

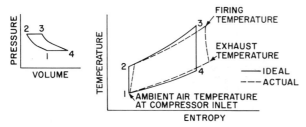

Figure 3.11. Cycle diagrams for the Brayton cycle.

The combustion turbine plant capacity for U.S. utilities is shown in Table 3.1. Combustion turbines are relatively small (most units are under 100 MW) and are used mainly for peaking power generation and for reserve capacity. Although they must burn expensive oil or natural gas, combustion turbines have low capital costs, short installation times, and rapid starting and loading capabilities, characteristics that suit it for peak operation.

The working fluid in combustion turbines is air, which is a mixture of gases: N_2, O_2, CO_2, CO, and so on. At the temperatures and pressures at which air is used in a gas turbine, it behaves very nearly as a perfect gas. As a result, the perfect gas relationships may be used to calculate the performance of the cycle.

The Brayton cycle is illustrated in the P-V and T-S diagrams shown in Figure 3.11. It is composed of the following four processes:

1–2. Compression of air.
2–3. Constant pressure heat addition or combustion.
3–4. Turbine expansion.
4–1. Heat rejection or exhaust to atmosphere.

The differences between the ideal and actual cycles are due to conditions such as turbine and compressor inefficiencies, inlet and exhaust pressure drops, and so on.

For the compression and expansion processes, an energy balance gives:

$$\text{Work} = h_{\text{in}} - h_{\text{out}}$$

where h_{in} and h_{out} are the inlet and outlet enthalpies, respectively. Note that this results in compressor work being a negative quantity, with turbine work having a positive value.

For the ideal gas:

$$h_1 - h_2 = C_P(T_1 - T_2)$$

C_P is the heat capacity at constant pressure, which is the amount of energy required to raise the temperature of a unit mass by one degree while at constant pressure. T is the absolute temperature. The units of C_P are Btu/lb–°R (kJ/kG–°K) and the units of T are °R (°K).

For an isentropic process with an ideal gas:

$$T_2 = T_1(P_2/P_1)^{\frac{\gamma-1}{\gamma}}$$

P is the pressure and γ is the ratio of specific heats, which equals 1.4 for air.

Finally, for the heat addition and rejection in the Brayton cycle:

$$Q = h_{\text{out}} - h_{\text{in}}$$
$$= C_P(T_{\text{out}} - T_{\text{in}}) \text{ for the ideal gas.}$$

Therefore, for the four processes we have:

Compressor work $= C_P T_1 (1 - (P_2/P_1)^{\frac{\gamma-1}{\gamma}})$
Heat addition $= C_P(T_3 - T_2)$
Turbine work $= C_P T_3 (1 - (P_4/P_3)^{\frac{\gamma-1}{\gamma}})$
Heat rejected $= C_P(T_4 - T_1)$

and the cycle efficiency is:

$$\eta_{\text{ideal}} = \frac{\text{Net Work}}{\text{Heat Input}}$$

$$= \frac{C_P T_3 \left(1 - \left(\frac{P_4}{P_3}\right)^{\frac{\gamma-1}{\gamma}}\right) + C_P T_1 \left(1 - \left(\frac{P_2}{P_1}\right)^{\frac{\gamma-1}{\gamma}}\right)}{C_P(T_3 - T_2)}$$

If there are no pressure drops, $P_3/P_4 = P_2/P_1$, and this equation reduces to:

$$\eta_{\text{ideal}} = 1 - \left(\frac{P_1}{P_2}\right)^{\frac{\gamma-1}{\gamma}}$$

Therefore, for the ideal cycle, the efficiency depends only on the pressure ratio of the machine.

For the more generalized case with 100% combustor efficiency and no pressure drops:

$$\eta = \frac{\eta_T T_3 - \dfrac{T_1 r_P^{\frac{\gamma-1}{\gamma}}}{\eta_c}}{T_3 - T_1 - \dfrac{T_1 \left(r_P^{\frac{\gamma-1}{\gamma}} - 1\right)}{\eta_c}} \eta_{\text{ideal}}$$

THE COSTS OF POWER GENERATION

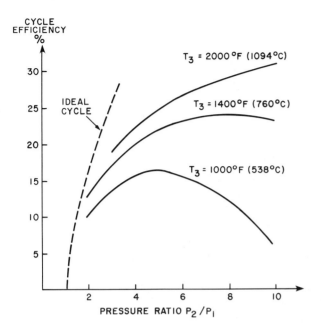

ASSUMES:
- TURBINE EFFICIENCY = 85%
- COMPRESSOR EFFICIENCY = 84%
- COMBUSTOR EFFICIENCY = 100%
- NO PRESSURE DROPS
- T_1 = 80°F (27°C)

Figure 3.12. Variation in Brayton cycle efficiency with cycle parameters.

η_T = turbine efficiency
η_c = compressor efficiency
r_P = pressure ratio, $P_2/P_1 = P_3/P_4$

Variation in efficiency with pressure ratio and firing temperature, T_3, is shown in Figure 3.12. The curves show that there is an optimum pressure ratio corresponding to each value of T_3, and that firing temperature has a significant effect on the efficiency level. Most heavy-duty combustion turbines used by U.S. utilities have values of around 2000°F (1094°C), and development work is being done on advanced turbines with temperatures approaching 3000°F (1649°C).

COMBINED CYCLE POWER PLANTS

Combustion turbines exhaust to the atmosphere at temperatures of about 1000°F (538°C), while the temperature in a condensing steam turbine is more nearly 100°F (38°C). As Carnot demonstrated, the way to improved efficiency is through higher source temperatures and lower sink temperatures. If a combustion turbine were to be used in conjunction with a condensing steam turbine, the resulting combined cycle would have the high combustion turbine firing temperature along with the low steam turbine exhaust temperature. A superior heat rate will result.

In a combined cycle plant, the exhaust from the combustion turbine passes through a Heat Recovery Steam Generator (HRSG), where the exhaust heat is used to produce steam. The steam then flows through a nonreheat steam turbine. Sizes of the plants range up to about 400 MW. All plants use a single steam turbine with from one to four combustion turbines, depending on the rated plant output.

Combustion turbines and heat-recovery steam generators may be used to replace old boilers in an application called repowering.

THE COSTS OF POWER GENERATION

The costs of power generation are usually separated into three categories (1) the capital or investment costs, (2) the fixed operating costs, and (3) the variable operating costs.

In the category of capital costs are all the land, equipment, construction, and interest during construction. The actual level of these capital costs is dependent on many factors and constraints, such as system reliability, reserve and energy needs, as well as air quality, water quality, noise level, and other environmental requirements. Capital costs are usually stated in millions of dollars or in dollars per kilowatt.

Capital costs are usually related to the utility's cost of doing business through the "fixed-charge rate," or revenue required to support a project. This is the ratio of the required annual revenue that must be obtained by the utility to support these capital-dependent costs to the installed cost of that equipment.

The fixed operating cost includes all those noncapital expenditures that result from the commitment to operate, but are largely independent of the amount of use, such as the fixed part of operation and maintenance (O & M).

The variable operating cost is for those expenditures, required to operate, which are directly dependent on the unit's use. Included in this category are

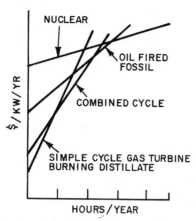

Figure 3.13. Relative power generation costs.

the variable part of O & M, and the fuel cost for a power plant.

In expressing costs or making an economic study, many factors must be known or assumed. For a general study, these factors must be assumed. Figure 3.13 shows the cost relationships between several different kinds of power generation.

POWER GENERATION RELIABILITY

The best available measure of reliability of generating units is the forced outage rate (FOR). The Edison Electric Institute, which collects and publishes outage data, defines FOR as follows:

$$\text{FOR} = \frac{\text{Forced Outage Hours (FOH)}}{\text{FOH} + \text{Service Hours}}.$$

In system reliability studies, FOR is taken as the best estimate of the probability that a generating unit will not be available to serve the load when needed. The impact of the forced outage rate on system design can be quite substantial as measured by reserve requirements. The amount of reserve required by a particular unit varies from one to five times the FOR. This depends upon unit size relative to system capacity.

Forced outage rates vary according to the type of prime mover, the unit size, and the age of the unit. In addition, it is usually expected that prototype designs will have higher than normal forced outage rates. Considering variations of all these parameters, values of FOR can be anywhere between 1 and 15%.

POWER PLANT SITING

Utilities today face critical problems when selecting a site for a power plant. Among the concerns that must be dealt with are the following:

1. Selecting a plant size that meets system load growth and reliability criteria.
2. Selecting the type of plant to be constructed (e.g., nuclear, coal-fired, or combustion turbine) based on an assessment of projected fuel costs, O & M expenses, and capital costs.
3. An assessment of any impact on the environment and demonstrating compliance with environmental laws and regulations.
4. Obtaining approval for a plant site.

Environmental factors now have a considerable influence on generation economics and site selection. For example, cooling towers, which are used to reduce the thermal discharge to a body of water, both add investment cost and reduce plant efficiency.

The burning of coal, oil, or gas causes some air pollution. All three release nitrogen oxides. Coal and oil release sulfur oxides, and coal releases some particulates. These emissions must be controlled to levels set by environmental regulation. The equipment necessary to control emission increases plant investment and operating cost.

The dispersion of the remaining emission into the air surrounding the plant is added to the emissions of other sources in the vicinity. These total emissions must also conform to the applicable environmental regulations.

Air Quality

The Clean Air Act of 1970 gave the Environmental Protection Agency its charter to establish ambient air quality standards for the United States, as well as provide emission standards as guidelines to achieve this ambient air quality. While the federal government provides the standards, the states provide enforcement. The National Ambient Air Quality Standards (NAAQS) are shown in Table 3.3.

Areas that are presently above the NAAQS must reduce emissions to achieve the NAAQS. Areas below are permitted to have an increase in concentration of pollutants only to an increment specified by the August 1977 Clean Air Amendments. This permitted amount of increase is called the prevention

TABLE 3.3. National Ambient Air Quality

Pollutant	Primary[a]	Secondary[b]
Sulfur oxides (sulfur dioxide)		
Annual arithmetic mean concentration	80 $\mu g/m^3$ (0.03 ppm)	60 $\mu g/m^3$ (0.02 ppm)
Maximum 24-hr concentration[c]	365 $\mu g/m^3$ (0.14 ppm)	260 $\mu g/m^3$ (0.1 ppm)
Maximum 3-hr concentration[c]		1300 $\mu g/m^3$ (0.5 ppm)
Particulate matter		
Annual geometric mean concentration	75 $\mu g/m^3$	60 $\mu g/m^3$
Maximum 24-hr concentration[c]	260 $\mu g/m^3$	150 $\mu g/m^3$
Carbon monoxide		
Maximum 8-hr concentration[c]	10 mg/m^3 (9 ppm)	10 mg/m^3
Maximum 1-hr concentration[c]	40 mg/m^3 (35 ppm)	40 mg/m^3
Ozone		
Maximum 1-hr concentration[c]	235 $\mu g/m^3$ (0.12 ppm)	235 $\mu g/m^3$
Lead		
Average 3 months	1.5 mg/m^3	1.5 mg/m^3
Nitrogen dioxide		
Annual arithmetic mean	100 $\mu g/m^3$ (0.05 ppm)	100 $\mu g/m^3$

SOURCE. Title 40 of the Code of Federal Regulations, Part 50.

[a] To protect the public health, figures at standard temperature and pressure.
[b] To protect the public welfare, figures at standard temperature and pressure.
[c] Not to be exceeded more than once per year.

of significant deterioration (PSD) increment and varies depending on the degree of industrial development desired by each state for each air quality control area.

Control of Pollutants

Emission of pollutants is controlled by fuel selection (e.g., low-sulfur oil), by postcombustion cleanup procedures, and, in some cases, by coal washing. Sulfur dioxide removal processes are in varying stages of development, with some, such as the wet limestone scrubbing process, commercially available. Control of nitrogen oxides has been achieved in combustion turbines with new combustor designs and with water injection to reduce flame temperature. NO_x control on large fossil-fueled boilers requires modification of combustion techniques and furnace redesign. Reduction of particulate emission can be achieved by electrostatic precipitators or fabric filters.

Thermal Dispersion

The total energy release (waste heat) from all sources in the northeastern United States, which has a relatively high concentration of energy production, is only approximately 1% of the absorbed solar energy and is projected to be about 5% at the beginning of the 21st century.

The concern is the effect of condenser thermal discharge on aquatic ecology. Water temperature standards have been adopted by some states and federal guidelines were included in the Water Quality Act of 1965. The standards are written in terms of maximum temperatures and allowable temperature rise for various types of receiving waters. Unfortunately, some of these criteria can be exceeded on some hot summer days without man-made temperature discharge, and some standards only refer to surface temperature.

Electric power plant cooling systems are classified as to whether heat-sink water is available directly to the condenser cooling intake, or whether the condenser cooling water is recirculated and the heat removed by some secondary heat exchange. The former system is called once-through and the latter is a closed cycle. Closed-cycle heat exchange systems may involve combinations of cooling ponds, spray ponds, mechanical and natural draft evaporative, or "wet" cooling towers. Wet cooling towers require make-up water to recover losses from evaporation, water droplets in the form of spray carried away by cooling air (called "drift" or "carryover"), and "blowdown" water used to prevent excessive deposits of minerals left by evaporation. Dry cooling towers are not advantageous from the aspect of having the highest cost, largest size, and most reduction of plant efficiency. Operating costs are a significant additional cost in the case of mechanical draft towers because they contain motor driven fans.

Water Discharges

Chemical waste water, total suspended solids in waste water, oil and grease discharges are also regulated. Specified discharge concentrations in milligrams per liter must be achieved. These concentrations are established by the Environmental Protection Agency under the Clean Water Act of 1977. Likewise, waste disposal is regulated by EPA pursuant to the Resource Conservation and Recovery Act and Toxic Substances Control Act of 1978.

Radioactivity and Reactor Safety

Nuclear power plants have none of the atmospheric pollution problems resulting from fuel combustion. The question of the effect of radiation caused by nuclear fission has received considerable public attention. It is usually part of the opposition to proposed nuclear plant sites, as well as to development of the breeder reactor. Many radiologists believe that the on-site radiation level standards issued by the Atomic Energy Commission* in June 1971 have been adequately supported by currently available data, and future study will conclude that the standards are actually conservative.

Current data supports the premise that the dosage from a nuclear power plant is far below what is judged as harmful and even far below the natural background radiation level. For example, cosmic radiation at sea level is equal to 40 mrem per year, and the average U.S. exposure to radiation through food, water and air is 25 mrem. In total, the average annual dose of radiation in the United States is 225 mrem. Meanwhile, exposure to radiation at the site boundary of a nuclear plant amounts to less than 5 mrem per year.

The units of radiation used here are millirems. A rem which is an acronym for Roentgen-equivalent-man, is a term for the per-unit energy absorbed per gram of exposed material, assuming a base of 100 ergs per gram, multiplied by the relative biological effect of a particular form of radiation. The base of

---

* Now the Nuclear Regulatory Commission (NRC).

relative biological effect is the biological effect from 100 ergs per gram of gamma radiation. In equation form,

$$\text{Rem} = \frac{\text{Ergs of given radiation type/gram}}{100 \text{ ergs/gram}} \times \text{relative biological effect}.$$

The following is a tabulation of relative biological effects.

Radiation	Relative Biological Effect
X-rays	1.
Gamma	1.
1 Mev Beta	1.
Thermal neutrons	5.
Fast neutrons	10.

Nuclear plants are designed to prevent accidental radiation release and contain any accidental radiation release in the extremely unlikely event an accident were to occur. The opinion has been expressed that no other technology has received comparable design for safety.

SUMMARY

The nuclear power plant is now recognized as a member of the power generation family that includes conventional coal, oil or gas-fired boiler steam plants, hydroelectric plants, combustion-turbine plants, internal-combustion engine plants, geothermal plants, and combined cycle plants. Each of these has different characteristics in the areas of performance, capital cost, generation cost, etc., but all have their place in the modern utility.

Although energy technology can provide answers to the questions "How can we produce electrical energy?" and "What will be its costs?", questions such as "What energy sources are acceptable?", "To what extent should we use these sources?", and "In what manner and where should these sources be located?" also need to be addressed. Answers to these question, and others, are possible, and are to be found from a number of different sources. The information that follows is not intended to answer all these questions, but rather to provide information on the energy technology aspects of this problem.

REFERENCES

1. Creager, W. P., and Justin, J. D., *Hydroelectric Handbook,* Wiley, New York, 1950.
2. Doolittle, J. S., and Zerban, A. H., *Engineering Thermodynamics,* International Textbook Co., Scranton, Pennsylvania, 1948.
3. Dusinberre, G. M., *Gas Turbine Power,* International Textbook Co., Scranton, Pennsylvania, 1952.
4. Faires, V. M., *Thermodynamics,* MacMillan, New York, N.Y. 1971.
5. Ferrar, T. A., Clemente, F., Uhler, R. G., *Electric Energy Policy Issues,* Ann Arbor Science Publishers, Ann Arbor, Michigan, 1979.
6. "Implications of Environmental Regulations for Energy Production and Consumption," Volume VI of *Analytical Studies for the U.S. Environmental Protection Agency,* National Academy of Sciences, Washington, D.C., 1977.
7. LaMarsh, J. R., *Introduction to Nuclear Engineering,* Addison-Wesley, Reading, Massachusetts, 1975.
8. Macklis, S. L., "Wind Turbines for Electric Utility Power Generation," *Electric Forum* (Magazine published by General Electric Company in Schenectady, New York), pp. 35–39.
9. Marsh, W. D., *Economics of Electric Utility Power Generation,* Oxford, New York, 1980.
10. *Sci. Am.* **224,** (3), a special issue devoted to energy.
11. Starr, C., "Choosing Our Energy Future," *EPRI Journal,* September, 1980.
12. "Steam Turbines," *Power Magazine* Special Report, June, 1962.
13. The Babcock & Wilcox Company, *Steam. Its Generation and Use*. The Babcock & Wilcox Company, New York, 1975.
14. *United States Energy Data Book,* GEZ7049C, General Electric Co., Fairfield, Connecticut, 1982.

4
POWER FROM STEAM AND COMBUSTION TURBINES

**THERMAL ELECTRIC
GENERATING STATIONS**

The energy delivered to an electric power system is generally referred to as "generation" and is measured in kilowatt-hours (kWh). Thermal generation, i.e., the energy produced by prime movers that are thermally powered (for example, steam turbines) produce most of the electricity used in the United States. Table 4.1 shows the percentage generated by the type of prime mover in the United States in 1977.

The essential elements of a fossil-fuel steam plant are shown in Figure 4.1. Fuel is burned in a furnace, providing heat. Pressurized water entering the boiler is converted to steam, which expands through the turbine causing it to rotate. The turbine, in turn, drives the electric generator.

Low-pressure steam exhausting from the turbine enters the condenser, where it is condensed back into water. Finally, this condensate is pumped back into the boiler and the cycle repeats itself.

To sustain the cycle, the following must be supplied continuously:

1. Fuel.
2. Air (oxygen). Note also that it is necessary to dispose of the combustion gases from the burning of the fuel.
3. A supply of cooling water for the condenser. This can be taken from a river, ocean, or lake. Cooling towers are commonly used today.
4. A supply of clean makeup water to keep the amount of water in the cycle constant, since some loss of steam is encountered in specific locations, e.g., boiler blowdown.

Simple power plants such as the one shown in Figure 4.1 are not used by utilities for power generation because there are a number of modifications that can be made to improve the performance of the cycle substantially. These are shown in the diagram in Figure 4.2. They are (1) the economizer and air preheater, which reduces the amount of heat lost up the stack, (2) the feedwater heaters, which increase the temperature of the feedwater entering the boiler and reduce the amount of heat rejected by the condenser, and (3) the reheater and superheater, which increase the efficiency of the turbine.

Figure 4.3 shows one possible arrangement of the major components in a coal-fired power plant.

Modern power stations must meet stringent requirements governing the amount of heat that can be rejected to surrounding bodies of water as well as the emissions of various pollutants. In addition, in light of the uncertainty in the availability of natural gas and the desire to reduce the consumption of oil, it is likely that most new fossil-fired plants will use coal as a fuel. A heat balance for 717-MW net output coal-fired plant with steam conditions of 2400, psig/1000°F/1000°F (16,7 MPa/538°C/538°C) is shown in Figure 4.4. (Note Figure 4.4a gives steam conditions in English units and Figure 4.4b gives steam conditions in SI units.) Wet scrubbers and electrostatic precipitators are used to reduce pollution emissions and cooling towers are used to reduce heat rejection to local bodies of water. This

PRODUCTION OF STEAM

TABLE 4.1. Generation by Type of Prime Mover in the United States

Prime Mover Type	Percentage
Total = 2,247,372 × 10⁶ kWh	100
Hydro	12.4
Fossil fuel—steam	74.8
Nuclear fuel—steam	11.4
Combustion turbines	1.2
Internal combustion—diesels	0.2

SOURCE. *Energy Data Report,* June 1980, U.S. Department of Energy, DOE/EIA-0049 (79).

heat balance shows various flows, pressures, and temperatures in the cycle, as well as some of the auxiliary power requirements.

The majority of the heat rejected from the cycle is lost through the condenser. These losses increase as the turbine exhaust pressure rises. Since cooling towers produce a higher exhaust pressure than a simple once-through condenser design, there is a loss in efficiency with their use.

Stack losses represent the second largest heat loss from the power plant. Exhaust gases from the boiler must be maintained at a temperature high enough to avoid condensation of the water vapor generated during combustion. Condensed water reacts with combustion gases to form acids that will attack the exhaust ducting and stack. Note that in the cycle of Figure 4.4, the exhaust gases are sufficiently cooled in the scrubbers so that steam must be extracted from the turbine to reheat them above the dew point. Stack gas reheating significantly increases the stack losses for this type of cycle.

INCREASE IN SIZE OF UNITS

History has shown that the size of new individual units purchased and installed tends to increase with time. This is motivated by economic considerations and larger-size systems. The equipment cost, the construction labor cost, the site cost, and operating labor cost for a large unit is less per kilowatt than for a smaller unit. Finding a suitable site for generating facilities becomes more difficult every year. Consequently, the available sites must be used as effectively as possible, which dictates larger units. The continued growth of load also tends to increase the size of units required to meet future loads. Units now being installed in the United States range from approximately 200 MW to 1300 MW, with the average unit being about 600 MW.

PRODUCTION OF STEAM

Fuel Handling

Coal-burning stations usually require extensive coal storage and handling equipment. Coal is first passed

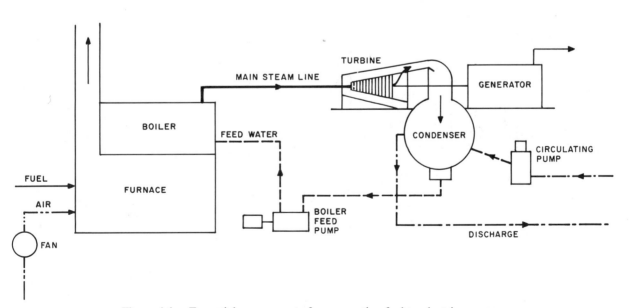

Figure 4.1. Essential components for converting fuel to electric energy.

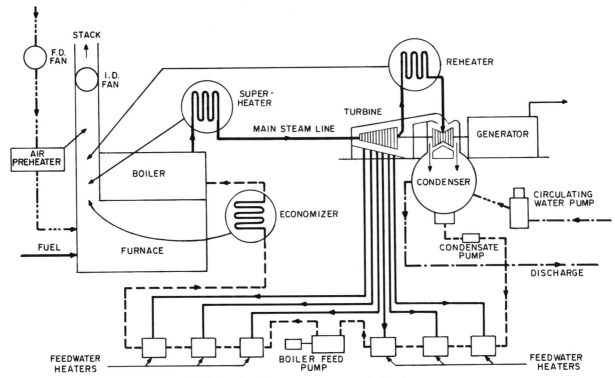

Figure 4.2. Additional major components needed to improve economy.

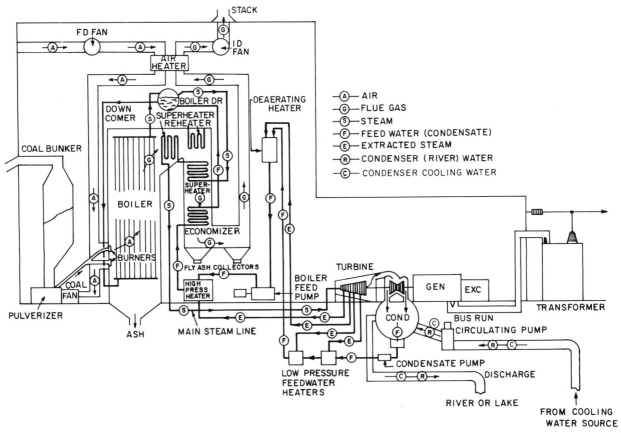

Figure 4.3. Physical arrangement of major components and flow patterns in a typical steam power station.

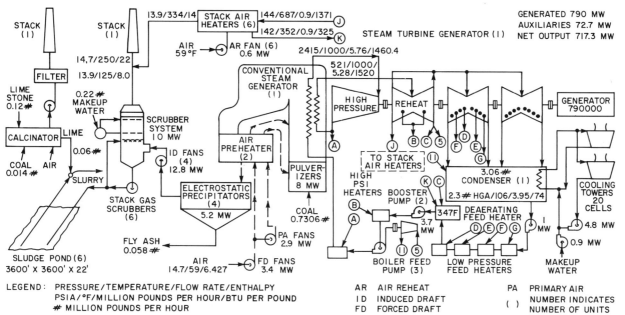

Figure 4.4.(a) Conventional steam cycle with wet flue gas desulfurization: English units. (For SI units see page 48.)

through a crusher to reduce its size to about 3 inches (7.5 cm) or less, and then is stored outdoors until required. Belt conveyors carry the coal from the point of unloading to the crusher house, from the crusher house to the storage pile, and from there to the station. In most installations, coal is pulverized as required; there is no provision for storage of pulverized coal. Coal is supplied to the pulverizers by variable-speed coal feeders controlled in response to changes in load demand. From the pulverizer, the coal is carried to the burners in the primary combustion air stream. Sufficient air is admitted in a secondary air stream to ensure complete combustion of the fuel.

In the case of oil-burning stations, the equipment required for fuel handling is much simpler. Most stations use a heavy-grade fuel oil which requires preheating to about 200°F (93.3°C) to reduce the viscosity before admission to the burners. For satisfactory combustion, the oil is atomized in the burners either mechanically or by steam. Miscellaneous fuel handling and fuel supply pumps are required, but their total power consumption is quite small.

Gas-burning plants have almost no fuel-handling equipment; only pressure-regulating valves and flow-measuring equipment are required. Since the availability of gas as a fuel for the generation of power has been restricted in many areas, many gas-burning plants also incorporate the equipment necessary to burn fuel oil.

Steam Supply: Fossil-Fired Boilers

In large steam generators, all of the steaming surface is in the waterwalls; and nearly all of the heat received by the waterwalls is transferred there by radiation. The steam is superheated in superheaters located in the exit gas path from the furnace. Reheaters are usually located between sections of the primary superheater. Steam temperature at the superheater outlet can be held within close limits over a load range from about 60% to 100% by the use of differential firing, tilting burners, or attemperation.

Steam generator efficiencies are in the range of about 82–90%, depending on the size of units, the steam conditions, and the type of fuel used. The largest heat loss in steam generators is that from flue gases. This loss can be reduced by the use of an economizer and air preheaters.

Nuclear Steam Supply System

Nuclear reactors provide an economic alternative to boilers using fossil fuels to supply steam for power generation. In the United States, light-water reactors (LWRs) predominate, although other reactor types are or have been used. Light-water reactors are so named because they use ordinary water, as opposed to heavy water (water in which the hydrogen isotope has a neutron as well as a proton), to slow down the fast neutrons produced in the nu-

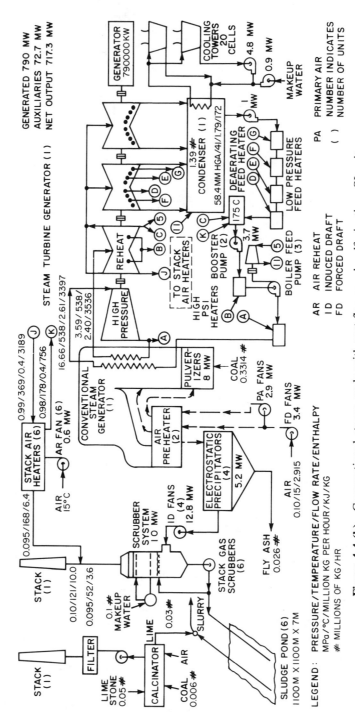

Figure 4.4.(b) Conventional steam cycle with wet flue gas desulfurization: SI units.

PRODUCTION OF STEAM

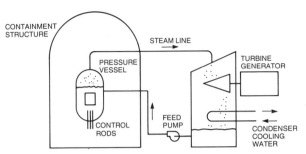

Figure 4.5. Principal elements of a boiling water reactor (BWR) power plant.

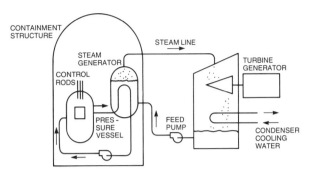

Figure 4.6. Principal elements of a pressurized water reactor (PWR) power plant.

clear reaction. The water is also used to carry away the heat generated, thereby cooling the reactor and generating steam.

There are two prevalent light-water reactor types used. They are the boiling-water reactor (BWR) and the pressurized-water reactor (PWR). Figures 4.5 and 4.6 illustrate the principal elements of the two types of plants. In the BWR, shown in Figure 4.5, a single loop is involved. Feedwater from the condenser is pumped, by the feed pumps, into the reactor pressure vessel where steam is generated and fed directly to the turbine. In the schematic of the PWR, shown in Figure 4.6, two loops in series are involved. Water at high pressure is pumped through the reactor pressure vessel and the steam generators by the reactor coolant pump in the primary loop. Feedwater from the condenser is pumped into the steam generators (large heat exchangers) and then to the turbine as steam by the feedwater pump in the secondary loop. Both LWR types supply steam to the turbine at or near saturated conditions and approximately 1000 psia (6.9 MPa).

The boiling water reactor system is shown in greater detail in Figure 4.7.

Among the major subsystems required in the BWR nuclear steam supply are:

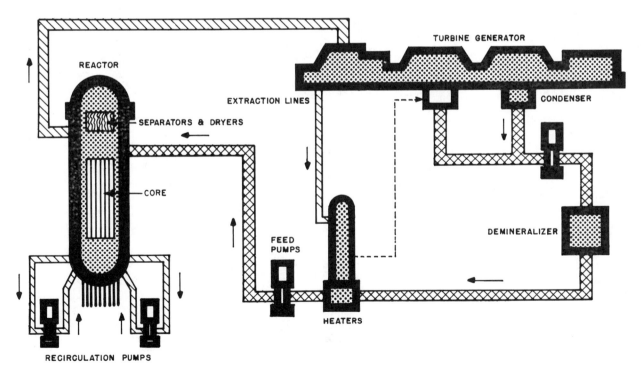

Figure 4.7. Boiling water reactor (BWR) power plant.

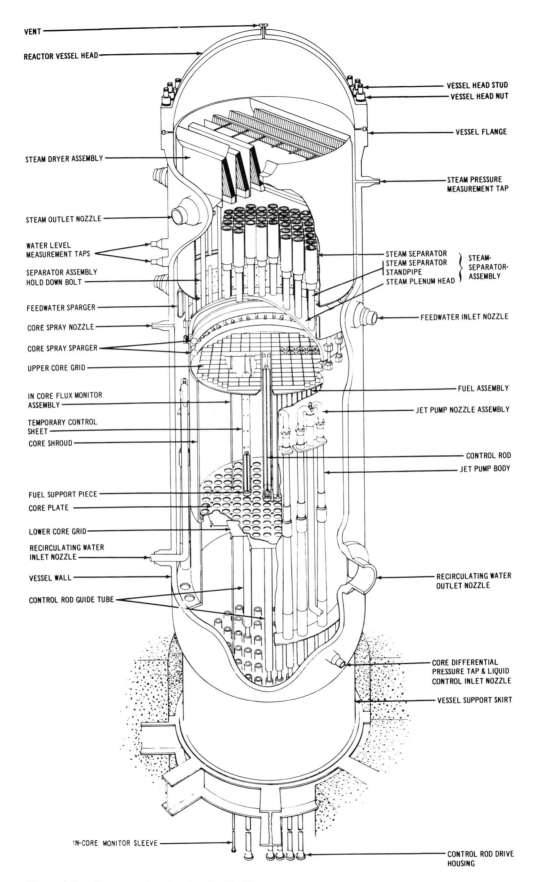

Figure 4.8. Cutaway showing detail of boiling water reactor (BWR) arrangement. In-core monitor sleeve

TURBINE

1. The reactor vessel and its internal equipment
2. The control rod drive system
3. The reactor water recirculation system
4. The reactor protection system
5. The reactor auxiliary systems
6. The radioactive waste treatment systems.

The reactor vessel supports and contains the nuclear fuel (or reactor core), and supplies the necessary flow paths for water entering the core and steam and water leaving the core. The vessel and its internal equipment are shown in Figure 4.8.

The principal elements of the BWR are shown in Figure 4.9. Feedwater enters the vessel at the top of the core and, joined by recirculation water, is driven downward around the core by the jet pumps. Part of the recirculation flow, which affords the primary control of reactivity, is fed from the vessel to external pumps, which return it into the jet pumps. These induce the main core flow. The generated steam and water mixture passes through the internal steam separators mounted over the core. The steam is dried in the vertical steam dryers, and then goes from the reactor vessel to the steam turbine. The core of fuel assemblies is centrally located in the reactor vessel.

TURBINE

In functional terms, the turbine is a device used to convert the stored energy of high-pressure and -temperature steam into rotational energy, which may then be converted into electrical energy in the generator. The steam is admitted to the turbine and passes through a series of stages, each of which consists of a set of stationary blades, called nozzle partitions, and moving blades attached to the rotor, called buckets. In the nozzles (the passages formed by the nozzle partitions), the steam is accelerated to high velocity, and this kinetic energy is converted into shaft work in the buckets.

Virtually all of the steam turbines used by the electric utilities are of the condensing type in which the steam is expanded to subatmospheric pressures, condensed, and returned to the boiler to repeat the cycle. For power systems operating at 60 Hz, fossil-fired units operate at 3600 r/min and nuclear units at 1800 r/min. On 50 Hz systems, these units operate at 3000 r/min and 1500 r/min, respectively.

Exhaust Loss

An important consideration in turbine design is the exhaust loss. This loss has two components: (1) hood loss, which is the loss in energy in the turbine due to the pressure drop from the exhaust annulus to the condenser and (2) leaving loss, which is the loss due to the kinetic energy of the steam as it exhausts from the turbine. The leaving loss is proportional to the square of the exhaust velocity and it becomes the predominant loss as the velocity is increased. Exhaust velocity is a function of three variables: (1) turbine exhaust annulus area, (2) exhaust flow, and (3) exhaust pressure.

Exhaust flow is determined by the unit rating, steam conditions, and feedwater cycle, while the exhaust pressure is fixed by the cooling scheme used as well as temperature of the cooling medium. As a result, the turbine designer is limited to changes in annulus area if he wishes to reduce the exhaust loss. The annulus area of a turbine is fixed by the maximum length of the last stage bucket that can be built to withstand the combined centrifugal and axial force resulting from the high rotor speed and steam flow. With the large steam flows common in modern steam turbines, multiple low-pressure turbines are used to obtain sufficient exhaust area to keep the exhaust loss to tolerable levels. The optimum annulus area can be determined only by balancing the additional cost of larger turbine and physical plant size against the fuel savings realized from a more efficient unit.

Various types of units are built for different unit sizes and steam conditions. A brief description of some of the more common types follows.

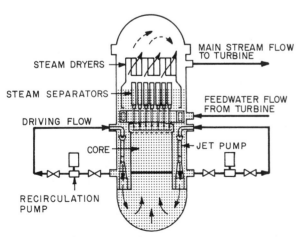

Figure 4.9. Principal elements of the boiling water reaction (BWR).

Compound Turbines

A compound turbine is one in which two or more turbine casings are coupled in series on one shaft (tandem-compound) or on two shafts (cross-compound).

Nuclear Turbines. Nuclear units are all of the tandem-compound design. Typical units larger than about 550 MW consist of one double-flow high-pressure (HP) turbine connected to two or three double-flow low-pressure (LP) turbines. After the steam expands through the HP turbine, it is sent to a moisture separator for removal of the water. Depending on the cycle, the steam may also be reheated before entering the LP sections. The purpose of moisture separation and reheating is to lower the moisture content of the steam in the LP turbines thereby reducing moisture losses and erosion rates.

Fossil Units. Fossil units can be of the tandem- or cross-compound design. These units normally consist of three sections: high pressure (HP), intermediate pressure (IP), and low pressure (LP). Steam reheating takes places between the HP and IP sections. These units are built in a wide variety of configurations which are described in the following sections.

Tandem Compounds. These units are built with either one, two, or three double-flow LP sections. In the smaller sizes, the HP and IP sections are housed in a common casing. These are known as opposed flow units since the steam flows in the HP and IP sections are in opposite directions. Larger units use single-flow HP and double-flow IP turbines in separate casing. There is some overlap in the ratings of units of these different design, but the general size progression is as follows:

1. Opposed-flow HP-IP with two LP ends (two flow).
2. Opposed-flow HP-IP with four LP ends (four flow).
3. Single-flow HP, double-flow IP with four LP ends (four flow).
4. Single-flow HP, double-flow IP with six LP ends (six flow).

Cross Compound. These units consist of two separate shafts, each connected to a generator. They are sold, designed and operated as a single unit with single control. Two basic designs are used, one in which both shafts rotate at 3600 r/min and the other in which one shaft rotates at 3600 r/min and the second at 1800 r/min. With the 3600/3600 r/min configuration, the HP, IP, and LP sections can be located on either shaft, but with the 3600/1800 r/min design, the HP and IP sections are on the high-speed shaft and the LPs are on the 1800 r/min shaft.

The cross-compound designs were utilized to obtain greater capacity than possible on a single shaft. Conductor-cooled generators were introduced during the last half of the 1950s. The resulting increase in generator capacity eliminated the need for two shafts for large power rating. Another advantage of the cross-compound units was a better heat rate or higher efficiency. Higher steam conditions were introduced in the 1950s and 1960s. The cross-compound designs were applied to keep the higher pressures and temperatures in the smaller 3600 r/min high-pressure turbine and obtain a better heat rate with the larger last-stage annulus areas available at 1800 r/min. Of course, the cross-compound units are more expensive. The higher investments are penalized when interest rates increase. Cross-compound units are seldom purchased now and most units being placed in service are of the tandem-compound design at 3600 r/min for fossil steam units and 1800 r/min for nuclear units.

CONDENSER

The function of the condenser and its associated hardware is threefold:

1. To produce a vacuum at the turbine exhaust.
2. To condense turbine exhaust steam for reuse in the closed system.
3. To deaerate the condenser to maintain a high vacuum and reduce the amount of dissolved gases carried into the boiler in the feedwater.

Approximately 30% of the turbine throttle flow is extracted for feedwater heating. Another 15% is extracted for stack gas reheating if the cycle is equipped with scrubbers. The remaining flow enters the condenser where the latent heat of vaporization of the steam is rejected to the surrounding environment. This heat represents the largest loss from the cycle. Typically, about 60% of fuel BTU (kJ) input is rejected in the condenser.

AUXILIARIES

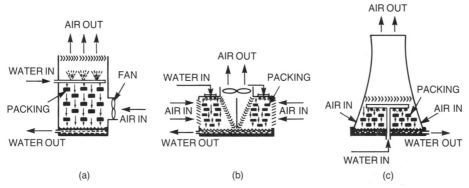

Figure 4.10. Cooling tower types: (*a*) forced draft; (*b*) cross-flow induced draft; (*c*) hyperbolic natural draft.

Noncondensible gases are removed from the condenser through the use of either mechanical vacuum pumps or multistage steam-jet air ejectors. This deaerating equipment is also used to establish vacuum when the unit is started.

HEAT REJECTION

In the past, condenser cooling water was usually taken from nearby bodies of water such as rivers, lakes, or the ocean. After passing through the condenser and picking up the heat rejected from the cycle, the warmed water was returned to its source. Today, in many instances, environmental considerations prohibit the use of such cooling methods. Other schemes, such as the use of man-made cooling ponds, canals, or cooling towers enable the electric utilities to meet restrictions on the amount of heat that can be rejected to natural bodies of water.

Cooling towers can be of three basic types: forced draft, induced draft, or natural draft. Forced- and induced-draft towers use fans to blow or draw, respectively, the cooling air through the tower. Natural-draft towers rely on a natural chimney effect to circulate the cooling air. Cross-sections of the three types are shown in Figure 4.10. They operate as follows: the water to be cooled is sprayed into the tower, splashed down over a series of baffles, and mixed with the throughflow of air. Cooling is achieved by the evaporation of a portion of the water and by the transfer of heat to the air. The water that drains to the bottom of the tower is then recirculated to the condenser. Because of evaporation and drift losses (some water droplets are carried off by the air flow), a source of makeup water must be available.

In locations where water is scarce, it may be necessary to use air cooled condensers. In this design, air, rather than water, is used as the cooling medium. A large number of fans is needed to circulate the cooling air.

In addition to the increased capital cost for the construction of the towers, there are other economic penalties that result from their use. The exhaust pressure obtained is higher than if a once-through cooling system were used, and this adversely affects the efficiency of the turbine. In the case of mechanical draft and air-cooled condensers, fan power adds to the units auxiliary requirements. Finally, turbines that use air-cooled condensers must be of a special design to allow them to operate with the very high-exhaust pressures that occur during the summer as well as the lower pressures that exist during cooler weather.

AUXILIARIES

Auxiliary functions are required to operate a steam power plant. The major auxiliary functions for a fossil-fueled plant include the feedwater system, combustion air and exhaust gas systems, condenser cooling water system, and fuel preparation and handling systems. The major auxiliary drives for these functions include boiler feed pumps (Figure 4.11), condensate and booster pumps, forced-draft fans (Figure 4.12), primary air fans, induced-draft fans, gas recirculating fans, condenser circulating water pumps (Figure 4.13); and coal crushers, and pulverizers for coal-fired plants. Other plant auxiliaries include air compressors, service and cooling water systems, lighting and heating systems, and coal handling systems. Auxiliary drives are usually pow-

Figure 4.11. Induction motors, in tandem drive, driving boiler feed pump.

ered by electric motors, with the large feed pumps and some fan drives powered by mechanical-drive turbines (Figure 4.14), using steam extracted from the main turbine. In a fossil-fuel-fired plant, the auxiliaries requiring the most horsepower are the boiler feed pumps and the draft fans.

The required auxiliary power expressed as a percentage of generating unit capacity has increased significantly in recent years due to the addition of pollution abatement equipment required to reduce the environmental impact of power plants. Air quality control equipment, such as electrostatic precipitators, baghouse filters, and flue gas desulfurization scrubbers, are often applied to new fossil-fueled plants, and have been recently added to many existing plants. Natural draft or mechanical draft cooling towers, spray ponds, cooling lakes, or cooling canals are added to both fossil and nuclear plants where thermal discharge regulations require auxiliary cooling.

Figure 4.12. Induction motor driving forced-draft fan.

AUXILIARIES

Figure 4.13. Vertical induction motors driving pumps for condenser circulating water.

Figure 4.14. Electric utility boiler feed pump turbine.

FEEDWATER SYSTEM

The condensed steam is removed from the condenser by condensate pumps. Condensate pumps are frequently known as "hotwell" pumps. These pumps operate with a high vacuum on the inlet. The discharge pressure must be sufficient to overcome the friction in the piping and low-pressure feedwater heaters, and any static lift required to pump the condensate to the level of the low-pressure heaters, particularly the deaerating heater. Some of the pumping power required to pump the feedwater to the boiler feed pump can be provided by booster pumps located between the condensate pumps and the boiler feed pumps.

The feedwater is usually heated in several stages, using steam extracted from the main turbine. The number of heaters in the feedwater string varies with the steam conditions and plant usage; a plant intended for use as a peaking plant and employing relatively low steam pressures and temperatures may have as few as one heater, while a plant intended for base loading and using high-pressure steam at 1000°F (538°C) may have seven or more heaters. Optimization of the feedwater string involves consideration of both the thermodynamics of the cycle and the costs incurred for the feedwater heaters and associated piping, balanced against the fuel savings resulting from a more efficient cycle. Feedwater heaters, with the exception of the deaerating heater, are of the closed type, where the steam extracted from the turbine does not mix with the feedwater passing through the heater, but is condensed and cascaded to lower heaters or added to the feedwater system at some point between the heater and the condenser. In the deaerating heater, the extraction steam mixes directly with the feedwater. The deaerating heater serves the double purpose of heating the feedwater and removing the noncondensable vapor entrained in the system.

The average steam plant requires some small amount of water to be added to the feedwater system to compensate for water lost due to blow-down in drum-type boilers and through leaks in pump seals, valves, turbine packings, and drains throughout the system. This water, termed makeup, may be 1–2% of the water in the system. The makeup is provided by demineralizers which give water of the required purity, and this water is admitted to the cycle in the condenser hotwell.

Main boiler feed pumps are usually multistage,

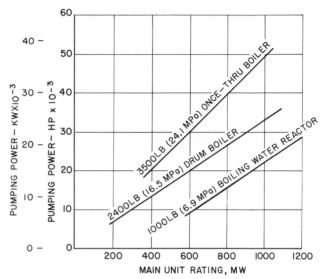

Figure 4.15. Typical boiler feed pump and reactor feed pump power requirements.

high-speed, centrifugal pumps. The boiler feed pump has to overcome boiler pressure, plus the friction in the high-pressure feedwater heaters, piping, and the economizer. These friction losses vary with flow, which varies with unit load. Flow control is achieved by varying the pump speed or by throttling the pump discharge. At full load on the turbine, the output pressure of the feed pumps must be about 25% above the boiler pressure. Since boiler feed pumps are essential auxiliaries, spare pump capacity is often provided to ensure that lost pump capacity will not result in decreased load capability for the turbine.

The power required by the boiler feed pumps depends on both the size of the unit and the steam pressure. The approximate power required is shown in Figure 4.15.

The boiler feed pumps are the largest auxiliary drives in terms of power required and energy consumed. This fact, combined with the importance of the boiler feed pump to the continuous operation of the plant, demands that the selection of a drive configuration involve careful scrutiny of the available options.

The pumps may be driven by one or more, or combinations of, squirrel-cage induction motors, mechanical-drive steam turbines, or gear drive from the shaft of the main turbine. Economic considerations that are evaluated when selecting a feed pump drive include the following.

1. Squirrel-cage induction motor drives require a variable-speed coupling for speed control, gears for high-speed pumps, and turbine-generator and auxiliary power system capability to provide the required power. The high-speed motor sizes available are limited as horsepower requirements increase.

2. Main turbine shaft drives require a variable-speed coupling for speed control, gears for high-speed pumps, and increased turbine capability to provide shaft power to the pump. A partial capacity motor-driven pump is required for plant startup.

3. Separate steam turbine drive has an inherent speed control advantage and requires the least main turbine capacity of the three options. Steam is usually extracted between the high- and low-pressure turbine sections. Additional steam and exhaust piping is required. An auxiliary steam supply for startup or a partial capacity motor driven startup pump is required.

4. Consideration of improved steam cycle heat rate due to the extraction of steam to drive feed-pump turbines and consideration of "released kW" available to the power system if the generator and turbine size are not reduced to coincide with the lower electrical auxiliary load when the feed pumps are not motor driven. Another consideration concerns main turbines that are last-stage flow limited. With feed pump turbine drives, the low-pressure sections of the main turbines do not have to pass the steam extracted for the feed pumps. Therefore, a correspondingly larger high-pressure turbine can be utilized to increase unit output due to the steam which must pass through the high pressure section to the extraction point. The mechanical drive steam turbine is the most frequently applied feed pump drive for large fossil-fueled generating units.

FORCED- AND INDUCED-DRAFT FANS

The combustion process in a furnace can take place only when it receives a steady flow of air and the combustion gases are constantly removed. The steam-generator draft system produces this air and gas flow. When only a chimney or stack is used, the system is a natural draft system; when this is augmented with a forced-draft fan or forced- and induced-draft fans, the system is a mechanical draft system.

Only the smallest boilers are natural draft; the boilers required by large turbines employ a mechanical draft system. The forced-draft fans supply the combustion air to the boiler and overcome the pressure drop through the air preheater and ducts. The induced-draft fans remove the combustion gases from the boiler and discharge them up the stack. The induced-draft fans overcome the pressure drop in the boiler, superheater, economizer, air heater, electrostatic precipitator or baghouse filter, and flue gas desulfurization scrubber and ducts.

The correct flow of air and gas is maintained by inlet vane and damper control, or by variation in fan speed, sometimes supplemented with vane and damper control. When speed control is employed, a squirrel-cage induction motor drive with a hydraulic coupling can be used to vary the fan speed. Other methods of flow control include multi-speed motors which can operate at two or more specific speeds, or adjustable-speed motor drives which operate with smooth control over a wide speed range.

During the late 1950s and the 1960s, many large steam units were built with pressurized furnaces. These units were built without induced-draft fans. For pressurized furnaces, large forced-draft fans provide the entire pressure requirement to drive the combustion air into the furnace, and the combustion products are driven out of the furnace through the economizer and precipitators and up the stack. Elimination of the induced-draft fan necessitates operating the furnace above atmospheric pressure so that the furnace has to be absolutely tight to prevent leakage into the boiler room.

Due to problems associated with leakage of combustion products from pressurized furnaces, many of these furnaces have been converted to balanced pressure operation with the addition of induced-draft fans. Another incentive to use the balanced pressure furnace is the increasing requirement for air quality control. The addition of larger precipitators and, particularly, the addition of large scrubbers has increased the pressure drop ahead of the stack. Large increases in induced-draft fan power requirements have resulted from these pollution abatement additions. The induced-draft fan horsepower is typically doubled by the addition of a wet limestone scrubber.

The forced-draft fans for some very large units are driven by mechanical-drive steam turbines. The economic considerations mentioned in the section concerning the selection of feed-pump drives are applicable to selection of fan drives. The increased induced-draft fan power requirements have resulted in consideration of steam turbine drives for in-

TABLE 4.2. Typical Fan Power Requirements

Unit Size (MW)	Pressurized Furnace	Balanced-Flow Furnace	
	Forced Draft Fan Horsepower(kW)	Forced Draft Fan Horsepower(kW)	Induced Draft[a] Fan Horsepower(kW)
1300	27000(20100)		
1100	22000(16400)	9000(6700)	18000(13400)
850	17000(12700)	7000(5200)	14000(10400)
700	14000(10400)	6000(4500)	11000(8200)
500	10000(7500)	4000(3000)	8000(6000)
400	7500(5600)	3000(2200)	6500(4800)
300	5500(4100)	2500(1900)	5000(3700)
200	3500(2600)	1500(1100)	3500(2600)
100		750(600)	1500(1100)

[a] May double where flue gas desulfurization scrubbers are used.

duced-draft fans. Consideration has been given to the application of noncondensing turbines for induced-draft fans where steam is required for stack gas reheaters; and condensers in the vicinity of the stack are objectionable.

Table 4.2 indicates the approximate fan power requirements for various sizes of units. The physical arrangement of the boiler greatly affects the required fan horsepower; certain furnaces may require up to twice the horsepower shown in the forced-draft column.

Either steam-jet ejectors utilizing high-pressure steam extracted from the main steam leads to "blow" the noncondensable gases out of the system, or motor-driven vacuum pumps are provided to extract the noncondensable gases from the condenser. This equipment is also used to establish a vacuum when the turbine is started.

The amount of condenser cooling water required is approximately 0.35–0.75 gallons (1.3–2.8 l) per minute per kW of generating capacity. The power required to circulate this amount of water varies with the condenser design and the length and cross-sectional area of the intake and discharge tunnels. An approximate figure is 10 hp (7.5 kW) per 1000 kW of generating capacity. Vertical low-speed, motor-driven pumps are used, more often than horizontal. Units up to 50,000 kW often use only one pump, while larger units use two or possibly three. Motor-driven traveling screens are commonly used to remove debris from the intake water which might otherwise damage the pumps or clog the condenser tubes.

POWER FOR THE AUXILIARIES

Most all large generating units built during the last quarter century have a "unit-connected" arrangement. Each generating unit has its own step-up transformer and its own auxiliary electric power system. Figure 4.16 shows a typical electrical one-line diagram for an auxiliary power system. There are two power sources available to the auxiliary buses, the unit auxiliary transformer which is connected to the main generator by means of isolated phase bus duct, and the start-up or reserve auxiliary transformer which is usually connected directly to the transmission system. The generating unit is started, using auxiliary power from the start-up source. When the unit is ready to accept load, the auxiliary power system is transferred from the start-up to the unit source. The auxiliary buses are normally composed of metal-clad switchgear. The larger motors are powered directly from these buses through circuit breakers. Load-center unit substations, where the power is transformed from the medium-voltage level to the low-voltage level, are also powered from the auxiliary buses. The load-center unit substations are composed of transformers and circuit breakers which distribute power to the medium-size motors and lighting and heating loads. Motor control centers, which supply power to and control the small motors and loads, are also connected to the load-center unit substations. Low-voltage dc power is also provided for control power and emergency power. The dc system is normally supplied from a battery charger system which is

POWER FOR THE AUXILIARIES

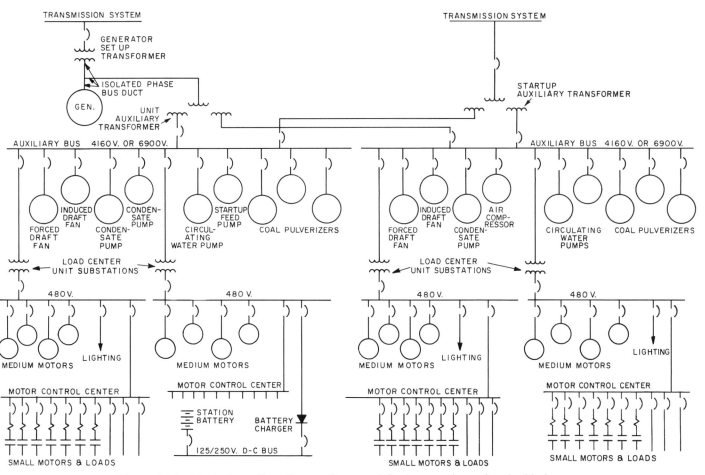

Figure 4.16. Typical one-line diagram for a coal-fired generating unit—double bus arrangement.

supplied from the low-voltage ac system. The battery chargers maintain a charge on the station batteries, which supply power to the dc system if the ac system fails. Uninterruptable power supplies are often provided to supply power to critical loads such as computers, controls, and monitoring systems.

Some units in recent years have been built without the unit source, with the auxiliaries being supplied from the start-up source for all operating conditions. Another approach which is being applied in a few plants is the use of either a high-capacity generator circuit breaker or a generator load-break switch located between the generator terminals and the isolated-phase-bus tap to the unit auxiliary transformer. The unit can be started with the breaker or switch open using the generator step-up transformer as a source of off-site power.

The amount of electrical power required to operate the plant auxiliaries depends upon the type of fuel, the type of boiler, the steam pressure and cycle, site considerations affecting condenser cooling water, environmental considerations, and whether any of the major auxiliaries, such as the feed pump, are mechanically driven. Prior to the strict requirements for environmental effects, the total auxiliary power required was approximately 5–10% of generating unit rating. With mechanically driven feed pumps, the electrical auxiliary power was approximately 3–5%. The addition of flue gas desulfurization scrubbers and cooling tower systems capable of cooling full condenser flow can result in auxiliary electrical power requirements as high as 12–15% for a coal-fired plant with mechanically driven feed pumps.

The typical auxiliary system one-line diagram shown in Figure 4.16 is a "double-bus" system. Each of the two medium-voltage buses is supplied from a separate auxiliary transformer secondary winding. Auxiliaries for small units requiring only a small amount of auxiliary power can be supplied from a single-bus auxiliary system. The auxiliary systems for the smallest units can be 480-volt or 2400-volt systems. For larger units, as the total auxiliary power becomes larger and the individual drive motors become larger, the coordinated application of system components requires more system design effort. Maintaining acceptable operating voltage levels under various auxiliary load conditions, minimizing voltage dip when large motors are started, and providing sufficient switchgear capability to protect the system from higher short-circuit current levels due to the increased loads are the governing design criteria. To provide for these criteria, the auxiliary system must either be divided into additional buses fed from additional auxiliary transformers, or switchgear with higher interrupting ratings must be used, often requiring an increase in system voltage level.

The industry has followed both of these approaches and also has used a combined approach. Up to four bus systems have been used at medium-voltage levels of 4160 volts and 6900 volts. Dual medium-voltage systems with 13,800 volts and 4160 volts, or 6900 volts and 4160 volts, are not uncommon. The largest motors are supplied directly at the higher voltage. The dual-voltage systems can be arranged in one of three ways:

1. Three winding auxiliary transformers, with one secondary winding for the higher voltage and one for the lower voltage.
2. Separate transformers for the two medium voltages.
3. A cascaded system with the auxiliary transformers supplying the auxiliary buses at 13,800 volts and 13,800–4160-volt transformers supplying 4160-volt buses from the 13,800-volt buses.

NUCLEAR PLANT AUXILIARIES

Auxiliaries and auxiliary systems for light-water-reactor nuclear units are similar to those for fossil-fuel plants. Only the major differences will be discussed in this section. The feedwater systems contain the same basic components with less pumping power required due to the lower feedwater pressure and low-pressure steam conditions. Reactor feed pumps provide the same function as the boiler feed pumps in a fossil unit. In contrast with large fossil units where almost all feed pumps are driven by steam turbines, motor drives are sometimes applied to reactor feed pumps in large nuclear units. The light-water reactors do not require large fan drives or large fuel preparation drives. There are some large pump drives associated with the reactors. The boiling water reactors require two reactor recirculating pumps approximately 7000 hp each for a 1200-MW reactor. The pressurized water reactors require four reactor coolant pumps of approximately 7000 hp (5200 kW) to 10,000 hp (7500 kW) each in the 1000–1300-MW reactor sizes.

The basic auxiliary power system for a nuclear unit is very similar to the system for a fossil unit. The same technical criteria apply with regard to system performance. Great emphasis, however, is placed on the engineered safety features (ESF) which are intended to ensure safe shutdown of the reactor under abnormal conditions. The auxiliary power system must supply power to redundant drives and loads which provide the safety features. Predominant among these drives are the emergency core cooling systems (ECCS). The auxiliary power system must provide separate paths to redundant ESF systems and must have two off-site sources of power available within the appropriate times after a shutdown is initiated.

A start-up auxiliary transformer and the power transmission system qualify as an immediately available source. Diesel-generator sets, capable of supplying the ESF loads, are provided in the auxiliary system as standby sources of power. The ESF loads and the standby generators can be isolated from the remainder of the auxiliary system. Auxiliary power requirements for nuclear units vary considerably depending upon whether the reactor is a BWR or PWR, whether the reactor feed pumps are motor driven or turbine driven, and upon the amount of condensing cooling water pumping power and auxiliary cooling power required. Considering these variables, the auxiliary power requirement is generally between 5–10% of unit rating for large units.

COMBUSTION TURBINES

Combustion turbines play an important role in utility system generation planning. By the end of 1979, over 50 million kW of combustion turbine capacity

COMBUSTION TURBINES

was in commercial operation for power systems in this country.

The combustion turbine possesses characteristics that make it attractive both for peaking applications and, where an exhaust heat recovery system can be used, for base-load operation. The peaking applications stress:

Low installed cost
Modular construction
Minimum transmission requirements
Short delivery times
No cooling water requirements
Minimum operation and maintenance costs
Complete automation
Low emission levels
Fast starting
Fast load pickup
Black start capability.

Combined-cycle plants utilizing combustion turbines, heat recovery steam generators, and a steam turbine provide most of the advantages of the simple-cycle peaking plant, with the additional benefit of very good heat rates. Also, combined-cycle plants require less cooling water than a conventional fossil or nuclear plant of the same size.

Simple-Cycle Combustion Turbine

The basic thermodynamic cycle involved in the simple cycle combustion turbine is the Brayton cycle. The thermodynamic cycle is implemented in a variety of configurations—single-shaft machines, two-shaft machines, heavy-duty turbines, and aircraft derivation turbines. This discussion will consider only the heavy-duty, single-shaft machine, a type often used in utility service.

In the single-shaft, simple-cycle combustion turbine, air at atmospheric pressure and temperature enters the compressor. The axial-type compressor uses blades mounted on the rotor to force the air through stationary blades, raising its temperature and pressure. The pressure ratio, or the ratio between the pressure at the outlet of the compressor to that at the inlet, is on the order of 10:1. The pressurized air leaves the compressor to enter the combustor, where part of the air enters the combustion liner to mix and burn with the fuel, and the rest of the air is used to cool the combustion liner. The two streams mix together before flowing into the turbine section. At present, the turbine inlet temperatures may range from 1650°F (900°C) to approximately 2100°F (1150°C), depending on the turbine and the type of fuel burned. In the turbine section, the hot gases expand through several stages, producing power, then exhaust to the atmosphere at a temperature of approximately 1000°F (538°C).

A cross-section of a combustion turbine is shown in Figure 4.17.

Heavy-duty combustion turbines may be fired using either gaseous or liquid fuels, or combinations of different fuels. The fuel system required for gaseous fuel is relatively simple: a valve controlled by the turbine control system opens to admit the proper amount of gas to the fuel nozzles in the combustion chambers.

The liquid fuel systems used on heavy-duty combustion turbine power plants can handle crude, dis-

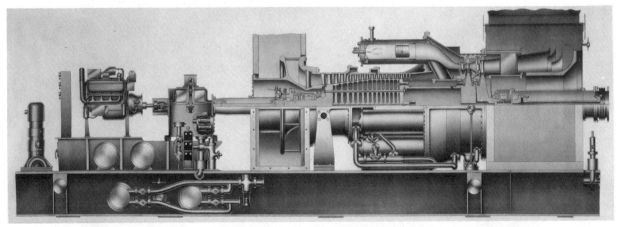

Figure 4.17. Cutaway view showing components of a simple-cycle combustion turbine.

tillate, and residual petroleum fuels. Distillate fuels utilize pressure atomization or air atomization depending on application needs; crude and residual fuels utilize air atomization. In all cases, fuel pumping, metering, and distribution are performed by a separate pump and gear flow divider fuel system. These systems:

1. Provide the proper amount of fuel for any operating condition, in response to control demands.
2. Provide the flow at the necessary pressure for continuous injection into the combustion chambers.
3. Divide the flow equally into as many portions as there are fuel nozzles and combustion chambers, there being one fuel nozzle per combustion chamber.

Dual fuel systems provide for on-line changeover between the two types of fuel, allowing the usage of the more readily available of the two.

Although combustion turbines will burn crude and residual petroleum fuels, the chemical makeup of the fuel as burned has a significant effect on the lifetime of the turbine parts in the hot-gas path. Many crudes and almost all residuals contain metallic trace elements which are highly undesirable for successful turbine operation. The two primary elements are sodium, usually in the form of water-soluble sodium chloride, and vanadium in the form of oil-soluble metal organic complexes. The effect of these two elements becomes more pronounced as firing temperatures are increased. Depending on the concentrations of these contaminants, fuel pretreatment may be required. Water-washing of the fuel is used to remove the sodium, while chemical inhibitors are used to counter the effects of the vanadium.

Since the combustion turbine is basically an air-breathing machine, its long-term performance and reliability are a function of the quality and cleanliness of the air. Inlet air conditioning, consisting of an inertial separator to filter out coarse dirt followed by high-efficiency filtration for the finer dirt, is provided. Additional pretreatment is available for marine environments or for especially dirty environments. Other optional equipment includes anti-icing packages for units that operate at very low ambient temperatures, evaporative cooling packages to increase the power output and efficiency of units that operate in dry, high-temperature environments, as well as additional silencing packages for the inlet and exhaust passages.

The combustion turbine, generator, and associated auxiliary systems are packaged in modules for quick installation. A typical power block of combustion turbines installed on a utility system is shown in Figure 4.18.

Combined-Cycle Plant

The gases exhausting from the combustion turbine are quite hot—in the vicinity of 1000°F (538°C). This represents a substantial heat loss from the cycle. In light of today's high fuel costs, it makes sense to

Figure 4.18. A power block of four combustion turbines.

GENERATOR DESIGN

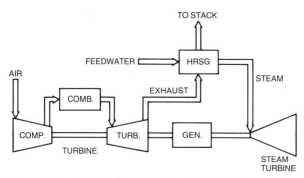

Figure 4.19. Block diagram of a single-shaft combined cycle plant.

consider the benefits of using this waste heat to generate steam for use in a steam turbine.

Modular heat recovery steam generators (HRSGs) have been designed to utilize the heat in the combustion turbine exhaust. The HRSG incorporates some of the features of a conventional fossil-fired boiler, such as an economizer, evaporator, and a superheater, but many of the design parameters differ from those applicable to a conventional plant. The HRSG must be designed to start and come to full load in a short period of time. It must be able to withstand the thermal shock when the exhaust gases are admitted from the combustion turbine. It must not severely restrict the exhaust flow from the combustion turbine, yet the boiler must efficiently utilize the heat in the exhaust gas and properly control the superheat of the steam produced.

Combined-cycle plants employ two basic designs, single and multishaft. In the single-shaft units, both the combustion and steam turbines drive a common generator. A block diagram of such a unit is shown in Figure 4.19. In the larger plant sizes, a multishaft design is used. These utilize a number of combustion turbines and HRSGs in parallel (usually four or six depending on plant size). The steam produced by all of the HRSGs drives a single steam turbine-generator. A block diagram of a unit capable of producing approximately 400 MW net at a net heat rate of about 7800 Btu/kWh (8230 kJ/kWh) is shown in Figure 4.20. The combustion turbine exhaust ducts are provided with a bypass system so that the hot exhaust gases may be routed directly to the stack without passing through the HRSG.

Although cooling water must be supplied to the steam turbine's condenser, the water requirements for the combined cycle plant are only one-half that of a fossil plant of comparable size. Most of the auxiliary systems required for conventional steam plants are required for the steam turbine portion of the combined-cycle plant. Careful evaluation of the cycle is required to optimize the feedwater string. Normally, the steam turbine operates with a relatively simple feedwater cycle.

GENERATOR DESIGN

Rotor

Forging. The basic requirement of the rotating field is to produce a given number of lines of magnetic flux, obtained by the product of the number of turns in the rotor winding and the current (amperes) in the winding. This product, known as "ampere-turns," is the driving force to produce the magnetic flux. Since the reluctance in the path of the magnetic lines affects the end result, the entire magnetic path must be designed for a minimum reluctance or highest permeability. The rotor body in which the field winding is located forms the path of the magnetic lines for part of the circuit, and the stator core and air gap provide the return path for the flux.

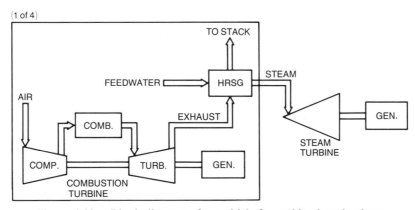

Figure 4.20. Block diagram of a multishaft combined cycle plant.

Figure 4.21. Milling coil slots in an 1800 r/min, four-pole rotor for a 687,500-kVA generator.

In relatively low-speed machines, separate field coils are attached to a rotating member, an arrangement known as "salient pole" construction. At 3600 or 1800 r/min, and at the diameters presently employed, the centrifugal forces involved will not permit salient pole construction. It is necessary to use a slotted integral forging (Figure 4.21) to hold the winding together at rated speed. The ampere-turn requirements for the field increase with an increase in rating, and this entails a combined increase in heating in the coil, more copper, and higher centrifugal load.

Rotor Windings. The winding used to produce the magnetic field is multi-coil, single circuit, energized with dc power fed through the shaft from the collector rings outboard of the main generator bearings. The fact that it is a relatively low-voltage, low-power circuit has been a major factor in building the generator with a rotating field instead of a rotating armature. It would be considerably more difficult to conduct the higher ac voltage (and power) through a set of collector rings and brushes.

Early generators were built with the excitation furnished by a 125-volt dc supply. This was increased to 250 volts, then to 375 volts. Today most large generators are being built with 500-volt or higher excitation systems. Earlier generators were built with a field coil arrangement, shown in Figures 4.22 and 4.23. The heat generated in the coil was conducted through the slot insulation to the field forging, and then to the cooling gas. The dielectric barrier forming the slot insulation was also the primary thermal barrier in the circuit; and, as current levels increased, this became a much greater obstacle. The answer to this problem was the "conductor cooling" arrangement in which cooling gas flowed directly through the conductors. This eliminated the thermal barrier of the slot insulation, allowing a substantial increase in the current carrying capability of a given size field.

Several different arrangements have been utilized by American and foreign manufacturers to accomplish the conductor-cooling principle. The arrangement presently used by General Electric in two-pole generators is referred to as "diagonal flow" cooling; this arrangement is shown in Figure 4.24 where six turns of a field coil have been separated for the purpose of photographing the individual gas passages. The gas flows down through a series of slotted holes, which are offset in each layer from those in the previous layer. The bottom turn is a channel which redirects the gas to another series of slotted holes upward in a diagonal progression to the top of the coil. The pumping action to provide gas flow is contained in the configuration of the slot

GENERATOR DESIGN

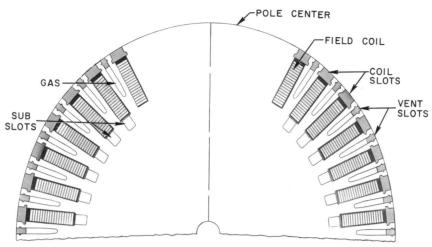

Figure 4.22. Winding and gas passage arrangement for two-pole filed.

wedges (Figures 4.25, 4.26, and 4.27). The holes for gas inlet are inclined in such a manner, that impact due to rotation of the field forces the gas through the wedge and down into successive turns of the coil.

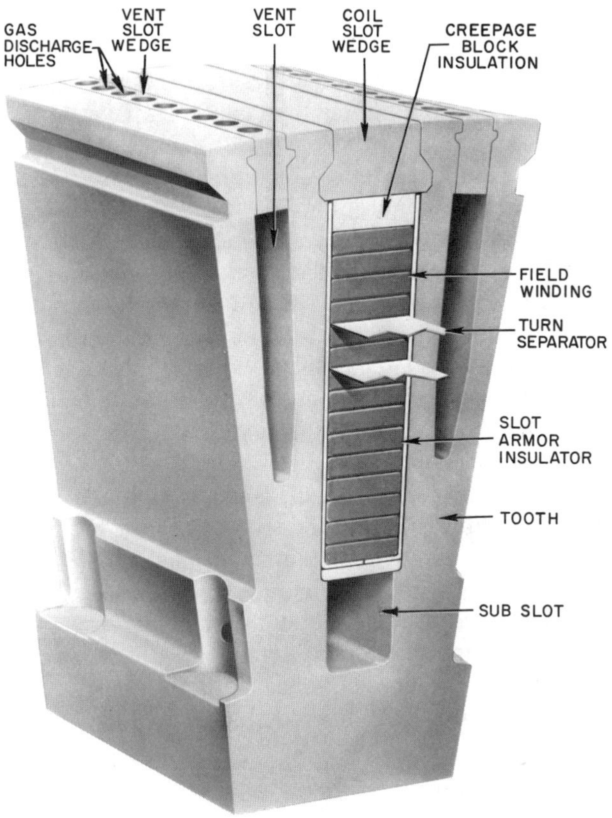

Discharge holes in the wedges have a raised section preceding the hole in the direction of rotation. This provides a pressure reduction at the hole, and the lower pressure induces gas flow by suction from the discharge end.

There are several alternate inlet and discharge sections throughout the length of the field which provide multiple parallel paths through the winding (Figure 4.28). The net result is shown in the temperature profile chart (Figure 4.29), which contrasts two conductor-cooling arrangements used in different types of construction. In the "axial" arrangement, gas flow through the conductor from both ends is allowed to discharge in the center part of the field. The resulting temperature, plotted as a function of the length of the field body, shows a cumula-

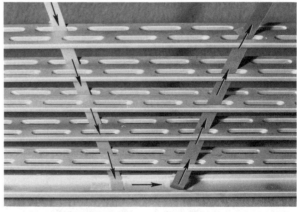

Figure 4.23. Model of conventionally cooled two-pole field.

Figure 4.24. Diagonal-flow, conductor-cooled field turns, showing gas-flow arrangement.

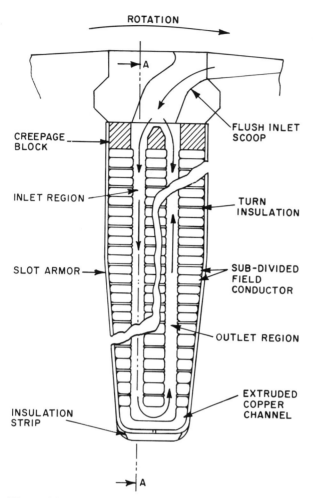

Figure 4.25. Section of a diagonal-flow rotor slot showing a gas passage.

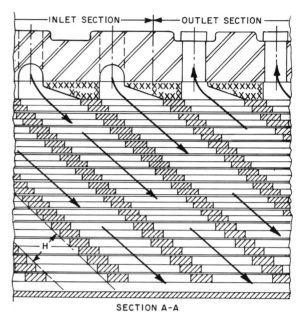

Figure 4.26. Diagonal-flow cooling of field winding showing stair-step pattern of cooling passages produced by offsetting the slots in successive layers of the winding.

tive build-up toward the center, with the hot spot at the center. In the "gas pickup" arrangement, gas which is fed into the field at several points traverses only a short section of the field winding before being discharged back to the gas gap. The temperature profile thus has a saw-tooth effect. In this arrangement, a ratio of hot spot rise to average temperature rise in the winding of about 1.25/1 can be achieved, contrasted to a ratio of about 2/1 in the axial flow scheme.

In operation, it is not practical to monitor hot spot temperatures. The operator has available only a reading of average winding temperature obtained from current and voltage. From an operating viewpoint, it is not desirable to have hot spot temperatures greatly exceeding the average winding temperature, which can be monitored.

Four-Pole Generators

The design of four-pole generators requires less duty on the rotor winding than two-pole generators, thus easing the problem of designing the rotor cooling circuit. As generator ratings increased, the transition to conductor cooling of the rotor winding was made on two-pole generators earlier than on four-pole generators. The four-pole generators which have been built with conventionally cooled rotor windings have much larger ratings than two-pole generators. The largest rated four-pole generators employ an arrangement where gas is fed from both ends through subslots under the coils, as in conventional cooling (Figure 4.30). A series of radial holes through the coils and through the slot wedges permit gas from the subslot to flow radially outward at a number of points in each slot. The gas is in contact with the conductor, eliminating the thermal drop through the slot insulation. This radial-flow conductor cooling arrangement is more effective than the conventional method, but less effective than diagonal-flow conductor cooling.

Stator

Stator Core. The function of the stator core is to provide a return path for the lines of magnetic flux

GENERATOR DESIGN

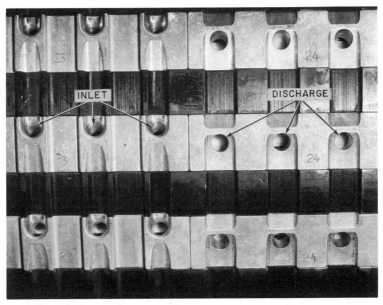

Figure 4.27. Section of air-gap pick-up conductor-cooled rotor.

from the field, and support to the coils of the stator winding.

The lines of magnetic flux generated by the rotating field are, in effect, a series of continuous loops which may be compared to a very large number of rubber bands, each of which tends to contract to the shortest path it is permitted to follow. Reluctance to this path is determined by the nature of the material through which the lines of flux pass. Air forms the highest reluctance, and magnetic steel (high permeability) forms a very low reluctance. The path of the lines of flux from one pole of the field to the opposite pole, outside of the field body, forms the longer part of the circuit. It is desirable to provide the return path with a high permeability (or low magnetic reluctance) material to minimize the magnetizing force that must be applied by the field winding. To accomplish this, the stator core is made in the form of a cylinder of high-permeability steel that fully encloses the field. Rotation of the field causes the flux to sweep around the stator. For a given point, this amounts to a rapidly reversing direction of the lines of flux in the stator core, dictating that the stator core be laminated, in contrast to the solid body of the rotating field.

The lamination is accomplished by stacking the core with a series of overlapping segments of steel, each of which is insulated with a layer of enamel on both sides. These segments are stacked on key bars at the outside diameter which maintain the align-

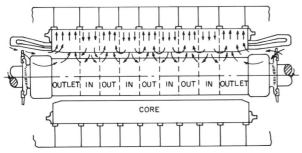

Figure 4.28. Line-up of rotor and stator cooling zones.

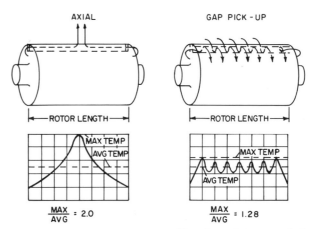

Figure 4.29. Temperature profile of conductor-cooled rotors.

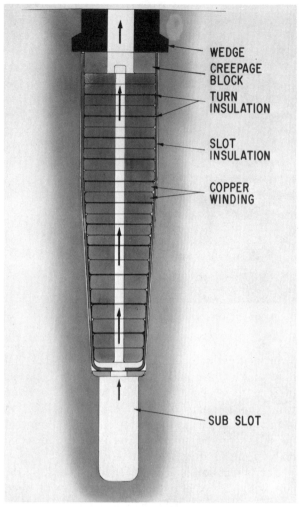

Figure 4.30. Radial flow ventilation.

ment of the outer edges (Figure 4.31). The inside diameter has slots in which the stator winding is later assembled. Alignment at this point is maintained during stacking by temporary slot aligners which are moved through the slots as the laminations are assembled, producing a straight slot for the stator coils. The complete core for a large generator may contain as many as 500,000 individual segments or punchings.

The effect of a two-pole field within the cylinder of the stator core is that of a very powerful magnet which tends to deflect the stator core toward the field in line with the poles (Figure 4.32). Although the stator core is very stiff, a deflection of two or three mils takes place due to the magnetic pull of the field. This is a rotating or traveling deflection which rotates in synchronism with the field. If the core were solidly mounted, the deflection would be transmitted to the stationary components and would appear in the station structure as a vibration of twice-per-revolution frequency. This annoying frequency must be suppressed within the machine. It is isolated by a series of spring-mounting assemblies between the core and the stator frame (Figures 4.32 and 4.33).

In a four-pole generator, the north and south poles are 90 mechanical degrees apart, instead of 180. The eight-nodal pattern that results, produces only a small deflection of the stator core, making it unnecessary to isolate a four-pole stator core as is done in a two-pole generator. The key bars can then be assembled rigidly in the stator frame without spring bars (Figure 4.34).

It is essential that the individual laminations in the stator core remain insulated from each other to prevent circulating current between laminations, due to the reversing nature of the magnetic flux. The laminations are shorted together at the key bars where the core is stacked. If the inside diameter of the core is damaged in such a manner that the punchings at the tooth tips are shorted together, a closed loop would be provided for circulating current between punchings. It is mandatory that care be taken to prevent damage to the inside diameter of the stator core, which would burr or short the punchings together. If this happens, repair work must be done to regain the separation or insulation between the individual punchings.

The generator stator is inherently a heavy component, due to the magnetic requirements of the core. The total number of lines of flux, and the flux density to which a given magnetic steel can be loaded, determines the required volume of the core. In some cases, the largest generators require a core that weighs almost 250 tons. The interleaved arrangement used in stacking the core and assembling the winding precludes the possibility of making a longitudinal split for convenience in handling. It is desirable to ship the factory-assembled core in one piece, including the winding. Shipment and handling are very critical and, as mentioned earlier, the design of the equipment must take into consideration the capability of the carrier's facilities and roadbed. The outside dimensions and total weight of the frame as shipped must not exceed the maximum that the carriers can handle, considering tunnels, bridges, roadbed, etc. There are many utility sites that could not be reached by railroad if optimum construction of the generator core and frame were used.

GENERATOR DESIGN

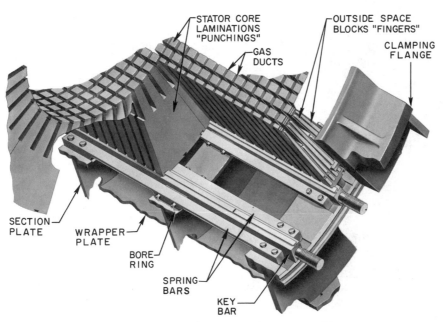

Figure 4.31. Spring-mounted core assembly.

Stator Windings. The magnitude of the voltage induced in the stator winding is a function of three factors: (1) the total number of lines of flux produced by the field, which were cut by the stator winding, (2) the frequency with which any given coil cuts the lines, and (3) the number of turns in the stator coil. An extreme simplification of the problem points out that the first factor is the field capability, the second is operating speed (fixed by frequency requirements), and the third is a function of the stator winding itself.

Schematic Arrangement. The three-phase stator winding must be so distributed that in each

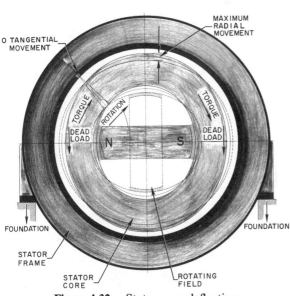

Figure 4.32. Stator core deflection.

Figure 4.33. Spring-mounting assembly for two-pole generator.

Figure 4.34. Stator frame for four-pole generator with key bars attached.

It has been customary to provide external connections by means of high-voltage bushings from each end of each phase for the purchasers' connections. The windings are normally connected in wye and grounded at the neutral point, through some

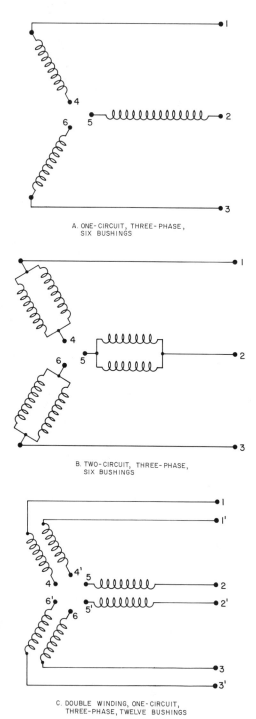

A. ONE-CIRCUIT, THREE-PHASE, SIX BUSHINGS

B. TWO-CIRCUIT, THREE-PHASE, SIX BUSHINGS

C. DOUBLE WINDING, ONE-CIRCUIT, THREE-PHASE, TWELVE BUSHINGS

Figure 4.35. Stator connections.

phase there is a uniform buildup and decay of voltage, 120 electrical degrees apart. The shape of the voltage wave is a sine wave. At a given instant, the phase that is at peak voltage would have one side of its winding being cut by flux from the north pole and the opposite side of the turns or coil near the center of a south pole. For a two-pole generator the sides would be approximately 180 mechanical degrees apart. For reasons of harmonics and telephone interference, it is not desirable to have the opposite sides of a coil exactly 180 degrees apart, so the turns are slightly less than full pitch. By arranging each phase in two sections, they can be distributed symmetrically with respect to the pole center positions.

To obtain the desired output voltage and kVA, the designer may vary the number of slots and the method of connecting the coils in the slots to achieve different winding patterns. Figure 4.35, A–C, shows some of the possibilities in a very elementary way, indicating one-circuit and two-circuit arrangements to obtain variation in voltage and current combinations. The two-winding arrangement has been used in the past but has not been widely applied. Designs of three- and four-circuit windings are used for highest-rated generators but are less common, due in part to the added complexity of manufacturing windings with a large number of bars and special connections.

GENERATOR DESIGN

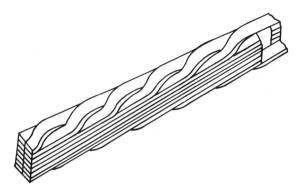

Figure 4.36. Transposition of insulated strands in stator bar.

form of external impedance. Delta-connected windings provide another alternative for the generator designer, but do not provide a convenient point at which the windings can be grounded.

Stator Bars. The armature is formed by insulated bars or half-coils assembled in the stator core slots, joined at the ends to form coils, and connected in the proper phase belts by connection rings at the end of the winding. The stator bars are composed of insulated copper conductors (strands) arranged in the form of rectangular bars. The bars are assembled so that each strand occupies every radial position in the bar at some point along the length of the bar. This is accomplished by a spiral progression of each strand around the bar, called transposition. The arrangement causes all strands to share the load current equally and minimizes circulating current losses within the bar, which would be caused by flux distribution within the coil slot. A schematic transposed section is shown in Figure 4.36. For many years, the 360° Roebel transposition has been in common use. Recently, other transpositions have been used, varying from 360° to 720°, depending on the flux distribution of the particular design.

The cross-slot leakage flux in the stator slot varies from top to bottom of the slot and, if the bar were solid, would prevent uniform current distribution in the bar. The insulated strands are transposed in such a manner that the voltage differential is balanced out from end to end; however, at intermediate points along the length of the bar, a voltage difference exists between strands. A strand short within the bar would permit circulating current, in addition to the normal load current, resulting in overheating and possible failure of the bar.

Bar Assembly. The copper strands, drawn to very close dimensions and protected with class-B strand insulation, are offset at the required spacing to produce the cross-overs, and are spiralled together to form the rectangular bare bar. This is molded to compact the bar, to give it accurate size, and to bond the strands together in the slot portion. The insulation of all stands is verified electrically and the bar ends are formed so that opposite coil sides can be connected with the design coil pitch and the strands in the end portion bonded for rigidity of the arm during taping.

Liquid-cooled bars have a header brazed to each end so that all strands are fitted into a hollow copper casting for the hydraulic and electrical connection (Figure 4.37). The electrical connection between top and bottom bars usually is made by brazed, interleaved laminations. Two bars are located in each slot, each bar being one-half of a different turn of the stator winding. In some slots, both top and bottom bars are half-turns of the same phase; in others the two bars are in different phases.

Cooling. The solid-strand, conventionally cooled bar used for many years (Figure 4.38A) has accounted for most of the total installed steam power generation, and is still being used for generators smaller than 200 MVA.

Heat transfer from the conventional winding through the insulation ground wall is retarded by the thermal barrier of the insulation. To circumvent this barrier and continue the progress in unit rating,

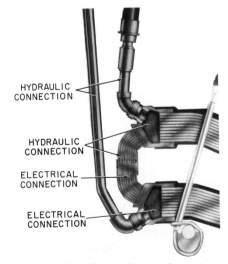

Figure 4.37. Stator bar end connections.

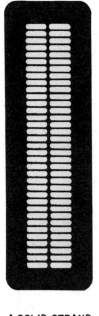 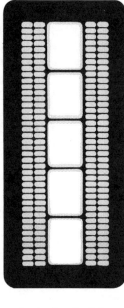 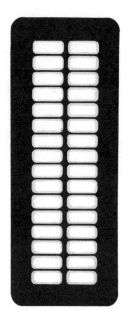

A. SOLID-STRAND CONVENTIONAL **B. GAS CONDUCTOR-COOLED** **C. LIQUID CONDUCTOR-COOLED**

Figure 4.38. Stator bar sections.

various conductor cooling methods have been devised. Functionally, conductor cooling involves the flow of a coolant inside the insulation ground wall for heat removal from within the bar. This has been accomplished by several arrangements.

Gas conductor cooling (compared with liquid) involves a relatively large mass flow of coolant and is accomplished by the interspersion of gas ducts within the bar. Increased circulating current losses in large strands prohibit making the strands large enough to accommodate the gas flow; separate, nonmagnetic, insulated gas ducts must be built into the bar. Figure 4.38B shows an arrangement previously used by General Electric in making such bars. Figure 4.38C shows a typical liquid-cooled stator bar in which all strands are hollow to provide room for coolant flow through each strand.

Continued refinements in the design take into consideration such factors as the difference in losses at the top and bottom of the slots. The top bars can be made a different size from the bottom bars to offset this difference, as in Figure 4.39. A design with a small number of slots would result in wide slots. The bars for the wide slot have a four-tier strand arrangement shown in Figure 4.40. Another variation of this is the mixed-strand arrangement (Figure 4.41), which provides higher utilization of slot space with copper. Strand loss,

due to induced circulating current, is a function of strand size. Additional interspersed solid strands can be made smaller without coolant space, thus minimizing strand loss. Temperature differential is only slightly higher than the all-tubular strand bar, and considerably less than the gas conductor-cooled bar.

Insulation. For many years, the insulation generally used on the high-voltage stator coils of large steam turbine-generators has been mica tape; this tape is applied over the slot portions and ends. After the tape is applied to the coil, it is impregnated with material to fill and seal the coil. The mica tape system previously consisted of three elements:

1. The main dielectric barrier, made of mica flakes roughly 1 inch (2.54 cm) square and 1 mil thick, built up in layers to a thickness of 4 or 5 mils.
2. A backing tape usually made of paper about 1 mil thick.
3. An insulating varnish or resin, normally of asphaltic or modified bitumen composition, used to fill up the interstices and to bond together the entire structure.

In 1947, progress made with new kinds of insulating materials—resins, varnishes, etc.—suggested that it

GENERATOR DESIGN

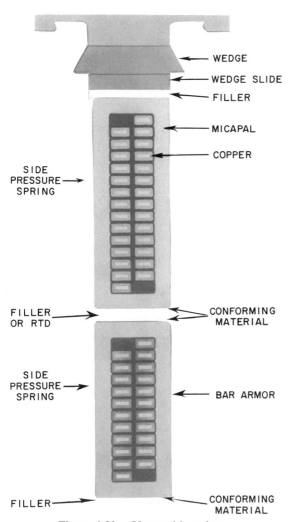

Figure 4.39. Unequal bar sizes.

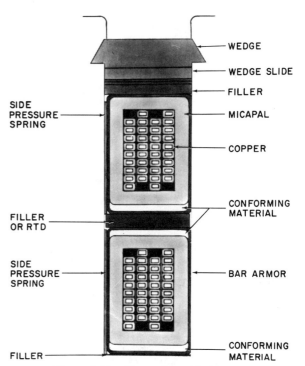

Figure 4.40. Four-wire wide bar.

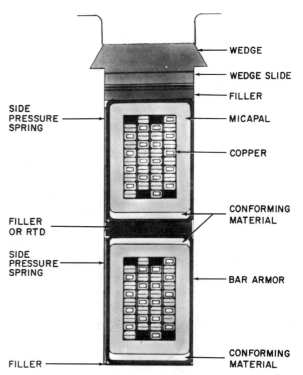

Figure 4.41. Mixed-strand bar.

should be possible to put together a new and better combination of materials in the three basic elements employed in the long-used mica tape insulation. An extensive development program was undertaken during which hundreds of base materials were investigated and evaluated and insulation systems employing various combinations of these materials were tried. Since no single material was found that was entirely satisfactory for the bonding resin, a new formulation of two resins was developed which had the desired properties. This new mica tape (Micapal*) system was adopted and is still used.

The three basic elements of the conventional tape have been changed. Instead of all layers of

---

* Trademark of General Electric Company.

mica flakes for the main dielectric barrier, the new tape employs a majority of layers of ultrathin mica platelets formed into a compact uniform structure known as mica mat, combined with some layers of mica flake. Replacing the old bituminous bonding resin is the specially developed duplex synthetic resin, composed of epoxy and alkyd resins blended to achieve the most advantageous combination of properties. Early Micapal tapes were made with paper backing reinforced with glass threads. This has been replaced by a special glass-cloth backing.

Each of these substitutions brought important advantages. Mica mat gave improved uniformity of tape thickness and dielectric strength, and also increased dielectric strength. The epoxy-alkyd resin combination contributed much better bonding, heat resistance, and endurance with greatly improved dimensional stability, chemical inertness, and resistance to moisture, oil, and other contaminants. The glass backing added great mechanical strength and permanence to the tape, facilitating its application to the coils, and eliminating the potential weakness of thermal deterioration of the paper previously used. There is also a procurement advantage in the use of mica mat, since it can be made from readily available domestic mica, instead of the much larger mica flakes that must be imported from India.

Another major advantage of Micapal tape is its resistance to mechanical deterioration due to expansion and contraction of the coils. In many full-scale laboratory tests where bars insulated with an asphaltic mica system were quickly damaged, similar bars with Micapal tape remained unaffected indefinitely. Another advantage of this tape is its corona-resistance properties. It also has a markedly higher corona-start voltage level and a far lower corona intensity at operating voltage than exists in the modified bitumen mica flake insulation. The latter has established an excellent service record.

The first Micapal insulated winding went into service in 1954, and has been followed by an increasing number of systems each year. Since 1959, all large General Electric generators have utilized Micapal insulation. It has established an excellent record and those few cases in which difficulty has been experienced have been due to mechanical damage to the insulation.

Magnetic Forces. One of the critical design limitations in large generators has traditionally been the temperature rise of the windings and insulation, due to losses. This is no longer the primary factor limiting the capability of a given generator design. Electromagnetic forces acting on various compo-

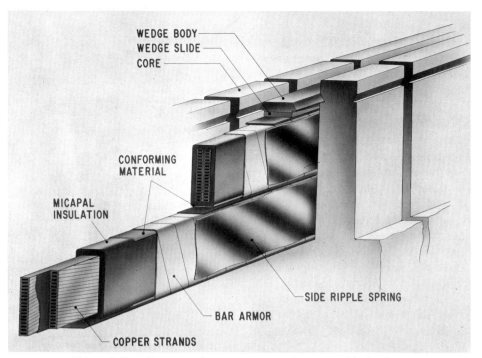

Figure 4.42. Cutaway of generator stator bar assembly in slots.

GENERATOR CONFIGURATION

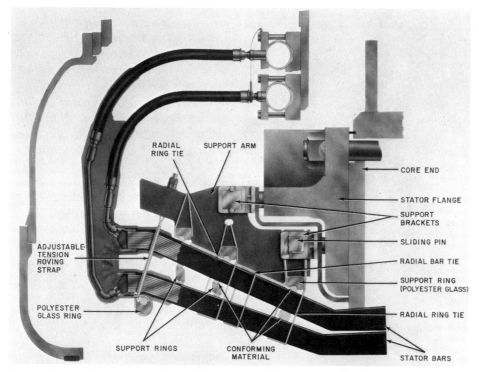

Figure 4.43. Arrangement of insulating connections and Tetraloc end-winding support system for a liquid-cooled unit.

nents of the generator are now of significant importance. This is true under steady-state running conditions, as well as transient or fault conditions.

Magnetic forces on the slot portion of the stator winding fluctuate at 120 cycles/sec. This tends to cause stator bar vibration if radial clearance exists. The bars are closely fitted to the slots, employing a conforming material at the top and bottom edges (Figure 4.42). This material is plastic at the time of assembly and is cured by heat with the bars in place. The bars are wedged in place under pressure and, as an added precaution against bar vibration, nonmetallic side pressure springs are assembled in the slot next to each bar.

The Tetraloc* end-winding support system (Figure 4.43) provides effective restraint against running frequency vibration as well as sudden short-circuit forces. Blocking between bars and at each support point utilizes a plastic thermosetting material. Glass fiber straps saturated with thermosetting resin secure the bar ends to the support structure, to prevent radial or circumferential movement. These plastic thermosetting materials permit accurate custom fitting and tying of each bar during assembly. The materials are then cured by baking the entire stator in a special oven to form a rigid high-strength assembly (Figure 4.44) The support brackets have slotted bearings to permit axial movement as required for thermal expansion and contraction during load changes.

GENERATOR CONFIGURATION

The functional requirements of the generator components have been described under separate headings. The mounting arrangement has varied through the years to accomplish the best match between design requirements and the shipping and installation limitations. These limitations do not permit a very wide latitude in variations of major dimensions of the stator core, although some flexibility in the design is possible.

Two-Pole Generators

The arrangement that has been most widely used on General Electric two-pole generators is shown in

---

* Trademark of General Electric Company.

76 POWER FROM STEAM AND COMBUSTION TURBINES

Figure 4.44. End view of liquid-cooled generator windings with one-pass cooling arrangement.

Figure 4.45. This design utilizes a single-piece stator frame with vertical hydrogen coolers, spring-mounted core, and conventionally cooled stator and field windings. The one-piece field forging also has a single-stage axial flow fan mounted at each end for circulation of cooling hydrogen. Prior to the development of conductor cooling, several other frame designs were used as generator ratings increased to accommodate the required gas flow to and from sec-

tions of the core with external "blisters," domes, etc.

Multi-piece frame construction was also used for several years on conventionally cooled generators, on which the outer part of the stator frame, including cooler housings, was built in sections for separate shipment, to be bolted together and seal-welded at the job site. Improvements in materials and advances in technology have made it practical to design and build conventionally cooled generators with one-piece frame construction to ratings in excess of 200 MVA. Conductor cooling is generally applied in ratings larger than 200 MVA.

The same basic arrangement of components is used in conductor-cooled generators (Figure 4.46) as in conventionally cooled, up to the limit of physical size of the frame for shipping. The limit for the single-piece frame, vertical cooler construction is presently well over 500 MVA. Continuing improvements in design, techniques, and facilities permit these limits to be revised upward occasionally.

Four-Pole Generators

Conventionally cooled four-pole generators in a wide range of sizes are built with a single-piece frame, vertical cooler arrangement, since it is the most practical configuration. Four, six, or eight hydrogen coolers have been used for units up to about 250 MVA.

With conductor cooling, the single-piece frame, vertical cooler construction can be used for genera-

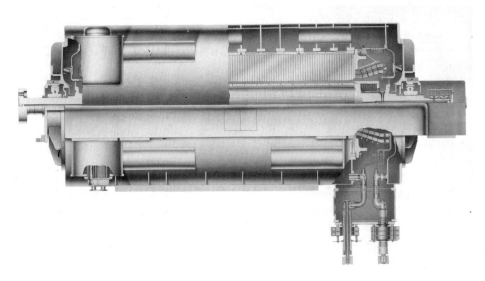

Figure 4.45. Axial section of 3600 r/min generator with hydrogen-cooled stator and rotor.

GENERATOR CONFIGURATION

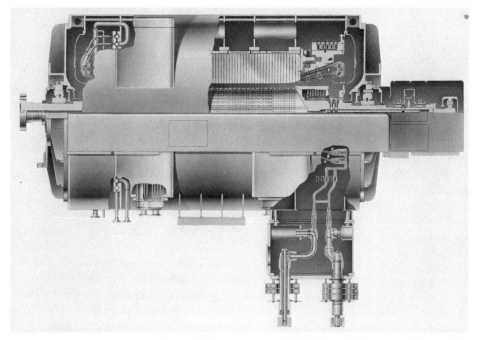

Figure 4.46. Axial section of 3600 r/min generator with liquid-cooled stator and gap pickup rotor.

tors with ratings well over 750 MVA; and the machine can be shipped as a single piece (Figure 4.47).

In the very largest 1800 r/min generators, it becomes necessary to utilize multi-piece frame construction for shipment, as described previously for two-pole generators.

Electrical Characteristics

Saturation Curve. Figure 4.48 shows the generator terminal voltage at no load as a function of field current. This curve is called the no-load saturation curve. For each design, the calculated no-load satu-

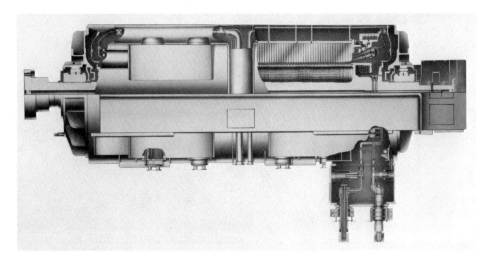

Figure 4.47. Axial section of 1800 r/min generator with liquid-cooled stator and conductor-cooled rotor.

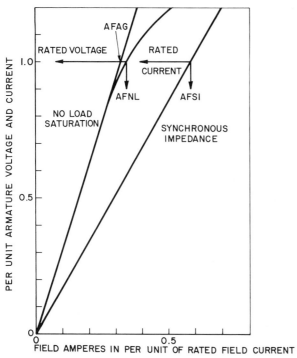

Figure 4.48. Typical saturation and synchronous impedance curves.

ration curve is substantiated by factory tests. On the curve of Figure 4.48, the term AFNL means amperes field no load. It is the value of the field current that would produce rated voltage at the generator terminals at rated speed and no load. AFAG means amperes field air gap, or the field current that would produce rated voltage at rated speed and no load if there is no magnetic saturation in the flux paths of the generator. This is used as the base field current or 1.0 per unit, in synchronous machine calculations.

Synchronous Impedance. In Figure 4.48, the synchronous impedance curve indicates the excitation required as a function of armature current, with the generator terminals short circuited. Excitation at rated armature current is labeled AFSI (amperes field synchronous impedance).

Typical "V" Curves. Figure 4.49 shows the correlation, at different loads, of excitation and power factor. At a given kVA level, the proportionate change in field current required to change from one power factor curve to another is shown. The right-hand limit of these curves is the 1.0 per-unit field

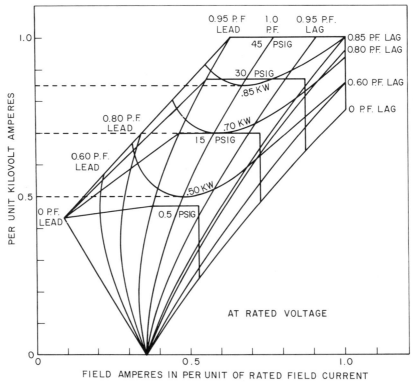

Figure 4.49. Typical V curves.

GENERATOR CONFIGURATION

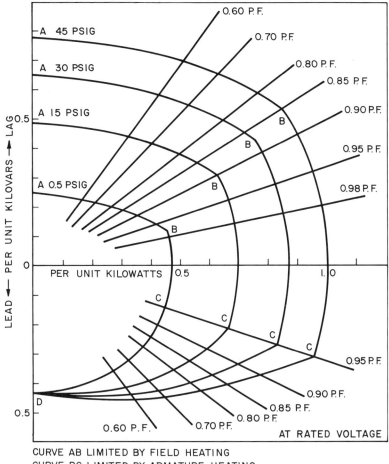

CURVE AB LIMITED BY FIELD HEATING
CURVE BC LIMITED BY ARMATURE HEATING
CURVE CD LIMITED BY ARMATURE CORE END HEATING

Figure 4.50. Typical reactive capability curves.

current line at which generator load is limited by field heating. A change to improve power factor at any load level shows an appreciable reduction in field current.

Typical Reactive Capability Curves. The available steady-state capability of a generator as influenced by the power factor is indicated in Figure 4.50, which is broken into three major components on each of the curves. These curves are divided so that the generator load is limited in each region as a function of the generator component most affected.

Operation on the upper portion of the curve (A–B) is from zero power factor lagging to rated power factor. In this region, the generator is said to be over-excited. On this part of the curve, the field current is at rated value. Since the field current should not exceed this rated value, this represents the capability limitation to prevent field overheating.

In the portion of the curve from B to C, which is from rated power factor lagging through unity to 0.95 power factor leading, the limit is on the stator current, and maximum nameplate stator amperes should not be exceeded.

In the region from C to D on the curves (leading power factor operation), the flux pattern in the generator is such that the end leakage flux from the core is at right angles to the stator laminations. This causes excessive heating in the stator-end iron and in some of the steel structural members. The end-iron heating imposes a limitation on the machine capability in this region. This capability is also reduced as the square of the voltage during reduced terminal voltage operation. In this region of under-excitation, the generator synchronizing torque is reduced. This can have an adverse effect on stability, so stability must be checked in this operating region.

GENERATOR PERFORMANCE

Generator characteristics that are influential in determining performance include the various reactances of the generator, such as synchronous, transient, and subtransient. These are of interest primarily in steady-state, normal response to regulator signal, and very short time characteristics under short-circuit or transient conditions, respectively. The field-winding inductance is also a factor influencing performance, since it governs the rate at which excitation current can be changed.

Short-circuit ratio (SCR) is another measure of the operating characteristics, the number itself being the ratio of field current at no-load-rated voltage to the field current required to produce rated armature current with the terminals short-circuited. In operation, to maintain constant voltage for a given change in load, the change in excitation varies inversely as the SCR. In other words, a greater change in per-unit excitation is required for a machine with low SCR than for high SCR for the same load change.

In general terms, a higher short-circuit ratio is a measure of greater stability, although this is not necessarily true under all conditions. Factors that must be considered are the speed of response of the voltage regulator and excitation system, the match between turbine and generator ratings, total turbine-generator inertia, "stiffness" of the system, etc. It is possible that a generator with 0.64 SCR, for example, in one system might be less stable than one with 0.58 SCR in another high-density system with quick-acting regulator and high-response excitation.

Several factors, such as the increased performance available in control equipment and closely connected systems, have permitted the industry to follow a trend of buying lower short-circuit ratio generators over a period of years. Design parameters involved usually permit a lower-cost generator at the lower value of SCR.

EXCITATION SYSTEM

The function of the excitation system is to provide direct current for the generator rotor windings, or field windings. In addition to this, the generator excitation system maintains generator voltage, controls kilovar flow, assists in maintaining power system stability, and provides important protective functions.

During changing load (or power factor), the excitation system must respond to maintain the proper voltage at the generator terminals. This may require rapid changes in field voltage due to disturbances on the power system. The excitation system should provide the appropriate field current at all times so that it is reliable and requires little maintenance.

Generator Excitation Power Requirements

Generator design considerations determine the field current, field voltage, and consequently, the power required for excitation. Figure 4.51 indicates approximate exciter ratings required for conductor-cooled turbine-generators from 200 to 1400 MVA. The exciter rating band was obtained by plotting the data for machines that have been built by General Electric Company, and there are several interesting points that this data reveals.

Generators of the same MVA rating may vary considerably in excitation requirements because of different power factor and voltage ratings, or short-circuit ratio. Although not shown, the exciter rating for a conventionally cooled generator can be estimated on the basis of 2.5 kW/1000 kVA (two-pole generators less than 200 MVA).

Conductor cooling increases the excitation requirements. An approximate rule-of-thumb for estimating exciter ratings for two-pole conductor-cooled generators appears to be 3.5 kW/1000 kVA. For four-pole conductor-cooled generators, the appropriate figure is 2.3 kW/1000 kVA. These estimates are likely to be accurate to within about 20%. As generator design progresses toward the use of better materials and more efficient utilization, excitation requirements will likely increase. Even with intensive conductor cooling, however, the excitation power requirements will probably not exceed 0.5% of the generator kVA rating.

Although the excitation power requirements are small when expressed as a percentage of the generator rating, the supply and control of the generator field current is important because:

1. It determines the generator terminal voltage and kilovar flow. (The extent to which the field current determines these variations depends on system connections, loads, and relative size of

EXCITATION SYSTEM

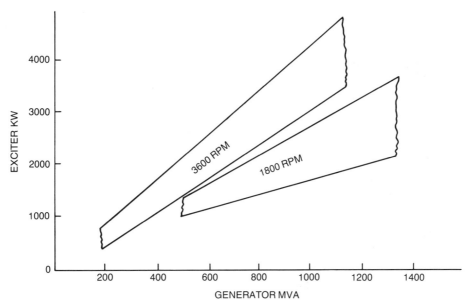

Figure 4.51. Excitation requirements for conductor-cooled generators.

the generator with respect to the remainder of the system.)
2. It enhances the ability of the machine to remain in synchronism with the system to which it is connected, and allows greater utilization of transmission system capacity.
3. It directly affects unit reliability and availability.

Components of an Excitation System

The generator excitation system consists of mechanically and electrically coordinated components. The components can be classified into five functional categories:

1. Power source
2. Rectification
3. Cooling systems
4. Control functions
5. Protective features.

Power Source. The excitation system must furnish the generator excitation power requirements. The excitation system must, in turn, derive its power from a reliable source. The common sources of power for the excitation system are the shaft of the main turbine-generator unit, the plant auxiliary bus, special windings in the generator, or the main generator terminals.

Rectification. The common power sources for the excitation system produce an alternating current; therefore, a means of rectifying the alternating current becomes necessary. A rotating commutator on the dc machine, used as an exciter, was a common means for achieving rectification in the past. Virtually without exception, today's systems use semiconductors such as silicon rectifiers (both controlled and uncontrolled).

Cooling Systems. In the past, air was the common cooling medium for components of most excitation systems. Today's trend toward more efficient utilization of material and space has added hydrogen gas and liquids for cooling where practical or necessary for optimum design.

Control Functions. One of the important basic qualities of a power system is the voltage of the supply to its customers, with a constant voltage at all times being ideal. The field current required to produce a constant voltage at the generator terminals varies over a wide range for various loads and power factors. Adjustment of the field current can be made manually by an operator, or automatically by a generator voltage regulator.

The generator voltage regulator provides the essential control of the generator excitation system. Fundamentally, it controls the ac generator terminal voltage within narrow limits under normal power system conditions and rapidly returns the

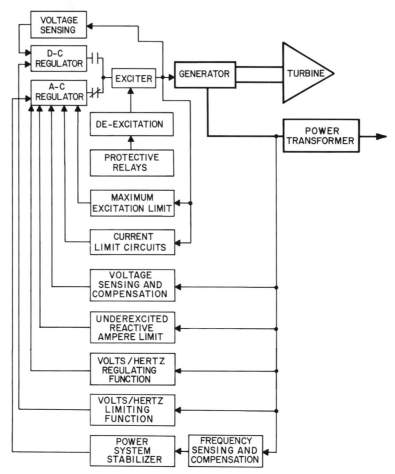

Figure 4.52. Control and protective circuits of a typical excitation system.

voltage to normal when abnormal conditions cause sudden system voltage changes. Additional control circuits are frequently provided for other purposes. Figure 4.52 illustrates the extensive nature of the control and protective circuits which may comprise the excitation system. The essential functions provided are:

1. **AC voltage regulator.** Increases both the steady-state and transient stability limits of the generator with respect to the power system by continuous, fast control of the exciter voltage to match the generator requirements.

    a. *Steady-state.* Adjusts the exciter voltage to maintain the ac generator terminal voltage at a preset value for all normal conditions of load and power factor on the generator. Increases the steady-state stability limit of the generator, allowing full power to be delivered to the power system under adverse operating conditions.

    b. *Transient-state.* Rapidly varies the exciter voltage during abnormal power system conditions to restore the terminal ac voltage to normal as quickly as possible. Allows the generator to continue synchronous operation under more severe disturbances. Generator over-voltage caused by sudden dropping of load is quickly reduced by the fast response of the excitation system. This reduces stress on the system insulation and may permit the use of surge arresters having lower-voltage protective levels.

2. **Power system stabililizer.** A power system stabilizer utilizing signals from shaft speed, frequency, power, or other sources may be used to supplement the voltage-regulating function.

EXCITATION SYSTEM

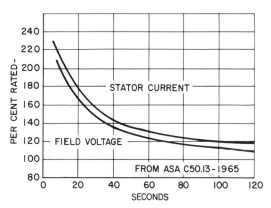

Figure 4.53. Generator short-time thermal capability.

Power system dynamic performance is improved by the rapid damping of system oscillations, and greater utilization of transmission network capacity is possible.

3. Reactive current compensator. Controls the distribution of reactive MVA between generators connected to the same bus so that the generators will operate in parallel with reactive MVA proportioned between the generators according to their rating, or some other predetermined ratio.

4. Line-drop compensator. Varies the generator terminal voltage in accordance with its load to hold constant voltage at some point electrically remote from the generator, such as the station high-voltage bus. This function is sometimes called active and reactive current compensation or line-drop compensation.

5. Underexcited reactive ampere limit (URAL). Limits the amount of underexcited reactive MVA that a generator can supply so that an adequate steady-state stability margin will be maintained with respect to the power system and generator; safe operating conditions will not be exceeded.

6. Maximum excitation limit. Limits the ceiling exciter voltage and the length of time that it can be applied to the generator field, consistent with the safe short-time over-load restrictions on the generator. These are defined in Figure 4.53.

7. Volts/Hertz regulators and protection. Control and limit overexcitation of power plant equipment. Overexcitation describes a condition of excess flux in an electromagnetic device that will saturate the core steel, producing high-eddy current losses in the core and adjacent conducting materials. A convenient expression of the flux is the per-unit excitation, defined as per-unit voltage divided by per-unit frequency (volts/Hertz). The control and protection must ensure that operation is confined within those restrictions defined by Figure 4.54.

Protective Features. Additional protective circuits within the excitation system serve as regulators, limiters, and protective functions. Proper coordination of these protective circuits with the protective relaying will permit the maximum permissible transient (short-time) operation of the generator and excitation system, yet provide proper protection for both normal and emergency operations. Protective devices may be provided to remove power from faulted sections of excitation equipment, thereby improving service continuity of the generator. If a short circuit occurs in the stator windings of the generator, the excitation must be removed as quickly as possible to minimize the damage. A generator field breaker, exciter field breaker, or other means energized by the generator differential relays, are used to provide this protection.

Excitation System Performance Requirements

Performance requirements for excitation systems are determined by considering the operation of the generator during normal steady-state conditions, for severe transient disturbances, and under less severe prolonged disturbances.

Steady-State Conditions. During normal continuous operation of the generator, the excitation sys-

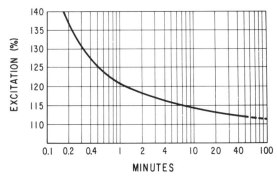

Figure 4.54. Permissible short-time excitation (general guide).

tem should automatically provide and adjust generator field current to maintain terminal voltage within close limits, and to divide MVA properly among machines. This means that the excitation system must be capable of supplying a wide range of direct current to the generator field windings.

In the past, parallel operation of excitation systems was a requirement, since many stations had spare exciters. Over the years, design philosophy has changed, and modern practice is to use a single exciter, depending solely on the reliability, redundancy, and maintainability of that source. This trend to eliminate spares is also applicable to other important power plant auxiliary equipment.

Transient Disturbances. A transient disturbance is an abnormal condition on a power system that may be caused by a system fault, switching, or load changes. The disturbance may have a duration of several cycles or perhaps a few seconds. When a transient disturbance occurs, the excitation system should rapidly respond to any dip or rise in generator terminal voltage. The excitation system must act to maintain stability during the abnormal condition and adjust excitation to meet the requirements of the conditions that exist afterward. There are two characteristics of an excitation system that are important in evaluating the effectiveness of the excitation system to assist in maintaining stability. Those characteristics are voltage response and ceiling voltage.

A need exists to define and specify such characteristics of the complete, operational excitation system. IEEE Standard 421-1972, "Standard Criteria and Definitions for Excitation Systems for Synchronous Machines," provides the framework for such specifications.

The historic definition for "Excitation System Voltage Response Ratio" was devised for response requirements within the first 0.5 second. This definition has been retained and is now supplemented by a new definition that covers higher-performance equipment with speeds capable of completing control action within 0.1 second. Figure 4.55 illustrates the concept of this basic definition of voltage response ratio. Curve AB represents the response of "conventional" systems. Curve AB' is for a "high initial response" system wherein ceiling voltage is reached almost instantaneously.

The new definition for a "High Initial Response Excitation System," was first proposed in IEEE Paper 69 TP-154 PWR, "Proposed Definitions for Synchronous Machines" and subsequently adopted

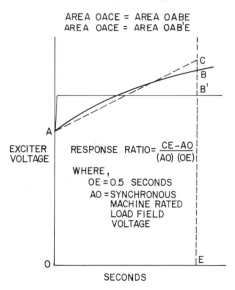

Figure 4.55. Excitation system voltage-response ratio.

in the previously mentioned IEEE Standard. The 0.1-second time interval was chosen as a practical value for any high performance, fast-acting system. This definition requires that 95% of the difference between rated-load field voltage and ceiling voltage be attained within this time period.

These two definitions will provide users sufficient means to specify excitation system performance and also give the manufacturers a means for competitively pricing specified systems. The time period of the definition of voltage response ratio is 0.5 second since stability following a power system fault is usually determined within this time span.

Exciter voltage response ratio has been used in the past to describe the response of the exciter portion of the excitation system. With some older types of excitation systems, this adequately describes the performance of the entire excitation system, but it is not sufficient to cover many systems in use today.

Another important consideration in determining the excitation system performance during a transient disturbance is the maximum or ceiling voltage output from the excitation system. The higher the excitation voltage during a transient disturbance, the more forcing action there will be for returning generator voltage to normal so that the power system voltages are brought back to normal. There is, however, the necessity to coordinate with the generator field insulation and to establish a practical limit to the ceiling voltage for which an excitation system can be designed without excessive cost. With present large generator designs, this maximum

EXCITATION SYSTEM

is approximately 200%. Likewise, it is desirable that a minimum ceiling voltage be established. The recognized minimum ceiling voltage in the power industry today is 120% of the generator-rated excitation voltage.

If the excitation system utilizes the generator terminals or the station auxiliaries bus as a power source, additional requirements exist. During power system disturbances, both of these voltage sources may be severely depressed. The average voltage level over the first few seconds following a severe fault may be as low as 70%. The excitation system must be capable of providing adequate field forcing voltage during such times of distress.

Prolonged Disturbances. A prolonged disturbance is an abnormal condition on a power system lasting several seconds and possibly longer. The disturbance is usually a consequence of some transient disturbance which has just occurred. When the transient disturbance has ended, the generator terminal voltage may be abnormally low due to loss of other generation, and the voltage regulator will attempt to bring it back to normal by applying maximum available excitation voltage to the generator rotor. These conditions may exist for an extended period. The generator has sufficient thermal capacity so that it can operate with ceiling voltage on its rotor and values of stator current greater than rated for a short time. Depending on the magnitudes involved, this short-time overload capability may extend from 15 to over 60 seconds as illustrated in Figure 4.53. A limiter is frequently provided in the excitation system to reduce the excitation before damaging generator temperatures are reached.

Types of Excitation Systems

Excitation systems have taken many forms over the years. The following is a brief historical review of those in common use.

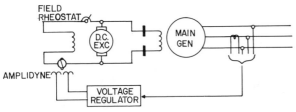

Figure 4.56. Dc generator-commutator exciter with GE amplidyne voltage regulator.

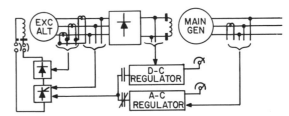

Figure 4.57. ALTERREX excitation system diagram.

Dc Commutator Exciter with Amplidyne Voltage Regulator. Figure 4.56 shows an elementary schematic representation. The dc exciter was either shaft or motor driven. When shaft driven, gearing was needed for the larger kilowatt sizes, as design limitations required reductions in speed.

This type of system was developed to a high degree of reliability. It has lost favor since 1965, mainly because of the large physical size of the equipment.

Alternator Rectifier System. In the ALTERREX* excitation system diagram (Figure 4.57), main turbine-generator excitation power is provided by the output of the alternator exciter. The ac output of the alternator exciter is rectified by stationary silicon-diode rectifiers. The dc output of the rectifier bank goes to the main generator field. A field breaker may be installed in the main generator field circuit or in the exciter field circuit as shown.

The self-excited alternator-exciter derives its field power through potential transformers (PTs) and current transformers (CTs). The voltage regulator uses thyristor rectifiers which establish proper alternator field voltage in response to either the automatic or the manual regulator signals.

The ALTERREX* excitation system consists of the following major components:

1. The alternator-exciter is a direct-coupled ac generator (Figure 4.58) driven from the main generator rotor. It has a stationary armature and a rotating field. The alternator is designed to permit handling as an assembled unit mounted on its own base. The bearings are supported in end shields and are pressure-lubricated from the generator system. The assembly is totally enclosed and is cooled by an integral air-to-water heat exchanger, which utilizes externally supplied cooling water.

* Trademark of General Electric Company.

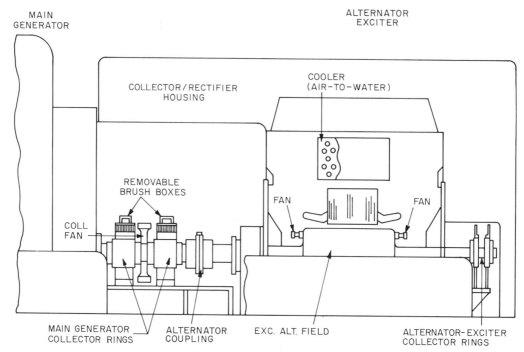

Figure 4.58. Direct-coupled alternator-exciter.

2. The power-rectifier system consists of multiple parallel three-phase diode circuit assemblies, the number depending on the rating of the system. Water provides an effective method of cooling these power-rectifier assemblies.

3. The semiconductor voltage regulator automatically regulates main generator terminal voltage. The regulator contains two switchable silicon-controlled rectifier (thyristor) bridges, either of which will supply rated alternator-exciter excitation. Two independent modes of regulation are provided: (1) ac automatic regulator to maintain desired generator stator terminal voltage and (2) dc regulator to maintain constant generator field voltage. The ac and dc regulators are also switchable. Either regulator can operate with either parallel bridge.

To conserve station space, the silicon-diode power rectifiers (Figure 4.59) are located in the sides of the exciter housing, and are fully accessible from outside the housing. This provides an assembly in which all the main power circuitry of the system is included in one compact unit, minimizing external power circuit interconnections. The voltage regulator with the air-cooled thyristor rectifiers and their associated stepdown transformer are mounted in a free-standing cubicle located elsewhere.

The rectifier circuits consists of water-cooled modules with bus connections along the top, and water connections at the bottom. The power-diode rectifiers are cooled by supplying deionized cooling water from the generator stator cooling water pumping unit.

Alternator Controlled-Rectifier (Thyristor) System. This is an alternator thyristor system. Functionally, this ALTHYREX* system is the same as the ALTERREX* system with the exception that the ac output of the alternator-generator is rectified by thyristors rather than diodes. The schematic diagram is shown in Figure 4.60.

Ac voltage from the exciter alternator is phase controlled by the thyristors to furnish dc voltage for the main generator field. The exciter is self-excited. Exciter terminal voltage is maintained by a static voltage regulator.

The thyristor power rectifier is constructed in a similar manner to the diode power rectifier used on the ALTERREX* system. The assembly consists of thyristors mounted on heat sinks, which are, in turn, connected to form parallel three-phase bridge circuits. Figure 4.61 illustrates a single thyristor mounted on a heat sink.

* Trademark of General Electric Company.

EXCITATION SYSTEM

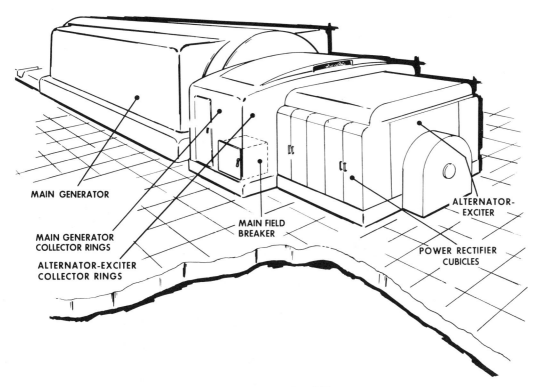

Figure 4.59. Compact ALTERREX arrangement.

When the main generator ac voltage is lower than normal, the ac voltage regulator advances the firing angle of the thyristors, causing a greater portion of the exciter alternator voltage to appear across the main generator field. When the ac voltage is higher than normal, the firing angle of the thyristors is retarded, causing a smaller portion of the exciter alternator voltage to appear across the main generator field. Since the exciter alternator voltage is maintained constant at a value of voltage corresponding to that required for ceiling dc voltage, it is available almost instantly.

Exceptionally high response ratios are, therefore, attainable with relatively modest exciter ceiling voltages. This system inherently provides high initial response.

This excitation system has been supplanted by the GENERREX* excitation system.

Potential Source Controlled-Rectifier. In this system, the power source is the generator terminal voltage, or the station auxiliary bus, rather than the generator shaft (see Figure 4.62).

Although this system also has a fast response, its

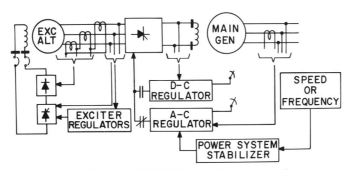

Figure 4.60. ALTHYREX excitation system diagram.

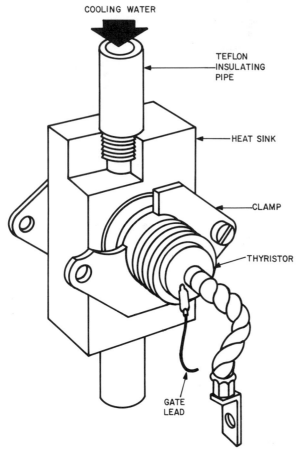

Figure 4.61. Water-cooled 500-ampere silicon thyristor.

ceiling and performance is limited by the bus or terminal voltage. During fault conditions, this voltage will be at a minimum when the regulator requires maximum output.

Compound-Rectifier System. The excitation system supplies field excitation to a generator by taking

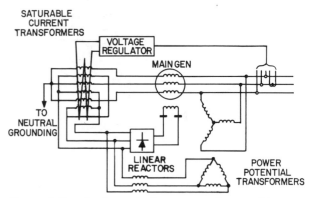

Figure 4.63. Static excitation system diagram illustrating the system applied to generators of approximately 240,000 kVA and less.

excitation power from the stator terminals of the generator through potential and current windings of the excitation transformer. The major components of this system include the excitation transformers, linear reactors, the power rectifier, the manual and automatic regulators with their auxiliary functions, and the main generator field breaker. These are illustrated in the schematic diagram (Figure 4-63) and in a typical station arrangement (Figure 4.64).

When the generator is not supplying current to a load, all field-excitation power is supplied by the potential source. When the generator is supplying current to a load, a portion of the excitation power is derived from generator-load current through current-transformer action. This current input also enables the exciter to supply ceiling excitation during system fault conditions, with severely reduced terminal voltage (or station-bus voltage).

The systems utilizing this concept are constructed in two forms, different in component arrangement, but functionally similar. For systems

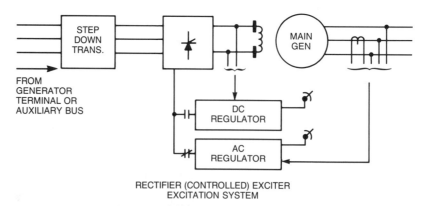

Figure 4.62. Potential source-controlled rectifier exciter.

EXCITATION SYSTEM

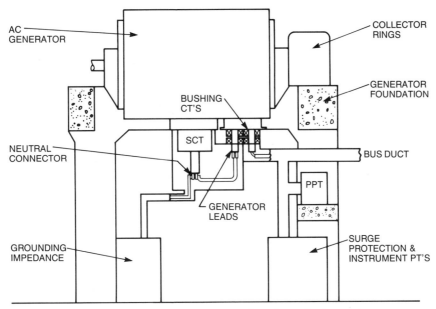

Figure 4.64. Station arrangement of static excitation system.

applied to generators rated up to approximately 240 MVA, the voltage source is obtained from separately mounted, power potential transformers (PPTs) and saturable-current transformers (SCTs). This is the system illustrated in Figures 4.63 and 4.64. An alternate configuration combines the voltage and the current sources into a single transformer. This single excitation transformer, referred to as a saturable-current potential transformer (SCPT), embodies the same basic functions as employed in the separately mounted configurations. The station arrangement for this alternate is shown in Figure 4.65. Designs have been accomplished accommodating generators up to 800 MVA.

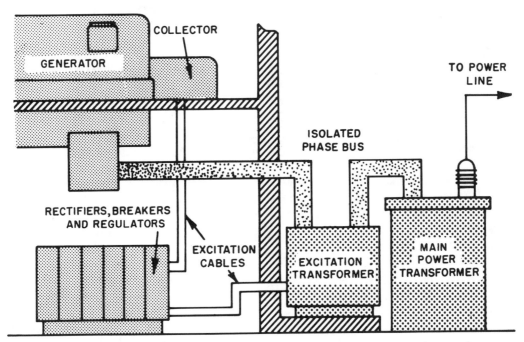

Figure 4.65. Typical station arrangement of static excitation system for large units.

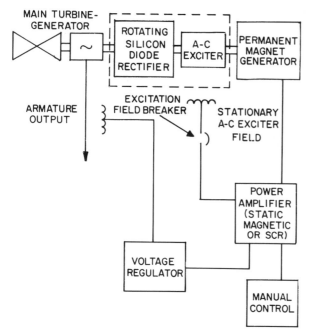

Figure 4.66. Rotating rectifier excitation system.

Rotating Rectifier Excitation System. A brushless generator-excitation system is accomplished by the rotating rectifier excitation system (Figure 4.66). It consists of an ac exciter and a rotating rectifier mounted on the same shaft as the turbine-generator field. Overhung from the ac exciter is a small permanent magnet generator whose stator output furnishes excitation energy, as controlled by the voltage regulator, to the stationary field of the ac exciter. The output of the rotating armature is fed along the shaft to the rotating silicon diode rectifier. The output of the rectifier is fed along the shaft to the field of the ac generator.

With special design of the ac exciter and larger amounts of excitation energy from the permanent magnet generator, high initial response performance can be accomplished.

Compound-Controlled Rectifier Exciter System. The GENERREX* is a highly integrated generator-excitation system. Utilizing all static components, this system promises a high degree of reliability.

The excitation power for the generator field is provided by a voltage-derived source and a current-derived source within the generator stator. A set of three-phase, water-cooled windings mounted in three generator stator slots and a series linear reac-

* Trademark of General Electric Company.

tor is the voltage source. Current is obtained through current transformer action directly from the stator windings. The neutral of the generator is made internally. These sources are combined through transformer action and the resultant ac output is rectified by stationary power semiconductors. This is illustrated in the one-line diagram of Figure 4.67, which shows the essential components.

A unique combination of diodes and thyristors in a "shunt" bridge provides the means of control. This control functions through periodic short circuiting of the ac source, and is made possible due to the "current transformer" characteristics of the source. Figure 4.68 schematically shows the nature of this bridge circuit.

The physical arrangement of the power source within the generator dome assembly and the power rectification equipment alongside the generator results in a compact station arrangement. An artist's conception is shown in Figure. 4.69.

The GENERREX excitation system is a high-initial-response system by virtue of the use of controlled rectifiers in the generator field circuit, and the compounding of action to provide excitation power during faults and disturbances on the power system.

The GENERREX excitation system is applicable over the full range of ratings suitable for electric utility application, including the largest now foreseen for the future. Voltage response ratios up to and including 3.5 per unit are being furnished; capability exists for even higher response ratios.

CONTROLS AND INSTRUMENTATION

Modern power plants are operated from some form of centralized control room. The centralized control room has the necessary controls and instrumentation to enable operating personnel to supervise and operate the boiler turbine-generator unit and its associated auxiliaries in a safe and efficient manner. The centralized control concept minimizes the operating staff but, of greater importance, it provides the operators with a compact arrangement of operating controls and a complete picture of the entire process from fuel to electrical output. When abnormal situations develop, complete information on the entire cycle permits faster and more accurate analysis of the problem, and the compact arrangement of operating controls provides the means to implement corrective action rapidly.

CONTROLS AND INSTRUMENTATION

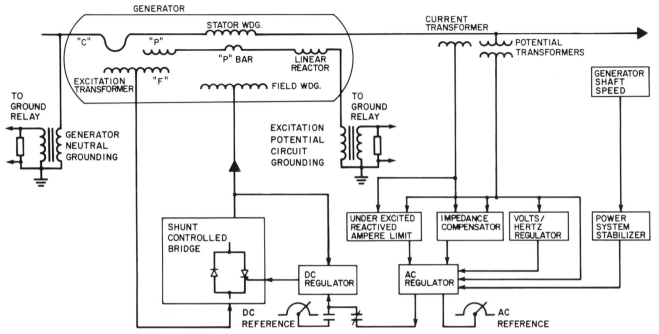

Figure 4.67. GENERREX excitation system—compound-controlled rectifier exciter.

The control functions in modern plants are accomplished manually by operators, by automatic controls, and by combinations of manual and automatic operation. While there has been increasing use made of automatic controls, they frequently require manual assists and in many cases, the control function must be accomplished manually during plant startup until the plant is at a load level that is in the operating range of the automatic controls. This is particularly true of controls relying on flow readings that are difficult to measure at low flow rates.

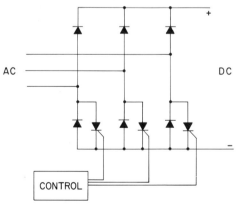

Figure 4.68. Shunt-controlled rectifier bridge.

Automatic controls constantly sense changes and take corrective action more quickly than a human operator and with greater accuracy and reliability. They also relieve the operator of control functions that would require virtually constant attention and would, therefore, be tiring. Control functions or actions requiring infrequent attention (such as positioning drain and vent valves, starting motors and pumps, and placing systems into automatic operation) are frequently left to manual operation. In general, all automatic control functions must be capable of being operated manually.

Instrumentation in modern power plant control rooms is provided by indicating and recording instruments, annunciators, indicating lights, and by digital computer-based data acquisition systems. Some minimum level of instrumentation is provided with the automatic control functions to permit manual operation.

The control and instrumentation in a fossil-fueled plant typically includes:

1. On-line, closed-loop, major automatic control systems such as boiler controls, turbine controls, flue gas scrubber controls, and many minor sub-loop controls, such as temperature controls for turbine lube oil and generator hydrogen, level controls for the deaerator and

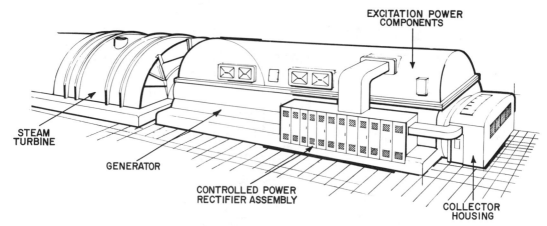

Figure 4.69. GENERREX excitation component arrangement.

condenser hot well, pump recirculation flow control, etc.
2. A variety of start-up and shut-down controls.
3. Monitoring, alarm, and protective tripping functions.
4. Indicating and recording instruments and annunciating devices.
5. Digital computer-based acquisition and control systems.
6. Man–machine interface, including the instrumentation and control means, and the extensive use of cathode ray tubes for data and graphic displays.

On-Line, Closed-Loop Automatic Controls

Boiler Controls. There are two basic types of boilers, the drum type and the once-through type. In the drum type, which always operates at subcritical pressures (less than 3208 psia), the steam is separated from the water in a drum which stores a large volume of steam and water. In the once-through type of boiler for either subcritical or supercritical pressure, there is no drum and the only storage of fluid is in the tube paths. Although the control of the two types differ, the primary functions of boiler control are the same for both the drum and once-through types of boiler. They are:

1. To supply heat to the boiler in accordance with the demand for steam.
2. To properly proportion the fuel and combustion air.
3. To supply the proper amount of feedwater.
4. To maintain the proper steam temperatures.
5. To protect against failures of the boiler and auxiliaries and to initiate tripping as required.

The controls utilize measurements of the various boiler parameters to be controlled (pressure, temperature, level, etc.) and operate control valves, dampers, vanes, and variable-speed drives through various forms of actuating devices. These actuators are driven either pneumatically or electrically.

Controls for a drum-type boiler generally include:

1. Combustion (pressure) control.
2. Feedwater (or drum level) control.
3. Temperature control.

Combustion Control. Combustion control controls the heat (or fuel) input to the boiler to maintain a desired steam pressure. To assure proper combustion, the air flow to the boiler must be coordinated with the fuel input. If the boilers were not subject to wide and rapid fluctuations in load, the fuel and air requirements could be determined by steam pressure alone. For example, when the demand for steam increases, the boiler pressure drops. This pressure drop can be sensed and the fuel and air flows increased to bring the steam pressure back to normal. Power plant boilers are subject to rapid load changes and a two-element control is commonly used. In addition to sensing pressure, the control also uses steam flow (or some equivalent signal) as a demand or feed-forward signal. Any change in steam flow will result in the fuel and air

CONTROLS AND INSTRUMENTATION

being adjusted rapidly to a value that is approximately correct and then trimmed by pressure-control action.

Maximum combustion efficiency requires more air than the theoretically correct value and this varies with load. As a result, the fuel-to-air flow ratio must be adjusted as a function of load. This adjustment is generally automatic. The characteristic of optimum excess air as a function of load is used as a reference, and the air flow is automatically trimmed to that value.

Another related air control is the furnace draft control, which adjusts the forced draft and induced fan air flows to maintain a slightly negative pressure in the furnace.

Feedwater Control. The water in the boiler drum is maintained at fairly constant level. During load changes, the water level in the boiler drum also changes. If the steam flow from a boiler is suddenly reduced, the pressure rises; this causes some of the new steam in the tubes to return to the liquid state. Also, the combustion rate quickly drops, and with it, the rate of evaporation. Therefore, the amount of entrained steam in the tubes is reduced so that the water level in the drum falls or shrinks. Conversely, when the demand for steam increases, the pressure drops, which increases the steam in the tubes and results in a higher-than-normal water level in the drum known as swell. These shrink and swell characteristics are the opposite of what would be expected when controlling water level. As a result, a feedwater control that senses only water level would be in error much of the time. The usual solution is a three-element control, two elements of which balance the flow of feedwater and steam, with a trimming action by drum level. Smooth regulation of the feedwater flow is obtained and the fluctuation of water level in the drum is kept to a minimum. This type of control system can be adjusted to maintain a high water level at high rates of flow and a low water level at low rates of flow (or start-up). This effectively increases the drum capacity to accommodate water level surges upon sudden load changes.

Steam Temperature Control. Because of the large amount of heat storage in the metal of the superheater and reheater sections of the boiler, there is considerable lag in the response of steam temperature to firing rate. Nevertheless, it is necessary to hold the steam temperature very nearly constant, especially in boilers that operate in the vicinity of 1000°F. The superheater is generally split into sections. The temperature of the steam leaving the superheater is frequently controlled by the introduction of a water spray in an attemperater located between superheater sections. The amount of spray water introduced is regulated by control valves. Other means of steam temperature control used with attemperation include gas recirculation, tilting burners, and dampers.

Many modern boiler control systems make use of coordinated control whereby changes in megawatt demand (increases or decreases) for the unit are sent to both the boiler and turbine. The rate of load pickup by the turbine is limited so that boiler pressure swings during load changes are within acceptable limits. Also, in many cases, interlocking protective-logic burner controls and boiler safety systems are integrated with the boiler control systems and can be equal to or greater than the scope of the boiler control.

Turbine Controls. Modern reheat turbines are furnished with a control and protective system that provides the speed/load control functions, the start-up/shut-down controls, the overspeed control, and trip protection and numerous other protective trip functions. Turbine controls also include a variety of minor control functions for auxiliary systems associated with the turbine. Present-day turbine control systems consist of hybrid analog and digital electro-hydraulic controls. The servos that control the main turbine valves are operated with fireproof hydraulic fluid with pressure in the neighborhood of 1500 pounds per square inch (10.3 MPa).

The speed control portion of the speed/load control function provides the speed governor characteristic with a 5% speed droop over the full load range. This 5% speed droop, or regulation enables turbine-generators to operate in parallel with stable division of load between units. The speed droop, or regulation, also provides the ability to respond to the power system load changes with changes in power system frequency. This is covered in more detail in Chapter 14. The load control function is provided by adjusting the speed-droop characteristic which sets the average position of the main control valves to control the admission of steam to the turbine and, therefore, the turbine output or load. During start-up, when the turbine is being accelerated to rated speed, the speed ramping is handled by an acceleration control.

The turbine speed control also limits the overspeed in the event the load on the unit is rapidly reduced or lost. This involves controlling the position of the main control valves admitting steam to the turbine and also the intercept valves which, during overspeed conditions, control the admission of the entrained steam from the reheater to the intermediate and low-pressure sections of the turbine. Modern units that have relatively low inertias may make use of additional control means of limiting overspeed after load rejection. These include fast closing of the intercept valves when necessary and a trip anticipator function which rapidly closes all of the turbine main valves, including the main stop, the control, the intercept, and the reheat stop valves. In all cases of overspeed control, the valves reopen automatically as the turbine speed returns to normal. If for any reason the overspeed is not being controlled to a safe level, overspeed trips will function to trip all turbine main valves. This overspeed trip function is provided by the mechanical overspeed trip and a backup overspeed trip.

Other protective trip functions are provided for:

1. High condenser pressure
2. High journal bearing vibration
3. High exhaust hood temperature
4. Low hydraulic fluid pressure
5. Low bearing oil pressure
6. Thrust bearing wear
7. Control system failures (power loss, speed feedback loss, etc.).

Turbine-generator auxiliary systems that include controls are the bearing lube oil system, hydraulic fluid system, steam seal system (which seals the turbine shaft), turbine exhaust hood water sprays for temperature control, turning gear system which rotates the main shaft at 5–10 r/min to prevent shaft bowing when the unit is shut down, generator hydrogen seal system, and generator stator and hydrogen-cooling systems.

Start-Up and Shut-Down Control. The turbine and the boiler undergo thermal and mechanical stresses in the change from ambient temperature and pressure to working temperatures and pressures as high as 1000°F (538°C) and 3500 psi (24.13 MPa). The turbine must be carefully started because of vibration, rotor thermal and centrifugal stresses, and differential expansions. The boiler also experiences thermal stresses and, therefore, requires a careful warm-up. These considerations require the start-up and shut-down to be closely controlled, in some cases, by a digital computer to avoid overstressing the unit.

To reduce thermal stresses in the turbine and the control valve chest, low-pressure, low-temperature steam is used to prewarm the turbine control valve chest and high-pressure rotor prior to accelerating the unit from turning gear to rated speed.

Initial warm-up of the boiler is accomplished with warm-up burners using oil or gas fuel. After the turbine-generator is at rated speed and synchronized to the power system, the main burners are placed in service. In the case of coal-fired units, this requires that the coal pulverizers associated with the main burners be started. The start-up of coal pulverizers and the light-off of burners in the furnace are critical functions and frequently make use of hybrid analog and digital control techniques. The pulverizer control may include sequencing for start-up as well as modulating control for operation. Burner control includes sequencing for light-off and overall burner management, as well as the flame monitoring function for alarm and/or tripping after burners are in operation.

Monitoring, Alarming, and Tripping. The operator cannot monitor the several thousand conditions in the power plant without automatic equipment. This generally includes a digital computer system which reads plant sensors, compares the values read with assigned limits, and notifies the operator when they are in an alarm condition. It also reads switch positions to check that valves, circuit breakers, and other plant devices are in the proper position. Abnormal conditions are displayed for operator use. Data may also be continuously recorded in memory for later analysis, or for quickly ascertaining the recent operating history of various plant equipment, should the operator require the information. Many devices are used as sensors: thermocouples and resistance temperature detectors (RTDs), pressure transducers, differential pressure transducers, vibration sensors, position transducers, etc. Back-up alarm devices of a simplified nature are used to ensure safe operation during digital system outage or maintenance.

Critical items are monitored continuously by protective tripping functions. The protective functions will initiate a unit trip in the event of an abnormal

CONTROLS AND INSTRUMENTATION

Figure 4.70. Typical control room with control and instrumentation bench boards.

condition which would jeopardize the safety of major equipment or personnel.

Indicating and Annunciating Devices. These are used to convey pertinent information to the operator. Since the most important function of the unit is to generate kilowatt hours, redundant systems and devices are commonly employed so as to be able to operate the unit while some systems and devices are inoperative.

A partial list includes:

1. A digital computer, which provides both indication and annunciation.
2. Indicating instruments for pertinent plant variables.
3. Indicating lights to display equipment status (on-off, open-closed, etc.).
4. Panel-mounted annunciators.

Recording Devices. Historical records are kept for plant reports, for analysis of problems, and for accounting purposes. These include the hourly and daily log of operation gathered and prepared by a digital computer or manually. Strip-chart recorders are commonly used for pertinent variables, such as feedwater flow, steam pressure and temperature, etc. Another recording function is to provide a printed record of the sequence of events that is associated with a trip or major disturbance. This function may be included in the computer system or may be a separate system.

Man–Machine Interface. The operator must be able to observe the operation of the plant, focus attention on any abnormal situations, and initiate necessary control action. These functions are accomplished through M/A (manual/automatic) stations of the boiler control equipment, through control and push-button switches, through the digital computer console and its input/output devices, and through the indicating, recording, and annunciating devices mentioned earlier.

Figure 4.70 shows a modern nuclear unit control room with the control and instrumentation bench boards. Fossil-unit control rooms do not differ greatly from this one. Much "human engineering" is involved in the layout of the BTG board (boiler-turbine-generator) to make the array of information as intelligible as possible to the operator and to provide controls that are easily used particularly during abnormal or emergency situations. Extensive use is made of video-graphics on CRT displays which, in Figure 4.70, are located on the near vertical section of the console. Video-graphics, bar charts, and other forms of video display are replacing many conventional instruments in modern control rooms. A trend that is developing is the concept of control "by exception," which emphasizes the use of the operator to take care of unusual or abnormal conditions in the plant, while automatic controls handle the normal within-limits adjustments built into the automatic control system. Because of the large quantity of information being transmitted from the plant to the control room, the technique is becoming more and more necessary.

As plants grow larger and control systems become more complex, it has become increasingly difficult to interface the interrelated subsystems properly to make a single overall system that functions properly. The development of more complex control and monitoring algorithms for implementation in computer systems has become an important aspect of plant and system design.

The proportion of plant cost for control and instrumentation has been steadily increasing, and this trend is likely to continue.

Nuclear Plant Controls

The controls required for the operation of a nuclear plant differ considerably in detail from those required by the fossil plants, but the functions to be performed are analogous. The control categories itemized in the previous section generally apply, bearing in mind the differences in the technology used to supply the steam for the turbine.

Nuclear plants are controlled by a combination of manual and automatic devices. The following brief discussion of the controls involved in the operation of a boiling water reactor plant will give an idea of the differences and similarities between nuclear and fossil plant operation.

Boiling Water Reactor (BWR) Plant Control. BWR plant output is increased or decreased by changing the reactor recirculating water flow and/or moving the control rods. As the reactor power output is changed, an initial pressure regulator automatically adjusts the turbine admission valves to maintain constant reactor pressure. The resulting change in reactor steam output thereby causes a corresponding change in the turbine-generator power output.

Plant Start-Up. Start-up of the plant is performed manually by the operator. Control rods are withdrawn according to a predetermined schedule to achieve a critical state in the reactor. Power is raised to the level that gives the desired rate of temperature increase based on specified allowable temperature differentials between parts of the reactor vessel. Several neutron-counting channels are used to monitor neutron flux constantly in the reactor core as the power level is gradually increased.

While the reactor temperature is being increased, the turbine is being rotated on turning gear. When reactor steam is available, and after a partial vacuum has been established in the main condenser, heating and loading of the turbine are accomplished by first establishing a flow of steam to the condenser through by-pass valves. This flow is gradually transferred to the turbine until rated speed is achieved, after which the unit is synchronized with the power system. Reactor power is then increased to pick up electrical load by withdrawing control rods and by controlling the recirculation flow. The time required to bring the reactor to a critical state and to pick up load on the unit will depend on plant size, operator proficiency, specific operating requirements, and the history of the reactor and the turbine.

Power Operation. After the generator is synchronized to the electric system and is producing a substantial output, the power output can be adjusted to meet the system requirements by manual adjustment of control rods, manual or automatic adjustment of reactor recirculation flow, or a combination of these two methods.

With a boiling water reactor, maneuvering capabilities in response to daily load fluctuations are provided solely by flow control, or in combination with control rod motion when circumstances warrant. Power change rates of up to 60% per minute are possible using flow control, but for normal operation, rates of up to 15% per minute are recommended. Power change rates of up to 2% per minute are possible with control rod motion.

Normal Plant Shutdown. Under normal shutdown conditions, reactor power and plant output are reduced by manual insertion of control rods. After turbine load is reduced to a minimum value, steam flow is established through the by-pass valve and the generator is disconnected from the system. If refueling or other functions requiring access to the vessel are planned, all control rods are inserted and the reactor is cooled down by release of steam to the main condenser.

Reactor Protection System. The reactor protection system initiates a scram (rapid insertion of control rods) for malfunctions that could damage the reactor.

The system performs a difficult task. First, it must have a very high probability of causing a scram when required. Second, it should not cause a scram when it is unnecessary. Thus, the system must be comprehensive in scope, and reliable in nature. Reliability of the system is assured by redundancy and testability in all vital systems. Some of the plant conditions that are monitored by the reactor protection system include: reactor pressure, water level, and neutron flux.

It should be noted that many of the control and protection features originally installed in nuclear plants are finding application in fossil plants as well. The improvements in plant protection and overall system reliability have made these features attractive to other large fossil stations as well.

SUMMARY

A modern steam-generating station contains furnaces, boilers, turbine-generators, and condensers; also, to improve thermal efficiency, feedwater heaters, economizers, air preheaters, superheaters, and resuperheaters. Fans are required to force the combustion air through the furnace, and pumps are used to force the condensate back into the boiler and through the system.

Nuclear plants are similar to fossil plants in that the turbine, condenser, feedwater string, and steam side auxiliaries are quite similar. The auxiliaries required by the nuclear steam generator, however, eliminate the auxiliaries required by the fossil-fired boiler in the conventional steam plant, while other different systems become necessary.

Combustion turbines are increasingly important sources of power on utility systems for peaking service. With exhaust heat recovery equipment as part of a combined cycle installation, combustion turbines can also serve as base load and midrange generation sources.

REFERENCES

1. Bartlett, R. C., *Steam Turbine Performance and Economics,* McGraw-Hill, New York, 1958.
2. Faires, V. M., *Thermodynamics, MacMillan,* New York, 1971.
3. Haywood, R. W., *Analysis of Engineering Cycles,* Pergamon, Oxford, 1980.
4. Haywood, R. W., *STAG Combined-Cycle Power Systems—Present and Future,* GE Publication GER-3115, 1980.
5. Haywood, R. W., "Utility Boiler Design/Cost Comparison-Fluidized Bed Combustion Versus Flue Gas Desulfurization," U.S. Environmental Protection Agency, Report EPA-600/7-77-126, November, 1977.
6. Rankin, A. W., "Auxiliary Turbine Drives For Boiler Feed Pumps," *Combustion,* January, 1964.
7. Salisbury, J. K., *Steam Turbines and Their Cycles,* Wiley, New York, 1950.
8. Woodruff, E. B., and Lammers, H. B., *Steam-Plant Operation,* McGraw-Hill, New York, 1977.

5
POWER FROM NUCLEAR FUEL

INTRODUCTION

In December of 1957, 15 years after the discovery and first demonstration of a controlled nuclear-fission chain reaction, the Vallecitos Boiling Water Reactor entered commercial operation, generating the world's first commercial electrical power using nuclear energy. This plant was designed and built by General Electric for the Pacific Gas and Electric Company in the San Francisco Bay area and had an electrical output rating of 5 MWe (megawatts electrical).

Today, the commercial nuclear power industry serves a growing market on a worldwide scale. Within the United States, the predominant reactor types supplying commercial electrical energy today are the so-called light-water reactors (LWRs), i.e., the boiling water reactor (BWR) and pressurized water reactor (PWR). This chapter contains a discussion of the fundamental physics concepts involved in the design and operation of LWRs, and of the BWR in particular.

The chapter is divided into three basic sections:

1. "Physics of Nuclear Processes," presents a brief review of the basic principles needed to understand the neutron behavior within a nuclear reactor core.
2. "Reactor Core Behavior," reviews the essential features of nuclear core operation.
3. "The Nuclear Fuel Cycle," reviews the physical processes and economics, with emphasis on the LWRs requiring slightly enriched uranium as nuclear fuel.

PHYSICS OF NUCLEAR PROCESSES

Nuclear Reactions

Radioactive Decay and Isotopic Transmutation. An element's chemical nature is determined by the number of electrons in its atom. The number of electrons in a neutral atom is exactly equal to the number of protons in its nucleus. As viewed from a distance, the positive charges of the protons are cancelled by the negative charges of the electrons. Thus, the number of protons in a nucleus determines the element; and the number of neutrons determines the isotope. For example, the uranium nucleus has 92 protons. If it has 143 neutrons, it is uranium-235. If it has 146 neutrons, it is uranium-238. There are 14 known isotopes of uranium.

The chemical behavior of an element is determined by the number of electrons needed to make the atom electrically neutral. The required number of electrons needed to do this is, of course, just equal to the number of protons in the nucleus.

Approximately 1000 distinct isotopes have been identified experimentally, and are grouped into 103 elements with names. There are four more elements shown on a chart of the nuclides, but they are unnamed. Of all identified isotopes, less than one-third (279) of the species are stable, that is, they are nuclei that do not experience natural decay or disintegration. The remainder have some intrinsic nuclear imbalance which eventually causes them to change their structure. Such changes always lead to a more stable isotope and involve the emission of some disintegration product. Typical examples of these emitted particles include electrons (beta decay), helium nuclei (alpha decay), positrons (like electrons, only they carry positive charges), and on rare occasions, neutrons.

All nuclei, except for the simplest of all, H_1 (consisting of a single proton), can be excited without altering their basic nuclear structure and will return eventually to their natural (ground) state by emitting one or more photons (energy). These photons are of very high energy and are called gamma rays. Nuclei can be excited artificially by bombarding them with photons or particles such as neutrons, protons, or helium nuclei.

Nuclei can also have their isotopic identification changed by bombardment and capture. Such induced changes are called isotopic transmutation and play a central role in the physics of nuclear reactors. For example, when a neutron of atomic number 1, charge 0 (denoted by $_0n^1$), is absorbed by a uranium-238 nucleus (containing 92 protons and 146 neutrons for a total of 238 "nucleons," denoted by $_{92}U^{238}$), a new isotope of uranium, $_{92}U^{239}$ is formed. Such a reaction is written schematically as

$$_{92}U^{238} + {_0n^1} \rightarrow {_{92}U^{239}} + \gamma \text{ (gamma ray)}, \quad (1)$$

where the symbol γ, denotes the fact that when the $_{92}U^{239}$ nucleus is created in this way (called a "compound nucleus"), it appears in an excited state which decays by emitting a photon (i.e., a γ ray). This preserves the energy balance.

The isotope $_{92}U^{239}$ is unstable and spontaneously decays by beta emission, to an isotope of the element neptunium:

$$_{92}U^{239} \xrightarrow{23m} {_{93}Np^{239}} + {_{-1}\beta^0} \quad (2)$$

where β denotes beta ray or the emitted electron. Here we have explicitly denoted the 23-minute natural half-life of the $_{92}U^{239}$ nucleus for decay to $_{93}Np^{239}$. The element $_{93}Np^{239}$ is also unstable. It decays by beta emission with a 2.3-day half-life to an isotope of plutonium:

$$_{93}Np^{239} \xrightarrow{2.3d} {_{94}Pu^{239}} + {_{-1}\beta^0} \quad (3)$$

Although $_{94}Pu^{239}$ is not stable (and for this reason does not occur naturally), it does have a long half-life (approximately 24,000 years). Its nucleus fissions readily under neutron bombardment. These characteristics establish $_{94}Pu^{239}$ as a nuclear fuel.

Fission in Heavy Nuclei. Two competing nuclear forces affecting the structural stability of a nucleus are the nuclear binding force between the nucleons that tends to hold them together, and the electrical Coulomb repulsion forces between member proton nucleons that tend to break the nucleus apart. For the very light- through intermediate-sized nuclei, the nuclear binding forces interact between all nucleons. In large nuclei, the force of attraction between nuclei is too short-range for interaction between all nuclei. Thus, the Coulomb force would break the nucleus apart if the same rule as for small nuclei were followed, i.e., the number of protons in a nucleus is equal to or nearly equal to the number of neutrons.

To increase the total attractive force, relative to the Coulomb repulsive force, the heavy nuclei have an excess of neutrons. Lead, with 82 protons, forms a dividing line between radioactive elements and stable elements. This is not the only measure since tritium (a very light isotope of hydrogen) is radioactive. Also, the famous carbon-14 is radioactive. Lead itself has stable isotopes and radioactive isotopes. However, no known isotope of an element heavier than lead (number of protons greater than 82) is stable. The well-known uranium (with 92 protons) fits this pattern, since it is radioactive.

It was discovered in 1939 that some of these heavy nuclei could be split into two large fragments by bombarding them with neutrons. This process is called fission. Two important facts emerged from this discovery. First, the fission process itself yielded a large amount of energy on the nuclear scale (approximately 200 million electron-volts [MeV], equivalent to about 0.33×10^{-10} watt-second for each fission). Second, aside from the major fission fragments, each fission produced an assortment of products including photons, electrons, and of great importance, additional neutrons. On the average, somewhere between two and three neutrons are ejected from each fission. This average number is denoted by ν; for U^{235} its value is about 2.5. Table 5.1 shows the average distribution of fission energies among the fission products for fissioning U^{235}. As shown in the table, the majority of fission energy

TABLE 5.1. Distribution of Fission Energy for Fissioning U^{235} with Thermal Neutrons

Fission fragment kinetic energy	165 MeV
Beta decay energy	5 MeV
Gamma decay energy	15 MeV
Fission neutron energy	4 MeV
Other (neutrino)	11 MeV
Total fission energy	200 MeV

release is carried off in the form of fission fragment kinetic energy.

Subsequent examination of the fission process revealed additional important features. For example, it was discovered that not all fission neutrons are released instantaneously at the time of fission, but instead a small fraction of them, less than one percent (for U^{235} fission ~0.75%), are emitted at times significantly later than the fission event. Contrasting with the "prompt" neutrons emitted when fission occurs, "delayed" neutrons come from radioactive decay of certain of the fission product fragments formed by the fission. They are born at slightly lower energies (in the range 0.25–0.7 MeV) than their prompt counterparts. Delayed neutron birth may be from a fraction of a second after the fission event to more than a minute, depending on the fission product involved.

Existence of these delayed neutrons is of great importance in stabilizing the nuclear fission process. Delayed neutrons cause the reactor response time to be such that the control of the neutron population resident in the reactor core, hence, the power level, is easily achieved.

The fission of a heavy nucleus does not always produce the same fission product fragments. However, the fragments formed do have a statistically determined basis; the mass distribution characterizing U^{235} fission fragments is shown in Figure 5.1. It shows the relative frequencies (called yields) of fission product isotopes as a function of nuclear mass number. The double-peaked nature of this mass distribution is a consequence of the fact that nearly all fissions produce two fragments (called binary fission). Examination of this distribution leads to identification of many isotopes that are primary fission fragments. Two of these isotopes, krypton-87 and xenon-137, are called "precursors" because they produce delayed neutrons.

Heavy nuclei fission occurs naturally without bombardment. This is called spontaneous fission, as opposed to the artifically induced reaction using neutrons.

Energy Dependence of Interaction Probabilities. Fission neutrons are born with kinetic energies 40–80 million times higher than that of an average free nucleus in thermal equilibrium at a room temperature of 20°C. The neutron speed is 6300 (or greater) times higher than the average speed it would have as a free particle in thermal equilibrium with its surroundings.

Subsequent to their birth, fission neutrons migrate spatially, interacting with whatever surrounding nuclei are present, and eventually distribute their excess energy among the other nuclei they encounter. Some general concepts of these postfission interactions will be reviewed in the "Other Neutron Interactions" section of this chapter.

Neutrons migrate and interact over a wide range of energy during their lifetime. They are born as "fast" or high-energy neutrons and, if they survive a sufficient number of interactions, lose energy until they enter thermal equilibrium at low energy. At this latter stage, they are called "thermal" neutrons.

For nearly all types of neutron interactions, the probability that a neutron will interact with a nucleus depends strongly on the relative velocity be-

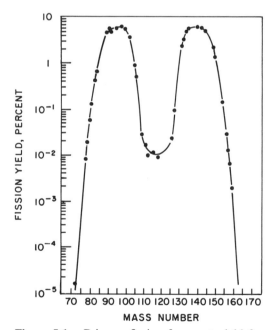

Figure 5.1. Primary fission fragment yield for uranium-235.

tween the two interacting particles. For certain heavy isotopes such as U^{235}, Pu^{239}, Pu^{241} (heavy isotopes with an even number of protons and an odd number of neutrons), the fission probability with thermal neutrons is much higher than with fast neutrons. The fission probability is so much higher that it is desirable somehow to slow down fast neutrons to enhance the probability they will cause fission in such isotopes. This is the basis for the so-called "thermal" reactor types, e.g., BWR, PWR, and HTGR (high-temperature gas-cooled reactor).

Those isotopes that will fission from thermal neutron interaction are called "fissionable" isotopes; isotopes such as U^{238} or Pu^{240} which do not thermally fission, but which transmute into a thermally fissionable isotope after absorbing a neutron (e.g., as shown by equations 1, 2, and 3), are called "fertile."

Essentially all fertile nuclei are fissionable if the neutron energy is high enough. This is called "fast" fission and occurs in U^{238} from fast neutron interaction in thermal reactors before the neutron has slowed down. The effect is small, both because of the small interaction probability involved and because in a thermal reactor the average neutron spends very little of its expected lifetime at high energies.

Other Neutron Interactions. Aside from fission, there are two other types of interactions of major importance to the neutron population in a reactor core. The first is the scattering reaction in which a neutron collides with a nucleus and emerges with a new direction and energy. The second is the so-called radiative capture reaction in which the neutron is captured by a nucleus of mass number A and charge Z, denoted by $_ZX^A$, according to the reaction

$$_ZX^A + _0n^1 \rightarrow \, _ZY^{A+1} + \gamma, \qquad (4)$$

where $_ZY^{A+1}$ is the new compound nucleus, usually formed in an excited state and possibly itself unstable. The combination of fission and capture reactions is referred to as "absorption."

Although there are several different types of neutron-scattering interactions, all of them yield the same important result, namely, they slow down fast neutrons. Each time a fast neutron collides with a scattering target, the target nucleus carries away some of the kinetic energy held initially by the neutron. Since energy must be conserved in such reactions, the neutron kinetic energy is decreased.

When two bodies collide, a maximum energy transfer occurs when the projectile and target are of the same mass. A ball hitting a "solid" wall rebounds with all its energy (assuming no losses). A heavy ball hitting a feather would lose negligible energy. Thus, to slow down neutrons, a target of neutrons or protons would be desirable. This indicates water to be a potential material, since each molecule contains two H' or protons. Of course, the oxygen contains eight protons, but they are bound together with eight neutrons giving an effective mass of 16 times the neutron mass.

The material used in a reactor to slow down the neutrons is called the "moderator." The moderator must not capture the neutron or it would be unavailable for fission and would stop the reaction. Boron is excellent at slowing down fast neutrons, but it will also capture them. Such material is named a "poison." The "goodness" of a moderator is in its slowing down power divided by the "cross-section" for capture. The larger the cross-section for capture, the more likely capture will take place.

Some typical moderators are: light water, H_2O, containing normal hydrogen, H^1; heavy water, D_2O, containing deuterium, H^2; graphite or carbon dioxide containing C^{12}, and so on. All of these materials are effective neutron moderators used in today's thermal reactor design types.

Along with these moderating materials, other materials are placed in the core for the express purpose of capturing neutrons. These are called "control" materials and include, for example, movable control rods, burnable poisons, and so on. In addition, some of the fission products that build up during reactor operation are strong neutron absorbers. Xenon and samarium are examples of such fission product neutron absorbers or poisons.

Nuclear Environment in the Fission Reactor

Nuclear Fuel: The Fissile Inventory. The objectives in assembling a nuclear core include creating an environment which will establish and maintain the nuclear chain reaction, providing a means for controlling the neutron population, and removing the energy released within the core.

Fissionable material can exist in different forms. The type selected depends upon the type of reactor being designed. For example, one might use uranium metal, uranium metal alloyed with dilutents, or uranium dioxide (UO_2). Light-water reactors use UO_2.

In light-water reactors, because H^1 is a neutron absorber as well as an efficient scatterer, it is necessary to enrich the uranium prior to fabricating the fuel. Heavy water is an excellent moderator and the uranium fuel need not be enriched. The heavy water, D_2O, already has a neutron, so it is not a neutron absorber, which makes D_2O a superior moderator. Natural uranium occurs with 0.711% by weight of thermally fissionable U^{235}. For application in LWRs it is necessary to increase the fissile content to values in the range 1.5–3.5% U^{235} by weight, using an isotopic separation process.

Construction of an LWR fuel bundle begins with fabrication of sintered cylindrical pellets of the slightly enriched uranium in UO_2 form. These fuel pellets are then loaded and sealed into a long tube of cladding. Based on today's technology, design objectives, and experience, zirconium is well established as the preferred LWR fuel-rod cladding material. Fuel bundles are then assembled into a square array of the Zr-clad UO_2 fuel rods, together with appropriate hardware components.

It is important to note at this point that this concept of fuel bundle design will result in a heterogeneous arrangement of fuel and other reactor materials. It is an intentional consequence, both from practical and theoretical points of view.

Other Essential Reactor Materials. A thermal reactor core has four essential components. These are fuel, moderator, coolant (working fluid), and control. LWRs are, by definition, characterized by their light-water moderator. In this case, light water is also used as the coolant. In contrast, the high-temperature gas-cooled reactor (HTGR) uses graphite (C^{12}) as moderator and helium gas as coolant.

Most nuclear cores employ boron-10 (B^{10}), in one form or another, for controlling the neutron population. B^{10} is a very strong neutron absorber. The control material, B^{10}, is placed in movable blades or rods which can be inserted or withdrawn from the active core by an operator. In some cases, such as the PWR, B^{10} is introduced in varying low concentrations into the moderator-coolant to augment movable control rods.

One other type of control material used in present day reactors is called a "burnable poison." The basic concept involved here is to introduce intentionally, in small amounts, in a fixed position (usually within the fuel itself), a neutron absorber or poison such as gadolinium (Gd). This poison material is distributed in such a way as to provide control of the neutron population for a predetermined period of time, on the order of months, and to eventually "burn out" via isotopic transmutation through neutron absorption (called "depletion" in this context).

The Neutron Life Cycle. To appreciate the essential factors involved in creating and sustaining a chain reaction, it is helpful to trace a neutron through a typical lifetime or generation. Figure 5.2 shows this expected neutron life cycle schematically, in a thermal reactor using U^{235}.

Starting at box 1, imagine N fast neutrons born from thermal fission. A few fast fissions ocur in a thermal reactor, so on the average, more neutrons are produced from fast plus thermal fissions. This ratio of total to thermal fissions is called the fast fission effect, and is denoted by ε. In thermal reactors ε is about 1.05: hence, at box 2 we expect to have, on the average, $N \cdot \varepsilon$ neutrons.

As the neutrons migrate in position, collide with nuclei and lose energy, there is always the chance that some will escape from the reactor core region. This is called neutron leakage; we denote the probability that a fast neutron will not leak out of the core by P_f, called the fast nonleakage probability.

There are $N \cdot \varepsilon \cdot P_f$ neutrons at box 3 to be slowed down. Some of these will be captured by nuclei having so-called resonance absorption behavior in the fast and intermediate energy ranges. The resonance escape probability, denoted by p, is defined in a fashion similar to P_f as the probability a neutron survives slowing down to thermal energies (i.e., to energies less than 1 ev) without being absorbed. Hence at box 4 there are $N \cdot \varepsilon \cdot P_f \cdot p$ neutrons in thermal equilibrium.

Not all thermalized neutrons will be available for fission since some will escape from the core. To account for these, P_{th} is defined as the thermal nonleakage probability to yield a total of $N \cdot \varepsilon \cdot P_f \cdot p \cdot P_{th}$ thermal neutrons left at box 5 to interact with available nuclei. The fate of the remaining neutrons, aside from intervening scattering reactions, is absorption. The ratio of thermal absorption in the fuel to total thermal absorption in all materials is denoted by f and called the thermal utilization. Hence, the total thermal absorption in the fuel at box 6 is given by $N \cdot \varepsilon \cdot P_f \cdot p \cdot P_{th} \cdot f$.

If the ratio of fission absorption to total absorption (i.e., the probability that an absorption leads to a fission) in the fuel is denoted by α, and the average number of neutrons emitted per fission by ν, then

REACTOR CORE BEHAVIOR

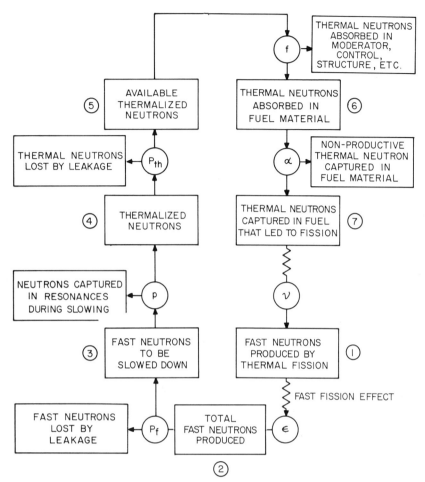

Figure 5.2. Thermal fission neutron life-cycle diagram.

the number of fast neutrons finally produced from fission due to the initial N fission neutrons with which we began is given by $N \cdot \varepsilon \cdot P_f \cdot p \cdot P_{th} \cdot f \cdot \alpha \cdot \nu$.

Each of these factors (except for N) represents one of the basic competing factors affecting the neutron population in the core. It should be noted that they have been carefully defined to separate, wherever possible, the physical phenomena into geometric, material, and nuclear processes. For example, the product $\alpha \nu$ is denoted formally by η, and to first order is solely a nuclear property of the fissile material. Likewise, $P = P_f P_{th}$ is a geometric nonleakage probability having principally to do with core size and shape.

Rewriting the final neutron count in the form $N \cdot P \cdot (\eta p \varepsilon f)$ expresses this separation in conventional form. The four-factor product $\eta p \varepsilon f$ is, by convention, denoted by K_∞, called the infinite medium neutron multiplication factor. This is expressed by equation 5,

$$K_\infty \equiv \eta p \varepsilon f, \qquad (5)$$

which is called the four-factor formula. Although an infinite medium lattice is not an exact description of any real lattice being represented, it can be made quite accurate using today's sophisticated theories, coupled with the availability of large computers. The product PK_∞, called the effective multiplication factor, is denoted by K_{eff}. K_{eff} differs from K_∞ because of the finite dimensions of real reactor cores.

REACTOR CORE BEHAVIOR

Nuclear Criticality

Measures of Criticality. Starting with N_n neutrons in generation n, the neutron life cycle produces

N_{n+1} neutrons in generation $n + 1$ where we have just seen that

$$N_{n+1} = N_n P(\eta n \varepsilon f) = N_n K_{\text{eff}}. \quad (6)$$

If we denote the mean generation time by τ, then

$$\Delta N = (N_{n+1} - N_n)\frac{\Delta t}{\tau} = (N_n K_{\text{eff}} - N_n)\frac{\Delta t}{\tau}, \quad (7)$$

where t is the time coordinate. In general, it follows that for a constant K_{eff},

$$\frac{\Delta N}{N_n} = \frac{(K_{\text{eff}} - 1)}{t} \Delta t. \quad (8)$$

The solution to this difference equation is (letting N_n be $N[t]$)

$$N(t) = N_0 e^{\sigma t} \text{ where,}$$
$$\sigma = \frac{K_{\text{eff}} - 1}{t}. \quad (9)$$

This is an interesting result since it contains the basic concept of criticality. For a given value of $N_0 > 0$ and τ, the neutron population $N(t)$ has the following behavior in time:

a. If $K_{\text{eff}} < 1$, $\lim_{t \to \infty} N(t) = 0$, "subcritical,"

b. If $K_{\text{eff}} = 1$, $\lim_{t \to \infty} N(t) = N_0$, "critical"

c. If $K_{\text{eff}} > 1$, $\lim_{t \to \infty} N(t) = \infty$, "supercritical."

K_{eff} is a measure of nuclear criticality. If a value of unity can be established and maintained, the chain reaction will just sustain itself; if values of $K_{\text{eff}} > 1$ can be achieved (e.g., by removing control materials), then the neutron population, and the power level, can be increased to any required level. Returning K_{eff} to 1 (e.g., by reinserting the control material) will cause the neutron population to be maintained at this new level.

The development of $N(t)$ given here is a gross oversimplification, both of reality and of the theoretical models currently in use. $N(t)$ is, in fact, a complicated function of core geometry as well as time and must be predicted with accuracy to describe the operating and performance characteristics of real power-reactor cores. Nonetheless, the fundamental concept of nuclear criticality is portrayed correctly by this simple model.

To examine the effects of small changes on the time behavior of the nuclear core, it is convenient to have a measure of criticality which is additive. Clearly K_{eff} as defined by

$$K_{\text{eff}} \equiv PK_\infty \equiv P\eta p \varepsilon f \quad (10)$$

is not additive with respect to changes, e.g., in p or f.

It is not difficult to develop such a quantity; it is obtained through partial variations of K_{eff} to yield

$$\frac{\Delta K_{\text{eff}}}{K_{\text{eff}}} = \frac{\Delta P}{P} + \frac{\Delta \eta}{\eta} + \frac{\Delta p}{p} + \frac{\Delta \varepsilon}{\varepsilon} + \frac{\Delta f}{f}. \quad (11)$$

Since equation 10 is valid only near critical, it is customary to define ρ, called the reactivity, as a fractional deviation from critical:

$$\rho \equiv \frac{K_{\text{eff}} - 1}{K_{\text{eff}}} = \frac{\Delta K_{\text{eff}}}{K_{\text{eff}}}. \quad (12)$$

As is seen by equation 11, values of ρ caused by changes in the individual components may be added together.

Core Reactivity Coefficients. Changes in core power level, fuel and moderator temperatures, material properties, among others, all have their effect on core criticality and can be expressed as changes in reactivity. These changes are normally treated in terms of reactivity coefficients, that is, differentials of ρ with respect to the variable experiencing the change.

For instance, the reactivity temperature coefficient is $\partial \rho / \partial T$, where T is the temperature. From equation 12, it is clear that the reactivity coefficient for changes in temperature can be written

$$\frac{\partial \rho}{\partial T} = \frac{\partial}{\partial T}\left(1 - \frac{1}{K}\right) = \frac{1}{K^2}\frac{\partial K}{\partial T} \simeq \frac{1}{K}\frac{\Delta K}{(K \Delta T)}, \quad (13)$$

where we have denoted K_{eff} by K. Therefore, from equation 11 the temperature coefficient can be written as

$$\frac{\partial \rho}{\partial T} \simeq \frac{1}{K}\left[\frac{1}{P}\frac{\Delta P}{\Delta T} + \frac{1}{\eta}\frac{\Delta \eta}{\Delta T}\right.$$
$$\left. + \frac{1}{p}\frac{\Delta p}{\Delta T} + \frac{1}{\varepsilon}\frac{\Delta \varepsilon}{\Delta T} + \frac{1}{f}\frac{\Delta f}{\Delta T}\right]. \quad (14)$$

In equation 14, the total temperature coefficient of reactivity, i.e., $\partial \rho/\partial T$ is made up of a sum of five terms, each of which contributes an amount equal to the fractional change in the corresponding factor appearing in equation 10 caused by a change in temperature. Implicit in these terms are all effects, such as density changes, nuclear behavior changes, geometric changes, etc., caused by changing the temperature, T.

In LWRs, there are three principal reactivity coefficients that describe effects caused by changes in core temperatures:

1. Void coefficient
2. Moderator temperature coefficient
3. Doppler coefficient.

Void Reactivity Coefficient. The principal reactivity feedback effect in the BWR is due to voids. This effect is of little or no consequence in a properly designed PWR. This is due to the presence of a boiling moderator coolant in the BWR and the absence of such steam voiding in the PWR.

Reduction of the moderator density within the BWR due to formation of steam voids has both local and full core effects on reactivity. These effects lead to changes in P due to increased neutron leakage as the moderator material density decreases within the core; changes in f, since the relative competition for thermal neutron absorption tends, by design, to favor radiative capture within the core lattice; and, to a lesser extent, changes in the remaining three factors in equation 14.

The BWR void coefficient is negative. As the coolant temperature approaches its saturation value and voids build up in the core, any addition of void content causes the neutron population, hence the power level, to decrease. The reverse is also true. This behavior leads to an extremely stable system with respect to effects that would tend to increase the void content. It is the void effect that permits the BWR to change power level using recirculation flow (i.e., the "flow control" power maneuvering capability).

Moderator Temperature Coefficient. For LWRs, the coolant and moderator are one and the same, light water (H_2O). An increase in the coolant's temperature causes it to expand, decreasing the mass of water in the core region. The attendant decrease in moderating capability, due to the presence of fewer H^1 nuclei, results in an inherently negative feedback effect on reactivity.

The BWR does have a moderator temperature coefficient of reactivity that takes into account nuclear behavior and density changes distinct from steam voiding. The effect is small compared with the void coefficient and is only of significance during start-up (i.e., when the reactor is taken from a cold to hot condition).

For the PWR, the opposite is true. With no steam voiding (by design), the moderator temperature coefficient is one of the primary reactivity effects. As is the case with the BWR void coefficient, calculation of the PWR moderator temperature coefficient is complex and requires advanced methods, experimental data, and experience. It is further complicated for the PWR due to including the effects of soluble B^{10} as a control material within the moderator coolant. As the coolant temperature increases, the density decreases and some of the strong B^{10} absorber is expelled along with some coolant from the core. This leads to a positive contribution to the PWR moderator temperature coefficient.

Doppler Coefficient. If, for any reason, the slightly enriched UO_2 fuel temperature in a LWR increases (e.g., due to a transient or due to intentional increase in thermal power output), there is observed a negative reactivity effect due solely to fuel temperature.

It is called the Doppler effect and arises from a temperature dependent increase in the U^{238} ability to capture neutrons as they are slowing down. In the context of equation 14 the value of p decreases (i.e., $\Delta p < 0$). Physically, the increased neutron absorption occurs, because, as the temperature increases, the U^{238} target nuclei move at higher velocities, on the average, relative to the incoming neutrons, and the fast neutron resonance interaction characteristic of heavy nuclei appears to shift and broaden the resonance absorption region. The net result of the broadening is an increased probability of neutron absorption in the slowing down part of the energy range.

This intrinsic negative feedback effect occurs very rapidly and produces an inherent safety feature in thermal reactors, thereby preventing rapid increases in local power levels within the fuel.

Although there are other LWR reactivity coefficients that must be considered in detailed design, the ones discussed here are the most important.

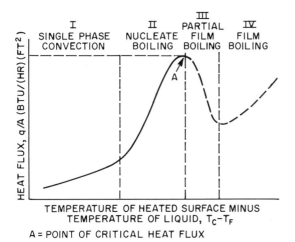

Figure 5.3. Diagram showing boiling regions (regimes).

Other effects important to the kinetic behavior of the core will be discussed later.

Nuclear Core Design

Energy Generation; Heat Transfer. It is necessary to consider how the energy generated in the nuclear fuel is to be removed from the core and used to generate electrical energy. Both BWRs and PWRs use light water as the moderator and coolant material, as well as the power-cycle working fluid.

To establish a consistent design, the heat generated within the fuel, together with the selected core geometry, must be properly matched to the coolant flow capabilities so that material temperatures are maintained well below damage thresholds. Normally, the limiting temperature from a thermal hydraulic point of view is that of the fuel-clad material. This single criterion limits the rate at which energy can be removed from the clad surface by the coolant.

Once the limiting characteristics of this heat removal rate (called the thermal hydraulic design basis) is known, the detailed fuel pellet and clad geometries can be designed on the basis of permissible stresses and strains in the clad occurring over the fuel rod operating life. Current LWR core designs use a heterogeneous lattice of vertical cylindrical fuel rods, containing slightly enriched UO_2 fuel pellets, clad in zircaloy tubing, and surrounded by the moderator (coolant). The UO_2 fuel is formed into pellets and is sintered to high density prior to loading into the tubing. A diametrical gap is provided between the pellet and the inside clad surface to allow for differential pellet-clad thermal expansion and to accommodate fuel pellet cracking and irradiation-induced swelling. Dimensions of the pellets and cladding are selected to yield acceptable clad stresses over the operating life of the fuel rod.

Determination of permissible heat fluxes at the fuel rod clad surface is a difficult task involving both theoretical and experimental work. This is especially true in the BWR where two-phase (liquid-vapor) flow occurs within the core. Figure 5.3 illustrates the prominent physical phenomena of two-phase heat transfer from an internally heated (fuel rod-clad) surface. These heat-transfer mechanisms are, of course, local phenomena. This requires the nuclear core designer to calculate the conditions at all points of clad surface within the core to ensure operation within stated limits. It is important to note here that, while these heat transfer phenomena occur locally, the conditions that induce their appearance in a forced convection system are predominantly nonlocal; this is the central issue, which makes the thermal hydraulic behavior a difficult technical problem to solve accurately.

As shown in Figure 5.3, several different mechanisms of heat transfer from the fuel clad surface to the water coolant-moderator can occur, depending upon the heat flux (i.e., the heat transferred per unit area), the coolant pressure, and the resulting difference in temperature between the clad surface and bulk liquid ($T_c - T_f$).

Single-phase convection means that heat is transferred to the liquid in such a manner that no local steam formation occurs on the clad surface. As the heat flux through the clad surface is increased, the temperature difference rises linearly, as shown in region I.

When nucleate boiling begins, the heat-transfer coefficient markedly improves as the heat flux is raised. In this mechanism (region II), steam bubbles form at points on the heated surface called nucleation sites; thus, the term nucleate boiling. The increased heat-transfer rate in this region is due to the increased mixing motion in the liquid, caused by the formation of isolated steam bubbles, and results in an extremely efficient heat transfer mechanism.

The peak point of heat transfer in the nucleate boiling region (A) is known as the critical heat flux (CHF). Attempting to attain higher heat fluxes results in departure from nucleate boiling (DNB) and the onset of film boiling. Water-cooled and moderated reactors are designed so that maximum surface heat fluxes are well below the CFH value for local surface conditions.

Partial film boiling occurs with attempts to attain a heat flux slightly higher than the CHF value, where unstable boiling conditions exist (shown as region III in Figure 5.3). This transition region will have violent nucleate boiling occurring at some points on the surface and film boiling at others. Rapid-clad temperature variations occur as local film boiling layers form and are swept away from the forced coolant convection.

Film boiling occurs with still higher heat fluxes when the steam bubbles coalesce and "blanket" the clad surface, preventing the liquid from coming into direct contact with the surface. This layer of steam film acts as an insulator between the clad surface and the liquid; its heat transfer coefficient is much less than that in nucleate boiling. Operation in this region (shown as region IV in Figure 5.3) causes the clad wall temperature to increase to high values and perforate (i.e., burn out). Clad burn out is prevented by keeping heat transfer rates well below those that produce film boiling.

Principles of Core Design Basis. Using a combination of experiment and theory, it is possible to develop a quantitative correlation between fuel lattice operating properties, coolant quality and flow, and the onset of the critical heat-flux event. Such a correlation between design parameters and critical heat flux is known as the thermal-hydraulic design basis.

Once the thermal-hydraulic and fuel-rod mechanical design bases are established, the designer can complete the detailed design of the nuclear core, using the remaining design variables to optimize performance of the system.

When the core coolant flow, pressure, and temperature values are selected, either the fuel rod mechanical or thermal hydraulic design limits the level of energy release within the fuel rod. This quantity is usually calculated in units of kilowatts/foot (kW/ft) of fuel rod. In most current LWRs, the kW/ft limit is produced nearly simultaneously by both design bases; this is a direct consequence of a well-designed nearly optimum system.

With the kW/ft limit set, the fuel rod maximum center line temperature is determined, and is independent of rod diameter. This is a direct consequence of the rod surface-to-volume geometry (cylindrical). Typical values in today's LWR designs fall into the 10–20 kW/ft range, with center line temperatures ranging from 3000–4500°F.

The number of fuel bundles to be included in the core depends on the power output required and the achievable peak-to-average power-peaking factors of the fuel design. Physical laws of nature prevent the power density from being absolutely uniform in position over the assembled lattice. Neutrons leak from the boundary surfaces of the core; hence, they fail to maintain the neutron population at the boundary surface that is achievable deep in the core's interior. This nonuniformity is taken into account by the power-peaking factors. Present-day LWR offerings are in the 800–1300 MWe output range.

Reactor Control

Operating Cycle Behavior. In addition to the two fundamental design bases developed so far (i.e., the fuel-rod mechanical strain and fuel assembly thermal-hydraulic CHF correlation), it is necessary for the designer to satisfy one more design concept. This concept relates to the nuclear behavior of the full core and simply states that the core must be equipped with sufficient reactivity control so that it may be shut down at any time and under any credible condition.

To appreciate this requirement more fully, it is helpful to list the principal components and factors that influence control of the nuclear core.

1. Movable Control. Includes control blades, soluble poison injection systems, and cluster-type control rods.
2. Burnable Poisons. Fixed-position neutron absorbers (e.g., gadolinium, boron, etc.), normally blended into the UO_2 fuel, designed to provide supplemental control early in an operating period and to deplete before the end of the period.
3. Fission Product Poisons. Neutron absorbers that accumulate in the fuel during operation as part of the fission product inventory. Some of these are relatively short-lived (e.g., Xe^{135} with a 9.2-hour half-life), leading to short-term transient behavior, while others are long-lived and "saturate" early in the operating life of the fuel (e.g., Sm^{149}), thereby contributing to a constant negative reactivity component.
4. Thermal Effects. Doppler moderator and void effects which, depending on the specific situation, may act as either negative or positive reactivity feedback effects.

All of these effects must be taken into account, their reactivity worths calculated, together with the positive initial reactivity worth of the enriched UO_2 fuel, and a reactivity balance established. It is this balance that must be accurately determined for each possible core state at each point of the operating cycle.

During the operating cycle, the fissile inventory initially placed in the fuel undergoes isotopic transmutation (in this context called isotopic "depletion" or "burn-up"), and the positive reactivity due to the fuel enrichment decreases with energy production. This factor has a direct bearing on the length of the operating cycle; an increase in the initial enrichment leads to a longer operating cycle capability and vice versa.

Because of the core shunt-down requirement and the finite reactivity worths of control components, however, one cannot increase the initial fuel enrichment indefinitely. There is a design-dependent upper limit of initial fuel reactivity (hence also operating cycle length) beyond which the shut-down criteria cannot be met. In current LWR designs, this upper limit is about 24 months.

Reactor Kinetics. Reactor kinetics is the name given to the study of transient behavior involving the response of the nuclear reactor neutron population to changes in core reactivity. All reactor systems operate in a rather complex physical environment where many reactivity feedback paths exist, some of which act very rapidly and others operate over longer periods of time.

Traditionally, the very slow kind of time behavior, such as isotopic depletion of fissile material, have not been included within the scope of kinetic analyses. Only the shorter time scale transients, such as the effect of xenon build-up and decay, which bear on the control of the reactor over a period of hours or less, are treated under the heading of reactor kinetics.

Figure 5.4 shows schematically a simplified version of the important feedback mechanisms involved in the dynamic response of an LWR to an applied reactivity insertion, ΔK_{CS}, from the core control system. The various reactivity effects discussed to this point are shown, together with the role played by the fuel and thermal hydraulic design.

Feedback systems such as those shown in Figure 5.4 consist of three basic phases. First, there is the input perturbation to the system (ΔK_{CS} in this exam-

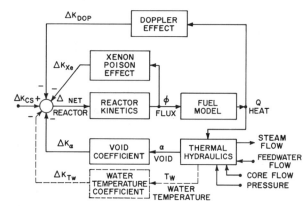

Figure 5.4. Light-water reactor dynamics response block diagram.

ple), which causes some change in the system. The second phase shown as a change in the neutron population, or flux, ϕ, in Figure 5.4, leads directly to altered system output variable levels. Taken by themselves, these first two steps constitute what is usually called the open-loop transfer function describing the system dynamic response.

In real reactor systems, such induced perturbations cause reactivity changes, which, in turn, feed back, closing the loop. Figure 5.4 portrays the closed-loop transfer function in schematic form.

The four feedback paths shown on Figure 5.4 come into play at different time points. The Doppler effect is very fast (with a time lag of 1 second or less), because it occurs within the fuel itself and is dependent only on fuel temperature. The void coefficient and water temperature coefficient are somewhat slower mechanisms (with time constants on the order of 8–10 seconds), because they both depend on the transmission of heat from the fuel into the water coolant. The xenon poison effect is much slower, depending on a radioactive decay chain with half lives of 6–10 hours.

The basic time behavior describing reactor kinetics was given in equation 9. In actual reactor operations, K_{eff} does not deviate very far from unity. For these small variations of K_{eff} from the critical condition, the effective value of τ, the mean generation time, is determined essentially by the delayed neutrons. This τ has a value on the order of 10 seconds or greater for normal operating reactor transients. The ensuing rate of changes of power level are easily monitored by instrumentation and controlled by operator actions.

Examination of Figure 5.4 indicates that the open-loop transfer function would be applicable, in an LWR, during start-up when there are no voids and the fuel and coolant are at relatively low temperatures. It is not until the plant is producing a significant fraction of its power rating (say 1%) that the closed-loop transfer function becomes important. At higher power levels, it becomes necessary to consider all the reactivity feedback paths shown in Figure 5.4.

For very slowly changing reactivity variations, the reactor power level stays in a quasi-steady state. One such slow change is the fuel reactivity depletion due to burn-up. In that case, the reactor control system is designed to compensate with the corresponding reactivity changes, inserting ΔK_{CS} either manually or automatically by:

1. Adjusting control (absorber) rod position,
2. Adjusting core flow (boiling water reactors only), or
3. Adjusting core soluble poison concentration (pressurized water reactors only).

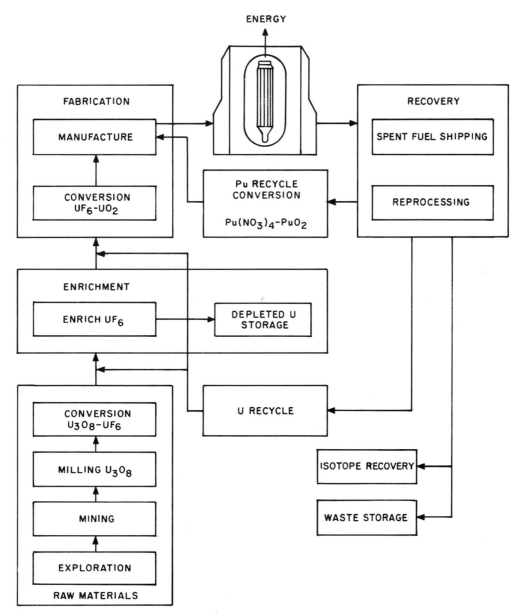

Figure 5.5. The uranium fuel cycle for light-water reactors.

Daily load variations also result in relatively slow changing reactivity situations, which are also accommodated by the reactor control system.

THE NUCLEAR FUEL CYCLE

Fuel Cycle Physics

Nuclear fuel has several characteristics that distinguish it from other power plant fuels. Notable among these are the high inventory costs and low energy generation costs compared with other fuels. Uranium raw material processing, uranium enrichment, fuel fabrication, energy release (irradiation), and fuel recovery constitute the major parts of the nuclear fuel cycle, and are illustrated in Figure 5.5.

The raw material (uranium) processing involves exploration for ore, mining, milling of U_3O_8, and conversion of the U_3O_8 to UF_6. A workable ton of ore will contain from 2 to 29 pounds (.9–13 kg) of U_3O_8. The concentrated form is called yellow-cake. This U_3O_8 is converted to uranium hexoflouride, UF_6, which, in gaseous form, enters the enrichment process. Through enrichment, the concentration of the fissile isotope U^{235} is increased, from the natural 0.711% (by weight) to between 1.5% and 3.5% for LWRs.

Fuel fabrication, as defined here, includes chemical conversion of the enriched UF_6 to uranium dioxide. The UO_2 powder is then pressed and sintered into a ceramic pellet a little less than $\frac{1}{2}$ inch (1.2 cm) in diameter and about $\frac{1}{2}$ inch (1.2 cm) long. The pellets are loaded into zerconium alloy tubes about 14 feet (4.2 meters) long. After being welded closed, the fuel rods are bundled into a square array ready for insertion into the reactor core.

Energy release from the nuclear fuel occurs during its residence time in the reactor. During this irradiation time, the fuel is partially depleted of the U^{235} isotope through fission, a buildup of the fissile isotopes Pu^{239} and Pu^{241} occurs, and a few other valuable fission products are produced. For LWRs, one-fourth to one-third of the reactor core fuel bundles will be removed at intervals of 12–18 months, and replaced with fresh fuel. The exact number of fuel bundles and the exact time of replacement depend on electric utility power generation requirements and the refueling strategy used.

Fuel recovery includes the transportation of the spent fuel elements from the reactor site, temporary storage, and the chemical operations to be performed at the remote reprocessing plant. It is expected that commercial reprocessing will involve chemical separation of the fissile plutonium and partially enriched uranium for recycling into LWR fuel. Valuable by-product isotopes will also be produced. Provision for temporary and permanent storage of the radioactive wastes is, of course, a very important part of the operation.

SUMMARY

The commercial nuclear power industry has now evolved into a mature, responsible industry that serves a needed market both domestically, as well as internationally. Nuclear fuel, as utilized in light water reactors (LWRs) supplies the vast majority of the electrical energy provided by this technology. LWRs are found in two variations: the boiling water reactor (BWR) and the pressurized water reactor (PWR). The fundamental attractiveness of nuclear power results from the availability and the relative inexpensive cost of nuclear fuel. In a LWR, fuel bundles can exist within the core providing energy for several years.

REFERENCES

1. Bonilla, C. F., *Nuclear Engineering*, McGraw-Hill, New York, 1957.
2. El-Wakil, M. M., *Nuclear Power Engineering*, McGraw-Hill, New York, 1962.
3. Lamarsh, J. R., *Introduction to Nuclear Engineering*, Addison-Wesley, Reading, Massachusetts, 1975.

6
POWER FROM WATER

INTRODUCTION

Hydroelectric power played a significant role in the development of the electric utility industry. As the industry grew, however, the percentage of the total energy generated by hydro plants decreased. For many years, hydro power accounted for 30–40% of the total energy distributed by utilities, depending on annual amounts of rain and snowfall. By 1965, this figure was under 20%. In 1980, water power, with an installed capacity of 76 million kW, accounted for 12% of the total generating capacity of the United States.

Hydroelectric power, since it uses no fuel, is regarded by some as a free source of energy. The initial cost to develop an available hydro site, however, can be significantly higher than the initial cost of an equivalent capacity base-load or peaking type thermal plant. The less expensive or "better" sites are naturally developed first so that future developments will tend to be more expensive. On the other hand, the rapidly escalating thermal fuel costs have made many formerly unattractive sites economically feasible.

Although the number of large new conventional hydro projects has been decreasing, many smaller sites are being developed, or existing sites are being uprated to increase their capacity. In addition, many hydro projects of the pumped-storage type are being planned or built. In pumped storage, water is raised to an upper reservoir by pumping during the off-peak hours. This is done by driving the "generator" as a motor from the power systems' thermal plants and operating the hydro turbine as a pump. During peak hours, the water is returned through the turbine, driving the generator to produce electric energy.

Regardless of the type of hydro project considered, factors other than economic considerations are becoming more significant in the analysis of the proposed project. For example, competing forms of land use, as well as the environmental impact of the project, are significant when analyzing a proposed site. In the final analysis it is often these factors that determine whether or not the proposed project is approved.

PLANT SIZE AND LOAD FACTOR

Few plants are operated at full capacity all the time. Sometimes a plant cannot carry full load continuously due to insufficient water. Use of the available water is based on coordination with other system generation. It was shown in Chapter 2 that system loads vary considerably throughout the day. If a major portion of the load is carried by hydroelectric plants, then the plants' combined output must vary similarly.

There are no longer many systems supplied entirely by water power. When there is both steam and hydrogeneration, the latter need not conform to the system load curve because the hydroelectric plants are operated so as to be of maximum value to the system, and consistent with the storage capabilities of the site. There is also an incentive to install sufficient plant capacity to capture all of the kilowatt hours available from the stream flow. The degree to which this is practical depends on the storage characteristics of the site and on the trade-offs of the plant capital cost and the value of system energy costs.

Where there is very little storage available, there is little opportunity to optimize the use of the hydro energy on the system. However, there is a trade-off to be made between the amount of capacity installed to capture the variable kilowatt hours of stream flow and the value of this energy to the system. In times of extraordinary increases in fossil fuel costs, this trade-off favors development of additional hydro capacity. Where there is some storage available, the additional opportunity exists to minimize costs by utilizing the energy at the best times in the daily load curves. Rather than generate at a relatively low output in a base load mode, it is likely that a much higher output for shorter periods would be justified. This requires additional hydro plant capacity which is justified by the much higher peak-load system energy costs that can be saved during peak-load periods. Typical variation in the incremental energy costs of a power system between base load and peak load periods may be 5 to 1 or more.

An example of stream utilization for peaking purposes is the Smith Mountain development of Appalachian Power Company on the Roanoke River in Virginia, utilizing both peaking generation and pumped storage concepts. Under normal "base-load" stream flow concepts, the plant rating was considered to be approximately 60,000 kW; however, the actual plant-generating capacity that was installed was over 400,000 kW, operating at a low load factor to replace system fossil-generated energy during highest cost periods.

Most streams are subject to variable flow. A few, like the St. Lawrence or the Pit in California, have comparatively moderate variations in flow, but most streams fluctuate rather widely from destructive spring floods to summer droughts. The Susquehanna River, for example, ranges from a flood flow of 35 times the average to a low flow of 10% of the average. Obviously, the lowest flow on record does not occur every year. In fact, a record extending over a great many years may show that the flow dropped to the minimum only once; the next lowest flow may have been five times this minimum, and this also occurred only once.

The day-to-day and month-by-month flow of a stream, in cubic feet per second, commonly called second-feet, is recorded by stream-gauging stations. There are thousands of these stations on many rivers. The flow during the period covered by the records can be summarized on a duration curve, also called a percent-of-time curve. This is a plot of the flow in second-feet, as ordinate, against the percent of the time that the flow was at least that much. The time used may be any arbitrarily chosen period

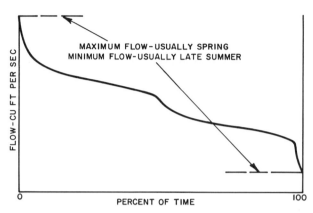

Figure 6.1. Typical stream flow duration curve.

covered by the records, though it is customary to use the entire period of record. The longer the period, the greater the spread between maximum and minimum flow is likely to be. Figure 6.1 shows a typical duration curve. From such a curve, an estimate of the probable kilowatt-hours that could be produced can be determined. In so doing, any effects that variations in flow might have on the available head must be considered.

VARIATION IN COST PER KILOWATT OF HYDROELECTRIC DEVELOPMENTS

There is a wide variation in the cost per kW of hydro plants. Plant size is one factor, as in any type of generating station. There are other items that have little to do with the size or kilowatt rating of the station. The proportion of the total cost represented by these items that are unrelated to size is likely to be higher for a low-head than for a high-head plant. These items include acquisition and preparation of the necessary real estate (this may be a great expense if the reservoir is to flood a large area); construction of the dam; building of highways and possibly a railroad spur to the site for transportation of construction materials and machinery; and finally, interest on the investment accumulated during construction, and legal, engineering, and other incidental expenses having no direct relation to the capability of the station. In a fairly typical development built some years ago, that part of the cost that varied directly with the rated output amounted to only 18% of the total cost of the project.

A significant effect on the cost of the turbine and generator is the operating head. Higher head plants utilize higher speed turbines and generators, which are physically smaller and less costly.

PUMPED STORAGE HYDRO

Pumped-storage hydro projects make use of electrical energy during off-peak hours to pump water from a lower reservoir to a higher reservoir for use in the generation of power during system peak periods. Due to losses in the cycle, only about two kilowatts are generated for every three kilowatts used during pumping, but the cost of energy for pumping may be less than half the cost of on-peak energy. For example, pumping energy is normally supplied on nights and weekends by large fossil and nuclear units. This improves the capacity factor of these units and reduces undesired cycling. The energy generated by this water is used during system peaks to reduce the other peaking units required.

Most of the modern pumped-storage plants in this country utilize a reversible Francis-type turbine, operating in one direction of rotation as a pump and the opposite direction as a turbine. This is connected to a synchronous generator/motor operating in one direction as a motor driving the pump and in the opposite direction providing power as a generator.

A pumped-storage plant has the same favorable operating characteristics as a conventional hydroelectric plant—rapid start-up and loading, long life, low operating and maintenance costs, and low outage rates. Its fast response allows it, when generating power, to follow system load changes. When not generating, the plant can be designed to operate as a synchronous condenser for voltage control. As a result of these benefits, the number of pumped hydro plants has increased rapidly since the early 1960's. In 1979, about 12,000 MW of such capacity was in operation, with another 5,000 MW under construction.

CONTROL OF STREAM FLOW

Pondage

Although we cannot store electric energy, we can store the raw material from which it is made. Sometimes this is true of water, as well as fuel. Part of the water coming down a stream at night, when the load is light, may be retained in the pond above the dam and used during the peak load period the next day. This is commonly called pondage. By this means, a development on a stream whose steady flow is good for only 10,000 kW, may actually carry a 20,000-kW load, provided the load factor is not over 0.50. This applies to an isolated plant—one that is the sole source of power for its load. But if that plant is part of a system having other generation, it can be developed to carry more than 20,000 kW during the system peak—at a lower plant load factor.

Pumped Storage

The stream flow requirements for a pumped-storage plant are very low compared to a conventional plant, since only water to make up for evaporation and leakage is required. For example, the Union Electric Company's Taum Sauk Plant is located on a stream with a minimum flow of only five second feet yet has a generating capability of about 410,000 kW. Many plants use a dam only to form the lower reservoir. An upper reservoir with no natural source is built and filled solely by operation of the pumps.

One problem that must be considered during the design of such a plant is that of minimizing leakage from the upper reservoir to maintain plant efficiency, since such leakage must be made up by additional pumping.

Seasonal Storage

Sometimes it is possible to build storage reservoirs to hold some of the spring flood water for use later on when dry weather comes. When this can be done, the firm rating of a development may be substantially increased. Consequently, seasonal storage offers two major benefits: flood waters are held back, and these waters are available for power production at a later time.

Multi-Purpose Developments

Over the past several years we have heard a great deal about multi-purpose dams—i.e., dams built to provide flood control and seasonal storage, to irrigate, to improve navigation, and so forth. To a degree, the concepts of flood control and of regulating stream flow for better power production are, as indicated above, mutually complementary; yet, to a degree, they are also mutually antagonistic. This is because a flood-control reservoir should be empty most of the time, ready to catch any flood that may come along. In the usual combination of storage reservoir and hydroelectric plant, however, maximum power production demands maximum head all the time so that the reservoir should be kept as full as possible. Water is drawn from storage only dur-

POWER FROM WATER

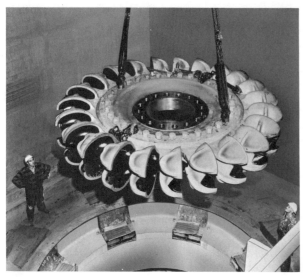

Figure 6.2. Impulse or Pelton wheel used in high-head plants.

ing times of low natural stream flow, or just before the approach of the expected spring flood or fall rains, so as to have reservoir capacity available to retain as much water as possible.

POWER AND ENERGY EQUATIONS

The power that can be developed at a site is a function of the available head and water flow:

$$\text{Horsepower} = \frac{(QH)e}{8.8},$$

where Q is flow in cubic feet per second, H is net head in feet, and e is efficiency.

Assuming an 88% efficiency, which is a reasonable number, a simple approximation results:

$$\text{Horsepower} = (QH)/10.$$

The energy available from a volume of water in a storage reservoir is:

$$\text{Kilowatt-hours} = \frac{(W\bar{H}V)e}{K}$$
$$\cong 20 \times 10^{-6}\,(HV),$$

where W is weight per cubic foot of water (62.5), $\bar{H}$ is the weighted average head of available water in reservoir, V is volume of reservoir in cubic feet and K is 2.65×10^6 ft-lb/k Wh.

ESSENTIAL ELEMENTS OF A HYDROELECTRIC DEVELOPMENT

It is much simpler to get kilowatt-hours out of water than out of fuel. The necessary elements are:

1. Some water at one elevation (probably, also, a dam to hold it there).
2. A suitable location at a lower elevation where the water from item 1 can be directed through a hydraulic turbine.
3. A hydraulic turbine connected to a generator.
4. A penstock to conduct the water from the upper level into the turbine at the lower level.
5. Controls and auxiliaries.

TYPES OF HYDRAULIC TURBINES

Turbines are of two basic types—impulse and reaction. The impulse, also known as the Pelton wheel, is shown in Figure 6.2. It is used in high-head plants—typically, where the head is 1000 feet (304.8 meters) or more. In this wheel, the head is converted into one or more high-velocity jets located around the periphery of the wheel and directed into spoon-shaped buckets. Energy supplied to the turbine in this manner is wholly kinetic, and the wheel is under atmospheric pressure.

Figure 6.3. Scroll case, by which water will enter wheel around its entire periphery.

TYPES OF HYDRAULIC TURBINES

The impulse, or Pelton wheel, usually has a horizontal shaft and a single horizontal jet. Large units are often of the double-overhung wheel construction with a generator between the two wheels, all on the same horizontal shaft. Manufacturers sometimes use two or more jets per wheel. The efficiency of the Pelton wheel is somewhat less than that of a reaction wheel at full load, but at part load the situation is reversed.

Reaction wheels are of two general types, Francis and propeller. In both types the water passages are completely filled with water. The water enters the wheel around its entire periphery from a scroll case (Figure 6.3) through guide vanes (Figure 6.4), which are also the gates that control the amount of flow. Except on the discharge side of the wheel, the pressure is above atmosphere. The energy which drives the wheel is in both kinetic and "pressure head" form.

The Francis is the most common type of reaction turbine. Figure 6.5 shows a reaction runner lying on its side, together with a scale model used for its design. Figure 6.6 is a side view of another runner, also a Francis type, but designed for a plant with a higher head. In the Francis wheel, the water flows from the scroll case to the runner through guide vanes which impart a tangential velocity. The water gives up most of its energy to the runner, then continues in an axial direction downward. This type of turbine is used for heads up to about 1200 feet (365.7 meters). Wheel efficiencies are often better than 93%.

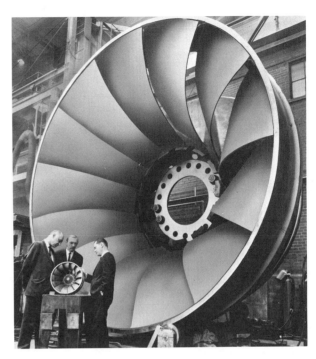

Figure 6.5. Reaction runner (on its side) as used in Francis-type reaction turbine.

There are two types of propeller wheels, fixed blade and adjustable blade. Both are high specific-speed runners (see following section) for use on low heads. Increasing the speed by means of propeller-type wheels reduces the size and cost of the generator and turbine, but there is a sacrifice in efficiency with the fixed-blade wheel.

The adjustable-blade propeller, usually referred to as the Kaplan wheel, automatically changes its runner-blade angle as the guide vanes are opened or closed, thus maintaining more efficient conditions

Figure 6.4. Guide vanes which double as gates to control amount of flow.

Figure. 6.6. Side-view of reaction (Francis) runner for use in high-head plants of about 1000 feet (304.8 meters).

Figure 6.7. Runner on Kaplan wheel used up to heads of 150 feet (45.7 meters).

of flow. This results in high efficiencies at all loads. It is used on heads up to about 150 feet. The runner of a Kaplan wheel is shown in Figure 6.7. Figure 6.8 shows an installation in cross-section. Except for the turbine runner and the blade-control oil piping, Figure 6.8 applies to any low-head, reaction-wheel installation.

EFFECT OF HEAD ON TYPE OF DEVELOPMENT

The arrangement and general character of a hydroelectric plant depends upon the head. The head of a low-head plant is created by a dam across a stream. Prior to building the plant, there was little or no change in water elevation at that point. Low-head plants are, therefore, characterized by a dam, a power house (which frequently is part of the dam), vertical shaft reaction-type wheels, and short water passages. High-head plants owe their head to nature, not man. They are usually found on mountain streams where the fall is very rapid. A low diversion dam diverts the stream, or part of it, through the necessary flumes, tunnels, and penstocks to the impulse wheels in the power house perhaps several miles downstream.

SPECIFIC SPEED

Electrical engineers have only an indirect interest in specific speed, but to the water wheel designer it is basic. Specific speed is defined as the speed at which the water wheel in question would operate if all its dimensions were scaled down proportionately until it would develop one horsepower at a head of one foot.

Numerically,

$$\text{specific speed} = N_S$$

$$= \frac{\text{r/min (hp)}^{1/2}}{H^{5/4}}.$$

Therefore,

$$\text{r/min} = N_S \frac{H^{5/4}}{(\text{hp})^{1/2}}.$$

Thus, at constant specific speed, the speed of the unit varies with the five-fourths power of the head, and inversely as the square root of the horsepower. Therefore, to keep the normal speed from being too high on a high-head plant, or too low on a low-head plant, it is desirable to use as high a value of specific speed as possible for low-head plants and a low value for high-head plants. Typical values of specific speed for different kinds of wheels are given in Figure 6.9. From this it can be seen why impulse wheels are used in high-head installations: they have the lowest specific speeds, and, therefore, permit using a wheel and generator speed that is within reason. Conversely, the high specific speed of the propeller-type wheel enables us to build, for very low heads, wheels with speeds that are not ridiculously low.

Water wheel designers like to be conservative in the selection of specific speeds for reaction wheels. Too high a specific speed, in combination with certain other unfavorable factors, often causes a phenomenon known as cavitation. This occurs at critical points where the contours create a partial vacuum and it may cause pitting or erosion of metal. Cavitation is entirely different from the wear due to scouring by abrasive material in the water.

SPEED REGULATION

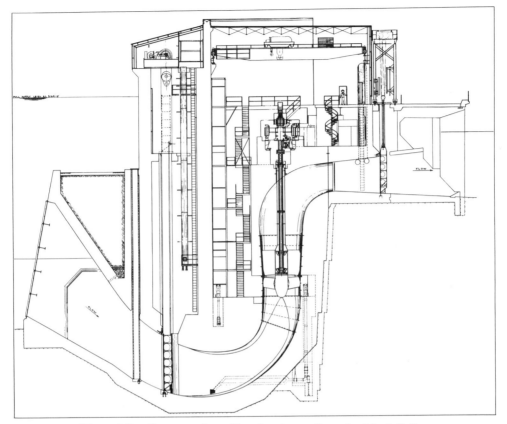

Figure 6.8. Cross-section of low-head, reaction wheel installation.

SPEED REGULATION

To change the power output of a water wheel, both the inertia and the relative incompressibility of water must be considered. It is not possible to make large and sudden changes in the quantity of water flowing through a hydraulic turbine, especially if the penstock is long. Neglect of this fact has sometimes been followed by unpleasant and perhaps astonishing consequences.

In low-head plants, where the power house is part of the dam and the water passages are short, the only concession to water hammer is to set a limit to the speed with which the gates may be closed, and then build into the generator whatever flywheel effect is needed to enable the turbine manufacturer to meet the specific speed regulation.

Some medium-head plants have surge tanks to absorb the energy in the penstock on sudden closing of the water wheel gates. A surge tank is merely a standpipe with a tank at the top. It is connected to the penstock, usually just as the latter enters the

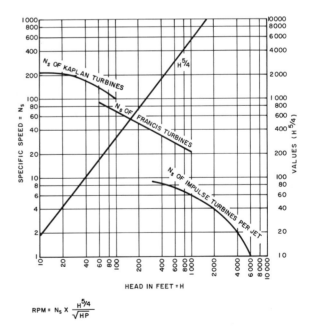

$$\text{RPM} = N_s \times \frac{H^{5/4}}{\sqrt{HP}}$$

Figure 6.9. Usual values of specific speed for different kinds of wheels.

power house. Surge tanks also help hold up the speed on sudden load increase.

Surge tanks are impractical on high-head plants, but fortunately the impulse type of wheel lends itself nicely to a different solution in either of two ways. One is to use a relief nozzle which is connected to the operating nozzle through a mechanism so that it opens whenever the working nozzle closes. The relief nozzles may remain open, or it may reclose, slowly, as determined by hydraulic conditions. The relief nozzle directs the stream harmlessly into a prepared pit, or possibly across the river. The second method is to deflect the jet from the buckets; then the needle valve can be closed, if desired—slowly, of course. While these schemes take care of overspeed on load reduction, obviously they are of no help in maintaining speed on sudden increase in load. Load pickup on an impulse wheel is slow because the penstock pressures will limit the rate of change of water flow that can be safely experienced.

RUNAWAY

Trash racks are usually provided to keep large pieces of debris from getting into the turbine and its water passages. However, it is recognized that sometimes logs or stones may get into the gates and keep them from closing and that load may be lost at the same time. In such a case, the unit will go to whatever overspeed is inherent in the wheel design—usually about twice normal. The rotating parts must be designed to stand this speed. It is impractical to prevent runaway by means of an emergency governor and a stop valve, as in steam turbines.

GENERATOR

The design of a hydroelectric generator is dependent on several external factors. Most of these factors directly influence the mechanical design, and affect the electrical design only indirectly. Usually it is possible to keep the electrical characteristics within generally accepted limits, although some mechanical requirements may necessitate an unusual electrical design to do so.

The approximate normal speed is determined by the turbine. The exact speed must be a synchronous speed for the frequency to be generated. The designers prefer some synchronous speeds over others, because they afford greater flexibility in design—for example, better choice of number of armature circuits, or of the number of armature slots per pole per phase.

Speed regulation or a need to limit penstock pressures during runaway, sometimes requires more than normal flywheel effect. If the necessary increase is considerable, usually the simplest way to get it is to increase the diameter of the rotor.

When it is necessary to overhaul or repair the turbine, it is a great convenience to be able to disassemble it and lift the parts through the generator stator without disturbing the latter. This procedure may require a larger diameter than would otherwise be used.

The overspeed requirement is also fixed by the turbine. It is on the order of twice normal—sometimes as high as three times normal on a Kaplan wheel. To build a rotor that can withstand the centrifugal stresses, it is sometimes desirable to go to a smaller diameter than that which would be chosen if there were no overspeed requirement. Obviously there may be times when the designer has to exercise his ingenuity to find a design that successfully meets the requirements of both overspeed and flywheel effect.

The thrust bearing of a vertical machine carries the weight of the generator rotor, the turbine runner, and (in the case of a reaction turbine) the thrust of the water on the runner. Under certain conditions, the water thrust may reverse sufficiently to lift the entire rotating element off the thrust bearing. This contingency must be factored in the design of the bearing and its support.

Sometimes the application dictates certain electrical characteristics. Years ago it was not unusual to have to design a hydroelectric generator to carry the charging current of a long, high-voltage transmission line, without losing control of voltage. This requirement has been replaced by an occasional demand for lower-than-standard reactance to improve the system stability.

The effect of these variable requirements is to discourage the development of a standard line by hydroelectric generators in all but the smallest sizes. All large machines are tailor-made. Any one station may contain two or more machines that are essentially duplicates, but rarely will there be duplicates between plants.

The standard short-circuit ratio (SCR) is 1.0 for 0.8 PF generators and 1.1 and 1.25, respectively, for

0.9 and 1.0 PF machines. Because of their low speed, hydroelectric generators have inherently high reactance. The transient reactance varies from 20% to 45%. Machines with amortisseur windings have lower values of subtransient and negative-sequence reactance than generators without these windings.

AMORTISSEUR WINDINGS

It has become common practice to equip hydroelectric generators with amortisseur windings. They afford the following operating advantages:

1. Reduction of overvoltage induced in the field winding by surges entering the stator windings, and by unbalanced conditions in the stator, such as coil failure.
2. Reduction of overvoltage in the stator winding caused by unbalanced faults on the machine, particularly on unloaded machines connected to capacitance, such as transmission lines.
3. Effective reduction in oscillation of generator output occasionally experienced on machines connected to their loads through high-resistance circuits.
4. Minor aid to system stability by reducing the magnitude of the machine rotor oscillations.

Hydroelectric generators normally have noncontinuous amortisseur windings where the amortisseur bars embedded in the pole-face are connected, but there is no connection between poles.

GENERATORS FOR PUMPED-STORAGE HYDRO

Pumped-storage generators are in most respects the same as generators used for conventional hydro plants. Both types are started in the generating mode by opening their gates to allow water to spin the turbine. When in the pumping mode, however, the pumped-storage unit requires other means to bring it to operating speed. Various methods have been used or proposed for this function, with the optimum for a specific project dependent on system and equipment considerations. The most frequently used techniques are briefly described here. These are the amortisseur-type start, wound-rotor induction motor start, synchronous start, and static start.

All of these methods generally unwater the pump/turbine by compressed air prior to starting. This limits losses at full speed to about 2–3% of the full unit rating.

Amortisseur-Type Start

Starting methods in this classification basically bring the unit up to speed as an induction motor, during which the amortisseur winding carries high-circulating current. Such machines are therefore equipped with heavy-duty "continuous" or connected amortisseurs. With an induction motor type start the total I^2R loss in the amortisseur winding is approximately equal to the rotational energy stored in the rotating parts of the unit. Since the starting time is relatively short, most of this energy is stored in the amortisseur bars during the start and becomes a limiting consideration as units become larger, especially at the higher speeds. On larger units, amortisseur-type starts are impractical and one of the other methods is used.

Several types of amortisseur starting have been used. The simplest of these is "full voltage starting," which is accomplished by simply closing the generator breaker. While simplest, this also results in the greatest voltage dips on the system. Frequently, "reduced-voltage starting" (by an autotransformer or by taps on the transformer) is used. "Reduced frequency starting" is another type of amortisseur start method that has been used at stations where both conventional and pumped hydro units are present. A generator is bused to the pump motor to be started and operated at a reduced frequency power source for induction motor starting of the pumping unit. Typically, at about one-third speed the motor and generator pull into step and are brought up in synchronism and together synchronized to the system.

Wound Rotor Induction Motor Start

This type of starting, utilizing a pony motor of about 5–10% of the rating of the generator/motor, has been widely used on larger units where amortisseur-type starts would not be practical. A control system is provided which maintains constant torque acceleration up to rated speed, then speed-matches the unit to system frequency for synchronizing. The starting motor is provided with one less pair of poles to permit bringing the unit up to full speed for synchronizing. Starting by this method is very smooth, and entails no significant voltage dip on the power system.

Synchronous Start

If the plant has conventional generators as well as generator/motors, the generator can be used for a synchronous start of the motors. The generator and the motor are isolated, excitation is applied to both machines while at rest, and they are then brought up in synchronism by admitting water to the turbine. The synchronizing torque available with this type of starting is much more dependent on the circuit resistance between units than on the reactance. Where the units are closely tied, a generator rating as low as about 10% of the motor rating may be used.

Static Start

Static starting is a form of synchronous starting in which a static variable-frequency source acts as the starting generator. The common arrangement utilizes a thyristor converter-inverter equipment rated at 5–10% of the generator/motor rating. A control system provides constant torque acceleration to rated speed, followed by synchronization to the power system.

AUXILIARY POWER

Auxiliary station-service power is generally obtained from the generator terminals and stepped down through a station-service transformer. Auxiliary power generally requires between 0.5–1% of the capacity of the unit. It operates the governor oil pump, sump pump, head-gate hoist motors, crane, air compressors, lights, heaters, and other miscellaneous motors and hand tools.

AUTOMATIC HYDRO STATIONS

Of the various types of plants used to generate electric power, the hydroelectric station lends itself most readily to automatic operation. Basically, stored energy in the form of water in the forebay can be released and converted to electric power by merely opening the turbine gates under the control of the governor, thus developing horsepower on the prime-mover shaft. The generator will produce real power under control of the governor and reactive power under supervision of the voltage regulator. The inherent design of the components involved makes the problem of obtaining automatic, unattended operation a relatively simple and inexpensive application.

Automatic hydro plants may best be classified as to the degree of automatic operation provided, namely: semiautomatic, fully automatic, and fully automatic with provision for control of the station from a remote location.

The degree of automatic operation provided at a hydroelectric station will depend on many factors, such as the size and importance of the station with respect to the power system, the expected load factor of the station, the relative location of the station with respect to other generating stations and to population centers, special hydraulic problems (including any required supervision of dams and other hydraulic works), and ice and debris control problems.

Semiautomatic Plants

The simplest and least expensive type of automatic control used in hydro stations is that found in semiautomatic stations. Stations of this type must be started and put on the line manually by an operator, but then may be left unattended.

This type of control is generally applied to stations that are small with respect to the system to which they are connected, and thus do not require that their output be coordinated with the system daily-load curve. Such stations are often operated for only an 8-hour period each day, and thus, once they are shut down, may not be available to meet emergency conditions on the system, unless an operator is available to go to the site to restart the machine.

Machines in stations of this type may be operated on free governor, load limit, or on load limit as controlled by a float-operated device installed to maintain constant forebay water-level elevation.

The control must provide means for shutting down the unit automatically on a signal from a time switch or float switch. Sufficient protective devices must also be provided to shut down the station automatically and prevent damage to the equipment under certain abnormal operating conditions. Other abnormal conditions may only result in an annunciator indication.

Fully Automatic Stations

The control for this type of station offers the advantage of being able to start the plant by means of a signal from a time, float, or system-frequency device without the presence of an operator at the site.

Automatic starting and synchronizing features must be added to the control scheme previously outlined for semiautomatic stations. Devices must also be provided to assure that certain operating conditions exist before the master control relay is energized—i.e., adequate governor oil pressure and relays in their normal reset position.

Fully Automatic Station with Remote Control

A fully automatic station controlled from a remote point offers most of the operating advantages of an attended manual station. The available water can be used in the most efficient manner by adjusting the output of the station to system load requirements. Added economic advantages can be obtained if several hydro stations and substations can be controlled from a single control station.

The functions to be performed from the control station should, in general, include most of the operations normally performed by an operator in an attended plant.

To carry out these functions, a communication channel must be set up between the control station and the outlying power plant. This channel may be direct wire, carrier, or microwave.

SUMMARY

It would be easier to get electricity from rain and snow water if the stream flow would conform to the system's needs for energy and power. But the uncertainty of spring floods and summer droughts raises two big questions: (1) How big a plant to build and (2) how to get the most out of the plant once it has been built.

These questions require the study of duration curves and other available stream-flow data. The prospects for seasonal storage also have to be considered. Finally, the decision has to be made whether the characteristics of the system dictate that major emphasis be put on energy production or on power production (for peaking).

In the eastern part of the United States, the conventional-type hydro plant has a low or medium head. Low-head plants are characterized by a dam across a river, which creates the head, and the power house is often part of the dam. The wheels are reaction-type units, to take advantage of their higher characteristic speed, and they have vertical shafts. In the mountainous West, some hydroelectric plants have high heads. These plants have impulse-type wheels, usually with horizontal shafts.

It is sometimes necessary to build extra flywheel effect into a water wheel-driven generator. The amount is dictated by hydraulic conditions. The purpose is to prevent excessive rate of rise in speed if load is suddenly lost.

It is also necessary to build both wheels and generators to stand runaway speed—about twice normal. This requirements is due to the possibility that full load may be lost at a time when the gate-closing mechanism is inoperative.

REFERENCES

1. Brown, P. G., and Otte, G. W., "Electrical Design Considerations in Pumped Storage Hydro Plants," *IEEE Transactions,* PAS-**S82,** 625–641 (1963).
2. Handel, R. D., "Design of the MICA Hydroelectric Project," *IEEE Transactions,* PAS-**96,** 1535–1545 (1977).
3. Harvey, III, C., "Concepts and Features of the Automated Unit Control System for the Racoon Mountain Pumped-Storage Plant," *IEEE Transactions,* PAS-**97,** 513–519 (1978).
4. Hydroelectric Power Subcommittee of the IEEE Power Generation Committee, "Bibliography on Pumped Storage to 1975," *IEEE Transactions,* PAS-**95,** 839–850 (1976).
5. IEEE Committee Report, "Survey of Pumped Storage Projects in the United States and Canada to 1975," *IEEE Transactions,* PAS-**95,** 851–858 (1976).
6. St. Onge, G. A., and Stewart, W. A., "Rock Islands Second Power House Uses Bulb-Type Hydro Units," *IEEE Transactions,* PAS-**96,** 1690–1696 (1977).
7. Stuart, R. H., Gesh, S. E., Busby, C. H., and Thom, R. H. P., "Electrical Features of the Churchill Falls Development," *IEEE Transactions,* PAS-**93,** 340–348 (1974).
8. Working Group on Protective Relaying for Pumped Storage Hydro Units, "Protective Relaying for Pumped Storage Hydro Units," *IEEE Transactions,* PAS-**94,** 899–907 (1975).

7

POWER TRANSMISSION

INTRODUCTION

Power transmission is the source of several of the more interesting phenomena and problems encountered in an electric utility system. Most of the problems are well understood today. However, they presented genuine concerns to the pioneers of the industry who first encountered them. For example, the early literature documents many references to corona—its effects, how to detect it, how to avoid it, and how to calculate or measure the radio (and television) interference it causes. Charging current also created a problem when a long high-voltage line was energized, and when it was switched off again. Later the problem of power system stability and the adverse effect of the high reactance of long transmission lines and their susceptibility to short circuits and interruptions was encountered. The last increase in transmission voltage to 765 kV has resulted in additional and continuing problems regarding audible noise produced by wet weather corona and ground-level 60-Hz electric field effects.

The increase in transmission voltage to the 765-kV level (765 kV is the highest level today) has called attention to two "new" problems. All phenomena may exist at all transmission voltages, and each time a substantial increase in voltage is used it reveals phenomena that were unimportant at the lower voltages. The "new" problems at the 765 kV level concern audible noise and ground-level field effects. Audible noise is a wet-weather corona phenomenon where the noise produced from the interaction of corona and water droplets is extremely objectionable. The second problem, ground-level field effects, produces an electrical shock to a person standing on the earth and touching a vehicle parked underneath the line. These concerns dictate the design of the transmission lines in the class often referred to as ultra-high voltage or UHV.

These new concerns are of major importance, and in many cases, are ruling design factors for upper-level extra-high voltage (EHV) and UHV transmission lines.

DEFINING TRANSMISSION

The definition of transmission depends on who uses the term. When water power was first developed at distant sites, the power was "transmitted" to the load area over high-voltage lines. When steam plants moved out of the load area, their output had to be "transmitted" too. Often steam power was also transmitted at high voltage—not primarily because of the distance, but because of the amount of power. "Transmission," in addition to its original function of moving energy long distances, has also come to mean the system backbone that ties together the important generating stations and primary substations. The term almost invariably designated the highest-voltage circuits on a given system.

The increase in transmission voltages over the years is shown in Figure 7.1. The highest voltages in commercial use in this country prior to 1953 (287.5 kV) brought power from Hoover Dam to Los Angeles, a distance of 270 miles (433 km). In 1953, the American Gas and Electric Co. (now American Electric Power Service Corp.) started operation of the first element of a new 345-kV backbone, superimposed on its existing 132-kV system in West Virginia, Ohio, and Indiana. In 1963, Hydro-Ontario started operation of a 500-kV line from Pinard to Manmer. In 1964, the first commercial 500-kV line

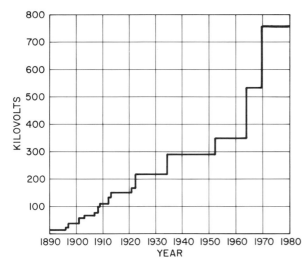

Figure 7.1. Highest transmission voltage in the United States.

HVTRF, complemented by work from other research centers, appears in the book, *Transmission Line Reference Book—345 kV and Above,* EPRI, 1974 (first edition), and the second edition published in 1982.

To reach 1100–1500-kV nominal levels, satisfactory ways must be found to limit switching surge potentials to, perhaps, 1.5–1.7 times the nominal operating voltages. Various arrangements of circuit switching equipment and other means of limiting surge crests are being studied.

Salt, industrial by-products, and other atmospheric contaminants can greatly reduce the insulating strength of exposed equipment. The addition of insulator elements and special designs with increased creep distance may not offer a complete solution of the problem for future UHV equipment. Long plastic insulators show promise for good contamination performance. However, their longevity under operating conditions has to be demonstrated.

in the United States was energized by the Bonneville Power Administration. This was a 110-mile (176 km) line from the Big Eddy substation to the Keller substation. This dramatic increase in higher-voltage usage is illustrated by the fact that from 110 miles of 500-kV transmission line in commercial operation in the United States in 1964, there were over 7200 miles (11,620 km) in operation by the end of 1970.

Several higher-voltage systems were commissioned in the United States and Canada. In Canada, Hydro-Quebec energized the world's first commercial 735-kV system in 1965. The first commercial 765-kV system in the United States was energized in the spring of 1969 by the American Electric Power Service Corp. Currently, Bonneville Power Administration has plans for a 1100-kV transmission level by the late 1980s.

This rapid upward trend in transmission voltages has been supported technically by a number of research test lines. One of these, General Electric's Project EHV, was under the sponsorship of the Edison Electric Institute. Project EHV developed an EHV transmission line reference book, incorporating the design technology developed through 1968. Project EHV, renamed Project UHV in 1967, and recently renamed the High Voltage Transmission Research Facility (HVTRF), is exploring innovations in line design and developing prototype designs for 1100-kV and 1500-kV systems. Figure 7.2 shows the surge generator used at these facilities.

All of the design technology developed at the

Figure 7.2. 5000-kV surge generator at the outdoor test area of the High Voltage Transmission Research Facility, Lenox, Massachusetts (Sponsored by the Electric Power Research Institute).

TABLE 7.1. Representative Conductors and Spacings in General Use or Proposed

Nominal Voltage (kV)	Maximum Voltage (kV)	Spacing (ft)	Diameter (in)	Conductors per Phase	Aluminum Cross-Sectional Area per Phase (MCM)	Number of Strands (Aluminum/Steel)
230	242	18	1.093	1	795	54/7
345	362	25	1.750	1	1414	58/19
345	362	25	1.245	2	2066	54/7
500	550	33	1.762	2	4312	84/19
500	550	33	1.602	3	5343	84/19
765	800	46	1.382	4	5088	54/19
1100	1200	60	1.602	8	14248	84/19

SOURCE. LaForest, J. J., Editor, *Transmission Line Reference Book*, EPRI, Palo Alto, California, 1982, p. 64.

COMPONENTS OF A TRANSMISSION LINE

A transmission circuit consists of conductors, insulators, supports, and usually shield wires, as shown in Figures 7.3 and 7.10. High-voltage transmission lines do not use the highways, as telephone and lower-voltage electric utility lines sometimes do; they are built on rights-of-way of their own.

Conductors

The most commonly used conductor material for high-voltage transmission lines is aluminum conductor, steel reinforced (ACSR) (The Aluminum Conductor Handbook, 1971). The reason for its popularity is its low relative cost and high strength-to-weight ratio as compared with other types of conductor material. Also, aluminum is in abundant supply, while copper is limited in quantity. The all-aluminum conductor appears to be gaining popularity in certain areas, especially where there is little or no snow, ice, or sleet loading problems. Typical conductors and spacings are shown in Table 7.1.

Insulators

Insulators are now of the suspension type (Figure 7.3) (LaForest, 1982, p. 470). Until transmission voltages reached 60 or 70 kV, however, only pin-type insulators were available. Any further increase in voltage appeared impossible because a longer pin insulator would be just too weak mechanically. Then E. M. Hewlett* developed the suspension in-

* For many years chief engineer, Switchboard Department, General Electric Co.

sulator. The first form was a chunk of porcelain having two cored holes at right angles to each other. Copper straps through these holes attached the required number of insulator disks to each other to form a string or chain, so that the porcelain was in compression. At first not even Hewlett dared try to work porcelain in shear or in tension. Appropriately, these insulators were known as "Hewlett insulators."

The Hewlett insulator showed that the suspension principle was practical. Before long a different kind appeared, in which the hardware was cemented to the porcelain, as shown in Figure 7.4. Known as the "cap and pin" insulator, it was cheaper and easier to use than the Hewlett, which it soon displaced.

It took several years for the manufacturers to learn how to make good cemented insulators. During this learning period, it was not unusual to have to replace 25% of the units on a line within a year after they were put up. Transmission lines were out of service so much of the time for testing and changing insulators that it was standard practice to build two-circuit lines to have one circuit available for operation.

Two new types of insulators for use in contaminated environments are the high-leakage distance and fog insulators. It is generally accepted by the industry that the insulation deterioration of suspension insulators in contaminated atmospheres, to a large extent, is a function of the leakage distance of the insulator. The skirt designs of these insulators provide this high leakage distance. The use of these insulators provides satisfactory transmission line insulation over longer periods of contamination

COMPONENTS OF A TRANSMISSION LINE

Figure 7.3. Suspension-type insulators.

build-up than the use of standard insulators. This provides the line designer with insulator application margins not available in standard suspension insulators. Figures 7.5 and 7.6 show two types of high-leakage distance suspension insulators, the ball and socket unit and the clevis-type unit, respectively.

Figures 7.7 and 7.8 show two arrangements of suspension insulator strings. The 765-kV single-circuit system shown in Figure 7.7 has two strings of insulators per phase, in a Vee arrangement. The 345-kV double-circuit system has only one vertical (tangent) string of insulators per phase. From a contamination standpoint, the V strings are more effective than vertical strings because of a "self-cleaning" possibility. Both sides of each insulator V string are exposed to the rain, allowing contaminants to be shed more effectively.

Supports

In recent years, there has been considerable effort to reduce the installed costs of transmission towers.

The cost depends not only on design and materials but also on the cost of transporting material and erection equipment to the tower sites, and upon the cost of labor for assembly and erection. This, coupled with the growing emphasis on environmental protection and aesthetic improvement, has led to the use of a variety of materials and design.

Galvanized steel, self-supporting towers, or wood H- or K-frame structures have been most commonly used for 345-kV lines. Guyed towers have the least weight, and this weight can be further reduced by the use of aluminum instead of steel. In some cases, these towers have been assembled in staging areas and flown to the site by helicopter. All double-circuit 345-kV lines have been built of self-supporting steel towers with the phases of each circuit arranged vertically to minimize the tower width, or, in the southwestern United States, on lower towers with each circuit in a triangular configuration. Figure 7.8 shows a modern 220-kV double-circuit tower with the vertical-phase arrangement. It also illustrates the effort to minimize environmen-

Figure 7.4. View of cap and pin insulator disk.

Figure 7.5. Pin and socket-type high-leakage distance insulator for use in contaminated environments.

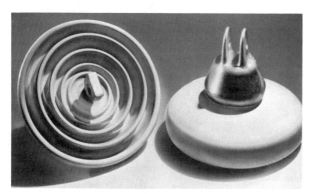

Figure 7.6. Clevis-type high-leakage distance insulator for use in contaminated atmosphere.

tal intrusion by changes in tower design and construction. Self-supporting steel towers have been most commonly used on 500- and 765-kV lines, and these tower structures are shown in Figure 7.17 and 7.7, respectively.

Shield Wire

Shield wires are used to protect the energized conductors from lightning. With a conservative design, practically all lightning strokes to the line will terminate on the shield wire rather than on a line conductor. Shield wires have been a part of nearly every high-voltage transmission line built for many years.

The two top wires (attached to the highest point on the tower structure) in Figure 7.7 and the one top wire in Figure 7.8 are shield wires. The other groups of wires attached to the suspension insulators are the line conductors that carry load currents. There are two conductors per phase shown in Figure 7.8 and four in Figure 7.7. For a discussion of multiple-phase conductors or "bundled conductors" see the section on bundled conductors.

Figure 7.7. Steel tower for 765-kV transmission system. (Courtesy of American Electric Power Service Corporation.)

Figure 7.8. Modern tower structure supporting two 220-kV bundle conductor transmission lines. (Courtesy of Southern California Edison Company.)

Right-of-Way

The right-of-way may be purchased outright or leased. If the latter, the lease will provide the right of such access to the line as may be necessary during construction and subsequent inspection, testing, and maintenance. The right-of-way has historically been selected on the basis of proceeding from the power source to the load area by the most economical route. The economic factors considered were expense of real estate, densely settled communities, bodies of water, or rugged, swampy, or inaccessible terrain. The cost of clearing and keeping the right-of-way free of trees and brush was sometimes a factor in deciding between alternate routes. Today, because of environmental and aesthetic considerations, utilities not only have trouble in getting an economical route, they have great difficulty in obtaining any right-of-way at all. In one particular system, a 115-kV interconnection, 5 miles (8 km) in length, was delayed for over 3 years by the opposition of over 15 environmental and conservation groups in hearings before a State Public Service Board, as well as by numerous meetings between utilities and the affected town and regional planning commission.

In some heavily populated areas, such as New York City, it is virtually impossible to construct overhead transmission lines, so that transmission system expansion has become limited to the use of underground circuits. As urban populations increase, this condition will become more widespread.

Underground Transmissions

In 1970, there were about 2000 miles (3200 km) of underground transmission lines in the United States—less than 1% of the nation's total transmission system. Currently, the most extensive system is that of Consolidated Edison Company in the New York City area. The highest-voltage installation, utilizing 525-kV oil-type cables, connects the power plant and the 500-kV switchyard at the Grand Coulee Dam. The first section of the system was installed in July 1973. The complete installation involves three circuits (nine cables) with an average length of about 6500 feet (2015 meters).

Although underground ac transmission would present solutions to some of the environmental and aesthetic problems involved with overhead transmission, there are technical and economic reasons that make the use of underground ac transmission prohibitive. A major difficulty of underground ac transmission lines results from a continuous flow of "charging current" between the conductor and the sheath. Its magnitude increases with voltage and varies inversely with the thickness of the insulation and directly with the length of the cable. At 345 kV, utilizing commercially available cable, practically all of the current carrying capability of the uncompensated underground conductor is utilized by this charging current in a distance of about 26 miles (41.6 km). Therefore, no capability is left for useful current to be transmitted, unless very expensive and bulky compensation equipment is installed.

Problems are also encountered in connections between underground lines and substations. Abrupt insulation changes cause very large concentrated electrical stresses. These stresses can be eliminated by terminating the cable above ground in an expensive device known as a "pothead."

Underground lines have the further disadvantage of poor accessibility. Overhead lines have more outages than underground per unit length, but the outages are usually shorter in duration. Because of the longer repair time for underground systems, expensive duplication facilities are sometimes installed to reduce the risk of overlapping outages. Overhead lines have greater flexibility because connections and repairs are relatively simple and they can be converted to higher voltage if necessary. Underground systems cannot be easily altered.

On the basis of present technology, it is estimated that in suburban areas underground lines cost, on the average, about 8.5–15 times more than overhead lines. As transmission costs represent between 10 and 20% of the utilities' total investment costs, the significance of these increases in the cost of transmission can be readily understood.

ALL-WEATHER TRANSMISSION

Transmission has to stand the heat of summer, the cold of winter, summer rains, and winter ice. Lines near the sea coast are subject to salt deposits, and those in certain inland regions are subject to deposits of alkali dust (IEEE Working Group on Insulator Contamination, 1979). Deposits of both salt and alkali on the insulator surfaces, when they are moist, tend to turn the insulators into semiconductors; in addition, the salt promotes deterioration of any steel in towers or hardware. Lines that pass near cement and chemical plants are subject to attack by whatever impurities are present in the atmosphere. Maintaining a line in good operating condition in the presence of these contaminants can be a challenge. For example, it is not easy to remove from line insulators the cement dust that has been intermittently moistened by rain and dew while it was accumulating.

The use of a silicone or petroleum grease on insulators lengthens the period between necessary washings and makes it easier to remove the contamination. This, in turn, aids washing while the line is energized. "Hot"-line insulator washing and other maintenance can be done even on lines up to 700 kV by the use of insulated buckets on extendable arms mounted on special trucks to elevate personnel to the necessary height.

Sometimes the heat and smoke from brush and grass fires under a line reduce the dielectic strength of the air so that it can no longer withstand normal operating voltage. Short-circuits and trip-outs follow. A line outage may result from any of a long list of natural or human causes. The list would include tornado, ice, wind, earthquake, switching surge, system instability, overload trip, personnel error, land or air vehicle, birds, mechanical failure, etc. Lightning is the most frequent cause of transmission-line interruptions in many parts of the country.

LIGHTNING

Lightning is an electrical discharge (Anderson, 1975; Anderson, 1979). It is the discharge of an accumulation of electrostatic electricity between cloud and earth, or between clouds. The only lightning that is of interest here is that which takes place between cloud and transmission line (or, rarely, between cloud and earth or some object very near a line).

Characteristics of Lightning

The lightning stroke is a high-current discharge that lasts only several millionths of a second. Many lightning strokes, however, contain two or more separate, individual discharges over the same path. One observer reports as many as 40 separate discharges, all occurring within less than a second.

The maximum lightning current measured in a transmission tower is of the order of 130,000 amperes; the maximum in a stroke, based on currents measured in towers and shield wires, is estimated to be on the order of 200,000 amperes. Reports of various investigations indicate that not over half the strokes exceed 15,000 amperes. The voltage between cloud and ground that initiates a lightning flash is not known, but is probably very high. Strokes have been observed up to 16,000 feet (4878 meters) in length, and it takes a very high voltage to spark across that distance. Half of those observed are over 4000 feet (1220 meters) long. When the stroke hits a line conductor, the resulting voltage build-up is almost invariably sufficient to cause flashover, even on the highest-voltage lines in service.

Although lightning is a short-duration phenomenon, it is not instantaneous. It starts at either cloud or earth and progresses at an observable and measurable rate toward the other. When the earth is fairly flat, the stroke is more likely to start at the cloud. On the other hand, very few strokes are from cloud to the Empire State Building; most of those observed start from the building.

Stroke to Tower or Shield Wire

When lightning hits a transmission line, it may terminate on the top of a tower, on the shield wire, or on a line conductor. If it hits a tower top, some of the stroke current flows into the shield wires, if there are any, and the rest goes down the tower to ground. If the impedance of the steel tower and the tower-footing resistance are low, and the stroke is moderate in terms of both current magnitude and rate of rise, the current passes harmlessly into the ground, and nothing happens. If these conditions do not exist, however, the flow of current through the tower raises the tower to a high voltage above ground. In extreme cases, the voltage between tower and ground may be enough to cause a flashover from the tower, over the line insulators, to one or more of the phase conductors.

In the past, the lightning performance of a line has been predicted on the basis of a lightning stroke of certain assumed magnitude and rate of rise. The tower was represented by its footing resistance and possibly by an inductance for the steel structure. The predictions obtained on this basis have been found to correlate very poorly with actual field experience, raising questions concerning additional factors that might affect the lightning performance of lines.

Because the voltage appearing across an insulator subsequent to a stroke to the tower or shield wire is affected by the electrostatic and electromagnetic field of its environment in the tower structure, the only accurate way to predict the relationship between stroke characteristics and insulator voltage is by test, using a geometrical model of the structure, or by analysis with a digital computer model. The response characteristic of the geometrical model is then fed into a digital computer together with other variables such as the statistical distribution of lightning-stroke magnitudes and rates of rise, point-of-stroke incidence, and phase-angle of the power system voltage on the phase conductor in question. The computer is programmed to analyze these variables statistically and to predict an outage rate for any proposed design.

Traveling Wave

When lightning strikes a phase conductor or shield wire, the current of the lightning stroke tends to divide, half going in each direction. If the shield wire is struck at the tower, current will also flow in the tower, as mentioned previously. The current of the lightning stroke will encounter the surge impedance of the conductor(s) so that a voltage will be built up. Both the current and the voltage will move along the conductor as "traveling waves." The velocity of wave propagation is a little less than the speed of light. If the line were truly "loss-less," the velocity would equal the speed of light. An approximate velocity of 1000 feet (310 meters)/μsec (a little high) is often used in approximate calculations. The voltage magnitude is equal to the current multiplied by the surge impedance. The surge impedance of a two-conductor per phase overhead transmission line is usually in the order of 300 ohms, and is almost pure resistance. Ultrahigh-voltage lines, with several conductors per phase have even lower surge impedance levels in the range of 200–250 ohms.

Stroke to Shield Wire

If the stroke hits a shield wire between towers, traveling waves along the shield wires are initiated. The lightning current flows harmlessly to ground at nearby towers, unless the tower-footing resistance is too high, or in the case of high towers, the tower inductance is too high. The tower top voltage is impressed across the line insulator strings and can cause a flash-over, resulting in a line "outage."

LIGHTNING-PROOFING A TRANSMISSION LINE

The only way to be sure a transmission line is lightning-proofed is to completely enclose it in grounded metal, a solution that is not very practical. The line, however, can be made practically lightning-proof.

To do so, shield wires are essential to intercept direct strokes. They must be properly located with respect to the phase conductors and have adequate clearance from the conductors, not only at the towers, but throughout the span. If tower-footing resistances are too high, they must be brought down to an acceptable value with driven grounds or with counter-poises. Then, by means of the described model tests and digital computer study, a decision is made on the most economical tower and insulator structure that limits the flash-over rate to an acceptable level. Assuming that the proper magnitudes of the quantities were used in the study, accounting for all changes that weather, aging, and so on might cause, and that the line was properly maintained, the line-outage rate should be approximately as predicted.

One important variable that should not be overlooked is the effectiveness of the shield wire or wires. A properly located shield wire intercepts a high proportion (probably above 95%) of the strokes that would otherwise hit a phase conductor. But lightning does not always follow a straight vertical path to the ground. Occasionally a stroke may pass the shield wire and hit the phase wire. This may be the result of the high wind accompanying the thunderstorm, blowing the phase conductor out beyond the zone of protection of the shield wire. Of special concern is the lightning performance of a double-circuit line on one tower where both circuits can be taken out by one stroke.

Design For Switching Surges

Insulation design for the transient voltages includes consideration of externally generated transients, lightning, and those generated by switching operations. In the low to midranges ($\leq$500 kV) of EHV transmission voltages, lightning surges may control the design in areas of high ground resistance, high lightning incidence, or where use of resistor preinsertion in the system circuit breakers minimizes switching surge amplitudes. In the higher EHV levels and UHV range, tower structures are by nature so massive that the lightning (backflashover) effects are considerably diminished and switching-surge effects rule.

Line design for switching surge focuses on a comparison of stresses and strengths (LaForest, 1982, Chapter 11). Switching surge stresses may be predicted through the use of a transient network analyzer or by a digital traveling-wave program, or maximum values may be assumed from past experience. Strengths for various kinds of tower and line gaps are the results of research efforts at the HVTRF or similar centers.

The traditional method of switching-surge design consists of first choosing a certain high value of surge amplitude and a representative combination of meteorological conditions; next the tower is designed to withstand this particular condition so that it will perform satisfactorily in service for most, if not all, of the other conditions that will appear. This method is best for preliminary calculations.

More refined design methods make the best possible effort to assess the relative frequencies of occurrence of all the combinations of electrical and meteorological events; evaluate how overall performance would change as the choice of insulation is changed; and determine how additional insulation investment would affect the outage rate (Anderson, 1964). These methods explore the problem in far greater detail, but usually require a digital computer program and a far more systematic evaluation of meteorological and transient voltage conditions.

Figure 7.9 illustrates the basic of the statistical design method wherein the stress and strength distributions are displayed along the line-voltage axis in the diagram. The numbers of flash-overs per set of switching operations may be obtained by integrating over the entire voltage range.

MECHANICAL DESIGN

For convenience, assume here that the problems of mechanical design and those of electrical design are unrelated and can be considered and solved separately and in series. Actually, some of these problems are interdependent and must be solved in parallel. For example, a more massive tower aids lightning performance, while a thin tubular steel tower aids switching-surge performance. The mechanical problems are those of supporting one or two three-phase circuits. Electrical design will have

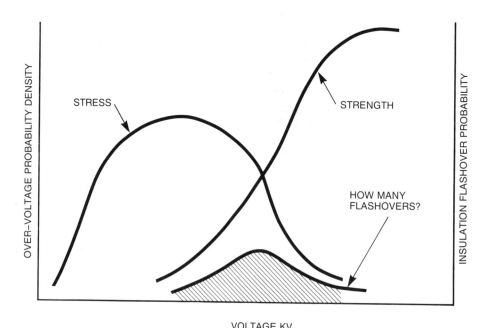

Figure 7.9. Statistical representation of insulation stresses and strengths.

specified the conductivity and outside diameter of the conductors; the number of units in the insulator strings; the minimum clearance from conductor to tower, and between conductors; the number and location of shield wires; and the maximum permissible tower-footing resistance.

Conductors

The conductors selected will be either ACSR or all aluminum. They must be strong enough to support an assumed thickness of ice, in addition to their own weight, in a wind of an assumed velocity.

Suspension Insulators

Standard suspension insulators are strong enough to support the conductor of any usual span length against the forces of gravity and wind, even when the conductor is covered with any probable thickness of ice. It may be necessary to use two strings of suspension insulators in parallel for the strain insulators at dead-end towers. Extra long spans—for example, at river crossings—may require several strings in parallel, with special hardware to equalize the pull among them.

Towers

All towers must be strong enough to support the conductors and shield wires, including their assumed ice loading during the maximum assumed wind loading. The structures that just meet these minimum requirements are commonly known as suspension towers and are used only on straight sections of the line. On many lines, the suspension tower is also required to withstand the unbalanced pull caused by the breakage of any one, or sometimes two, of the wires, either phase-conductor or shield wire. Every mile or so, there is a dead-end tower. A dead-end tower is expected to withstand the breakage of all conductors and shield wires on one side of it. Angles in the line require angle towers, whose strength is intermediate between suspension and dead-end towers.

Tower design must recognize that a suspension-insulator string is somewhat flexible. This flexibility allows the conductor to swing with a transverse wind. When the insulator string is at the maximum assumed angle from the vertical, the distance from the nearest tower member to the conductor or to any metal parts connected to it should be enough to withstand at least as much impulse voltage as the insulators themselves. This requirement fixes the minimum tower-top clearance (Figure 7.10) and the minimum spacing between the phase conductors.

In areas subject to ice, it is usual to build double-circuit towers with the middle conductor offset from the top and bottom conductors, as shown in Figure 7.10. The purpose of this arrangement is to reduce the chance that one of the phase conductors will interfere with another because of the accumulation of ice, or because of the sudden dropping of the ice load by one conductor only.

To achieve a minimum cost line, it is necessary to study the effect of variation in the span length. Several factors must be considered. Towers cost money. For this and other reasons, it is desirable to get along with as few as possible. But fewer towers mean longer spans and longer spans means greater sags. To maintain minimum clearance to ground at midspan, the towers must be higher.

The weight increases much more rapidly than the height, and therefore so does the cost. So a balance must be sought which gives the optimum number of towers per mile. The solution, once attained, can be used for the whole line, if it traverses level country. In rough, hilly, or mountainous country, the height, location, and probably the foundation design of each tower becomes an individual problem.

Computer programs have been written both to locate (spot) the towers along the terrain in the most economic manner, and, given the mechanical load requirements, to optimize the structural design of the tower itself.

Conductor Vibration

There are four basic types of vibration of transmission lines:

1. *Aeolian vibration.* This vibration is due to Von Karman vortex shedding on the leeward side of a cylindrical object in the wind. This is the kind of vibration that makes wires hum in the wind and the whistle of wind in the rigging of ships. It also causes the very noticeable vibrations sometimes observed on flag poles and street standards. This vibration can be lessened on transmission lines by adding vibration dampers (usually Stockbridge) which are a compound-pendulum type of arrangement that detunes the vibrating conductor and absorbs enough of the energy to stop or greatly lessen vibration. Other means of lessening this vibration include re-

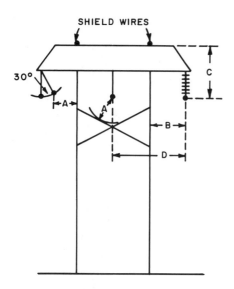

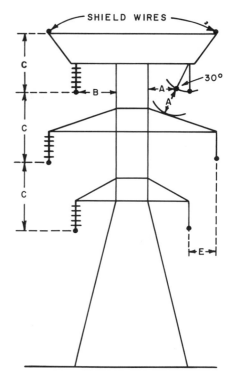

SINGLE-CIRCUIT TRANSMISSION LINE WITH HORIZONTALLY ARRANGED CONDUCTORS

DOUBLE-CIRCUIT TRANSMISSION LINE WITH VERTICALLY ARRANGED CONDUCTORS

A – CONDUCTOR TO NEAREST TOWER STEEL

B – CONDUCTOR TO VERTICAL TOWER MEMBERS

C – BETWEEN VERTICAL CONDUCTORS AND BETWEEN CONDUCTORS AND SHIELD WIRES

D – BETWEEN HORIZONTAL CONDUCTORS

E – OFFSET OF MIDDLE CONDUCTOR (VARIES FROM 0.5 A FOR ANTICIPATED FAVORABLE WEATHER CONDITIONS TO 1.5 A FOR ANTICIPATED UNFAVORABLE CONDITIONS

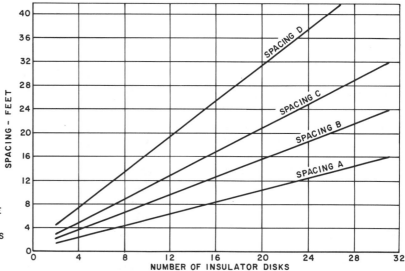

Figure 7.10. Tower-top clearance.

ducing the conductor tension and constructing the conductor so that it has a high internal friction. Aeolian vibration has sometimes been the cause of fatigue failures in transmission line conductors, particularly near suspension points where the conductor is flexed back and forth day after day.

2. *Subconductor vibration.* This type of vibration is only possible on bundled conductor arrangements. On a bundled conductor, the windward conductor has a wake that spreads out its leeward side similar to the wake a motor boat makes going through the water. One or more of the leeward con-

ductors is riding in a wake that has shear flow, and different velocities of wind cross over the top and bottom of the conductor. Depending on the conductor position, this can cause either negative or positive lift in a manner similar to the wind around an airplane wing where different velocities are also present on the upper and lower surfaces. In addition to this lift there is also a lessened drag on the leeward conductor compared with the windward conductor, tending to displace the conductors horizontally with respect to one another. This combined wind condition can be sufficiently unstable to force the leeward conductor into an oscillation condition. This oscillation is transmitted to the other conductors through the spacer that mechanically connects them together at intervals. Subconductor vibration, usually occurring only in open country where there is a continuous laminar flow of wind, can cause severe vibration on a bundle conductor arrangement, leading to the ultimate breaking of spacers and, in some cases, destruction of suspension points at the insulators. Subconductor vibration occurs at lower frequency (2–4 Hz) than aeolian vibration and is more difficult to control. Vibration dampers have been developed to help this problem, and considerable research is still in progress. Other means of lessening this vibration include the spacing of subconductors as far apart as practical, orienting the subconductors so that they are at advantageous points in the wind wake, and more frequent use of spacers within each span.

3. *Corona Vibration.* Corona vibration usually occurs in wet weather when water drops clinging to the underside of the conductor are forced off by an expulsion action due to the electrostatic field forces at the bottom of the conductor. Vibrational displacements of a few inches can appear between vibration nodes on any span. Corona vibration, because of its infrequent occurrence and relatively low amplitude, does not yet appear to be a major problem, but is of increasing concern as transmission voltages approach the UHV levels.

4. *Galloping.* The vibration known as galloping is usually created by a nonuniform air-foil surface formed around the conductor by ice. This type of vibration can be quite severe, is of a very low frequency, and is extremely difficult to control. Cases have occurred where this galloping has displaced the phase conductor all the way up to the ground wires, causing multiple trip-outs. It has also resulted in damage to conductors, spacers, and towers. Fortunately, it occurs very infrequently. Another form of galloping has occurred on conductors, even without ice, when the wind blows against the helical strands in such a way that it moves along the strands on the bottom of the conductor and across them at the top. This has occurred at river crossings, particularly at the Severn River in England. The obvious cure for this ice-free vibration condition has been to tape the conductor so that it represents a somewhat smoother surface to the wind at both top and bottom.

ELECTRICAL DESIGN

Two aspects of the electrical design of a transmission line are important: (1) its performance as a link in the 60-Hz system and (2) its vulnerability to electrical failure resulting from lightning, switching surges, temporary overvoltages, or normal service voltage under contaminated conditions.

As a link in the 60-Hz system, long-distance, high-voltage transmission has one characteristic different from nearly any other series impedance (Dunlop, 1979). That characteristic is the distributed shunt capacitance that must sometimes be taken into consideration. This capacitance carries current whenever there is an alternating voltage on the line. This so-called charging current is independent of the load current so that it continues to flow into the line even at no load.

In the pioneer days of long-distance, high-voltage transmission, charging current created a problem because it tended to increase the field flux in any generator in which it flowed. If the generator was too small with respect to the line, the charging current caused the generator to self-excite, thereby creating a difficult situation. Charging current also proved much harder for the circuit breakers of those days to interrupt than short-circuit current.

The flow of charging current through the line reactance creates another phenomenon. This is the "rise of voltage along an open circuited line," or Ferranti effect. The amount of rise along an open circuited line is given by $E_S/E_R = \cos rl$ where E_S and E_R are sending-end and receiver-end voltages, respectively, and $r = 0.12$ degrees per mile of line with l the length of line in miles. This is about 2% for a 100-mile (160 km) and 8.5% for a 200-mile (320 km) line. Capacitance may often be ignored without much error if the lines are less than about 50 miles (80 km) long, or if the voltage is not over 69 kV.

CORONA

When the surface potential gradient of a conductor gets so high that the dielectric strength of the surrounding air is exceeded, ionization occurs (Peek, 1911). This is known as "corona," and is manifested by power loss, an audible hissing sound, radio frequency noise, and the appearance of localized glow point and streamers (visible in the dark) that appear first at rough or sharp spots on the conductor and its hardware.

The major problems that may be caused by corona are radio and television interference, audible noise in the vicinity of the line, and power loss. For lines of 345 kV and higher, the conductor itself is the major source of audible noise, radio noise, and corona loss. Audible noise (LaForest, 1982, Chapter 6), is a relatively new environmental concern, and its importance has grown with increasing voltage level. The audible noise from the corona process primarily occurs in foul weather. In dry conditions, the conductors normally operate below the corona-inception level and very few corona sources are present. In wet conditions, however, water drops impinging or collecting on the conductors produce a large number of corona discharges, each of them creating a burst of noise. At UHV levels (≥ 1000 kV), in this country, audible noise will be a limiting design factor. Attempts at line compaction for EHV and UHV lines are blocked by this environmental design factor.

Radio interference (La Forest, 1982, Chapter 5) occurs in the AM band and not in the FM band. Noise energy is usually generated at FM frequencies by metal-to-metal sparks. Television interference is caused not only by metal-to-metal sparks, but rain and snow may produce moderate TVI, in certain cases, in a low-signal area. Metal-to-metal sparks sometimes occur between pins and caps of insulators, which may form a poor connection if the insulator assemblies become slack. These sparks can occur across any small dielectric gap and the conducting path need not be the EHV line structure. Particularly troublesome TVI has occurred from induced voltages in the hardware of nearby lines and from other paths such as fences and metal objects in or near the right of way. This interference must be limited to acceptable levels during normal atmospheric conditions. During rain and snow, both radio noise and corona loss may increase by as much as 100:1. Choice of conductors must, therefore, be a balanced consideration of fair-weather noise levels (frequently encountered) and higher foul-weather levels (rarely encountered). Power loss (LaForest, 1982, Chapter 7) is a major problem only if it occurs in sufficient amount during peak load, because then it adds to peak demand and affects system investment costs. In this case, the prospect of even an occasional short duration of excessive power loss may economically justify an increase in conductor size. The relation between the time when peak load occurs and when adverse weather conditions prevail will vary geographically.

Other things being equal, corona problems increase with higher voltage and these could assume large proportions. The corona problem may be averted by the correct choice of conductor size and the use of conductor "bundling," often made necessary by other requirements.

BUNDLED CONDUCTORS

There is now extensive application of bundled conductors, especially at voltages above 230 kV. Instead of one large conductor per phase, two or more conductors of approximately the same total cross section are suspended from each insulator string. Figures 7.7 and 7.8 illustrate the use of bundled conductors. The single-circuit 765-kV system in Figure 7.7 has four conductors per phase, while the double-circuit 220-kV system shown in Figure 7.8 has two conductors per phase. This increases the effective diameter of the conductor and provides these advantages:

1. Increased corona critical voltage and, therefore, less corona power loss, audible noise, and radio interference.
2. Reduced line reactance.

Disadvantages of bundled conductors include:

1. Increased wind and ice loading.
2. Increased complexity and cost.
3. Increased clearance requirements at structures.

Ground-Level Electric Fields

All overhead transmission lines at any voltage level have an associated 60-Hz electric field near the ground (IEEE Working Group, Part I and Part II, 1972 and Deno, 1978). In the past, at lower-transmission voltage levels, the significance of this electrostatic field has been minimal. At EHV and projected UHV levels, this parameter has become an important design factor. There are two general areas of concern regarding this parameter: electric shock and possible short or long term biological effects on people. An example of "electric shock" would be where a truck insulated from ground is parked under the line and a person touches the truck. If the ground-level field is high enough, the truck potential could be sufficient to overcome the electrical insulation provided by the person's skin and shoes, causing the current to flow. The amount of current flow is of primary interest and, to a large extent, is a measure of the severity of the shock. The 1977 edition of the National Electric Safety Code (ANSI-C2) addresses this question and presents a limit of 5 milli-amperes for the current for the largest anticipated trucked vehicle, or equipment shorted to ground under the line. When this current is greater than the limit, the clearances must be increased or the field reduced by other means to achieve the limit. Biological effects at current levels of operation, in the short term, have not proved to be harmful. Long-term effects are the subject of several current research efforts. No significant effects have been found to date regarding humans under transmission lines.

ECONOMICS

The objective of a transmission line is to move electric energy from one place to another at a minimum overall cost per kilowatt-hour, commensurate with acceptable reliability. This is easy to say, but often it is not easy to translate into definite decisions.

The cost is primarily made up of fixed charges on the investment, operation, and maintenance, and the cost of I^2R. The fixed charges vary directly with line cost. Important factors in the cost of a line are its voltage and conductor size.

Selection of the proper conductor size is mainly a process of weighing I^2R losses, audible noise, and radio interference level against fixed charges on the investment. As conductor size is increased, loss goes down, but costs of both conductors and towers go up. I^2R losses depend on both maximum load and load factor. There are many ways of evaluating losses, and each utility has its preferred methods. For EHV and UHV lines, the conductor diameter in the bundle must be large enough to limit audible noise and radio interference to acceptable values.

The cost of erecting the line is subject to wide variation because of the nature of the country through which it passes. Where the right-of-way is level and accessible to railroad and highway, delivery of materials and erection can be organized so as to cut erection costs to a minimum. This is not the case when the country is very rough. For example, a 165-kV line of the Pacific Gas & Electric Company follows a river canyon for several miles as it leaves the Caribou Plant. It zig-zags back and forth across the top of the canyon, with a tower first on one side, then on the other. Some of the towers are at such inaccessible spots that the steel and insulators had to be brought up the floor of the canyon and then hauled up to the tower site by block and tackle. This line was built a number of years ago. In terms of present-day dollars, the Caribou end of it would probably have cost from $150,000 to $200,000 a mile. Helicopters would be used in building such a line today.

Transmission-line costs change with geography as well as with time. Most power companies make their cost estimates and comparisons based on their local conditions and using their own unit costs.

Many generalized studies have been made to determine the effects of voltage, distance, load, and other factors on the cost of transmission. Figure 7.11 is based on one such study for 50-mile (80 km) transmission. The purpose of this figure is to demonstrate the feasibility of extra-high-voltage trans-

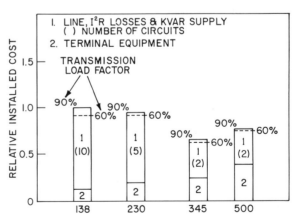

Figure 7.11. Installed cost in $/kW for 50-mile (80-km) transmission and 1500-MW power transfer at 138, 230, 345, and 500-kV.

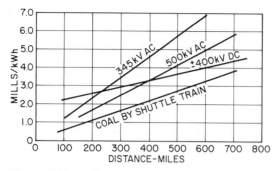

Figure 7.12. Effect of transmission distances on transmission costs.

mission wherever large bulk power transfer is required, even if the distance is only 50 miles (80 km). It will be noted that for a 1500-MW power transfer, the lowest cost is obtained with a transmission voltage level of 345 kV. It will also be noted that 10 circuits would be required at the 138-kV level to accomplish this 1500-MW transmission task. If the 50-mile (80 km) comparison is made for a 200- or 300-MW power transfer, the cost will change considerably and the lower transmission voltage levels represent the economic solution.

Figure 7.12 indicates the effect of transmission distance on transmission cost for voltage levels of 345-kV and 500-kV alternating current and ±400-kV direct current. A comparison with the cost of transporting coal by shuttle train is also indicated. For distances of 100 miles (160 km) and beyond, where circuit loadings are limited primarily by transient stability considerations, the cost of intermediate switching stations and series compensation is also included in the transmission cost. Although coal by shuttle train has a lower average cost for a fixed distance, transmission lines may also be used in terrain which prohibits the installation of rail lines.

DIRECT-CURRENT TRANSMISSION

Introduction

In recent years, HVDC (high-voltage direct-current) transmission has been employed in several special applications. Twenty-four systems are now in operation and nine are under construction (see Table 7.2). The HVDC system has certain advantages over the ac system that make it worthy of consideration when special characteristics are needed. Some examples would be underwater transmission, or where very long distance or asynchronous ties are required.

Description of a HVDC Transmission System

The circuit diagram of a HVDC transmission system is shown in Figure 7.13 where the principal items of electrical equipment in the dc terminals are identified.

Essentially a dc terminal consists of a number of converter units with dc outputs connected in series so that their combined outputs produce the full transmission line voltage. Each converter is an operating unit consisting of an ac switch, a converter transformer with a load tap changer, the dc valve, and a bypass switch. An ac breaker connects the converter unit to the ac system, and a set of ac filters is connected to the system to limit the flow of harmonic currents. In addition, a shunt capacitor bank is usually connected to the ac bus to provide the reactive requirements of the converters and limit the kvar demand on the ac system. The converter unit is connected to the dc line through a reactor that provides smoothing action (it is a high impedance to the high-frequency components of the ripple currents), and acts to limit fault currents in the dc system. The dc switching is performed by the valves using grid control. Control equipment for the converter units provides for the entry of current settings at one terminal, and supervisory equipment with a telecommunication link to coordinate the operation of both terminals.

A converter terminal capable of changing ac power to dc is referred to as a rectifier, and a terminal where dc is changed to ac is called an inverter. Most dc terminals are designed to operate as either rectifiers or inverters. A six-pulse, full-wave, double-way converter circuit is the basic building block for a dc terminal. In most modern installations, two six-pulse bridges are connected in series to form a 12-pulse converter. Several 12-pulse converters may be connected to achieve high-voltage and high-power ratings. For low-power ratings, a single converter unit may be used.

HVDC System Control

The normal control of Figure 7.14 shows the behavior of the rectifier voltage characteristic as a function of load current. A similar drooping characteristic is obtained in the inverter. To obtain a stable

DIRECT-CURRENT TRANSMISSION

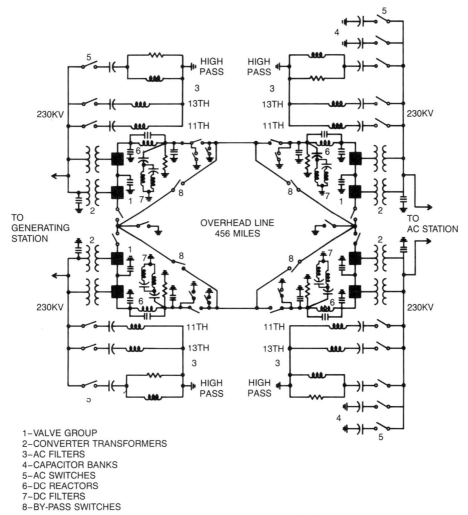

1 – VALVE GROUP
2 – CONVERTER TRANSFORMERS
3 – AC FILTERS
4 – CAPACITOR BANKS
5 – AC SWITCHES
6 – DC REACTORS
7 – DC FILTERS
8 – BY-PASS SWITCHES

Figure 7.13. Circuit diagram of a HVDC system.

operating point, the rectifier is given a current regulation characteristic so that it intersects with the inverter voltage characteristic at point A. Under normal control, the inverter determines the voltage of the dc system, and the rectifier controls the current supplied to the dc system. To do this, the rectifier must develop the voltage determined by the inverter, plus line drop. To provide stable operating points under contingencies, the inverter is given a current control characteristic at some value lower than the rectifier denotes by ΔI in Figure 7.14. In the event of a drop in rectifier voltage, as would be expected with a drop in sending-end ac voltage, there will be an intersection of the two characteristics.

The intersection of the control characteristic determines the load carried by the dc system. Changes in these characteristics may be made to obtain the desired behavior of the dc system. Rapid changes of dc power level may be obtained by changes in the current settings with changes of $\pm 10\%$ load in 50 msec or $\pm 50\%$ in 200 msec being possible. The control characteristics may be utilized to control dc power flow in response to any desired input. The effect of system faults near generation is acceleration of the generator units, and speed signals can initiate increased dc power flows to decelerate the machines and improve system transient stability.

Consideration has been given to controlling the power flow as a function of inverter receiving-system frequency drop, resulting in dc system behavior similar to that of an ac system, except that the amount of response is controllable and the dc link is nonsynchronous. It would also be possible to tele-

TABLE 7.2. Installations in Service and under Construction

System	Year Operational	Capacity (MW)	Dc Voltage (kV)	Length of Route (km)	Valve Type
Moscow–Kashira, USSR (experimental)	1950	30	200	113 (overhead)	Mercury arc
Gotland, Sweden	1954	20	100	98 (cable)	Mercury arc and thyristor
	1970	30	150		
Cross-Channel, England–France	1961	160	±100	65 (cable)	Mercury arc
Volgograd–Donbass, USSR	1962	750	±400	472 (overhead)	Mercury arc
Konti-skan, Denmark–Sweden	1965	250	±250	95 (overhead)	Mercury arc
				85 (cable)	
Sakuma I, Japan	1965	300	2 × 125	0	Mercury arc
New Zealand	1965	600	±250	567 (overhead)	Mercury arc
				38 (cable)	
Sardinia–Italy	1967	200	200	290 (overhead)	Mercury arc
				120 (cable)	
Vancouver Stage III, Canada	1968–1969	312	+260	41 (overhead)	Mercury arc
				32 (cable)	
Pacific Intertie Line I, United States	1970	1440	±400	1354 (overhead)	Mercury arc
Eel River, Canada	1972	320	2 × 80	0	Thyristor, air-cooled and insulated
Nelson River Bipole I, Canada	1973	810	±150	890 (overhead)	Mercury arc
			−300		
	1973–1975	1080	±300		
	1977	1620	±450		
Kingsnorth, England	1975	640	±266	82 (cable, 3 substations)	Mercury arc
Cabora-Bassa, Mozambique–South Africa	1975	960	±266	1410 (overhead)	Thyristor, oil-cooled and insulated
	1977	1440	+266		
			−533		

Vancouver Stages IV and V, Canada	1976	552	+260 −140	41 (overhead)	Thyristor, air-cooled and insulated
	1978	792	+260 −280	32 (cable)	
Stegall, United States	1976	100	50	0	Thyristor, air-cooled and insulated
Square Butte, United States	1977	500	±250	745 (overhead)	Thyristor, air-cooled and insulated
Skagerrak, Norway–Denmark	1977	500	±250	100 (overhead) 130 (cable)	Thyristor, air-cooled and insulated
EPRI Compact Substation, United States	1979	100	100 (400 kV to ground)	0.6 (cable)	Thyristor, freon-cooled and SF$_6$ insulated
CU Project	1979	1000	±400	656 (overhead)	Thyristor, air-cooled and insulated
Inga-Shaba Stage I, Zaire	Unknown	560	±500	1700 (overhead)	Thyristor, air-cooled and insulated
Nelson River Bipole, II, Canada	1978	900	±250	920 (overhead)	Thyristor, water-cooled and air-insulated
	1981	1800	±500		
Shin–Shinano, Japan	1978	150	125	0	Thyristor, oil-cooled and insulated
Hokkaido–Honshu, Japan	1979	150	±250	124 (overhead) 44 (cable)	Thyristor, air-cooled and insulated
Itaipu	1983	6300	±550	785 (overhead) 805 (overhead) 2 lines	Thyristor, water-cooled and air-insulated
Chateauguay, Canada	1984	1000	146	Back-to-Back	Thyristor, water-cooled and insulated
Oklaunion, United States	1984	200	82	Back-to-Back	Thyristor, air-cooled and insulated
Eddy County, United States	1984	200	82	Back-to-Back	Thyristor, air-cooled and insulated
Walker County, United States	1985	520	450	157 miles (overhead)	Thyristor air-cooled and insulated
Madawaska, Canada	1985	350	130.48	Back-to-Back	Thyristor, air-cooled and insulated
Miles City, Canada	1985	200	82	Back-to-Back	Thyristor, air-cooled and insulated
Intermountain Power Project, United States	1986	3200	±500	507.5 (overhead)	Thyristor, water-cooled and air-insulated
NEPSCO-HQ, United States/Canada	1986	690	±450	65 miles (overhead)	Thyristor, air-cooled and insulated

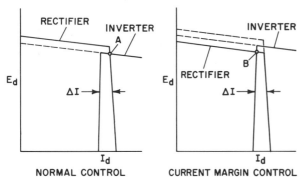

Figure 7.14. HVDC operating characteristic curves.

meter phase angles from various buses in the system to a central location or monitor power flow in specific ac ties to control the power flow in the dc link for the desired performance in the connected ac systems.

Conversion Equipment

Converter transformers are the link between the ac system and the HVDC valve. They are specifically designed to withstand the insulation requirements of dc application and to step up or down the ac voltage to appropriate levels for optimum valve design. During operation, the associated tap-changing equipment adjusts the valve voltage according to loading. The converter transformers also provide the required phase shift for 12-pulse operation through proper connection of the valve windings.

In Figure 7.13, the valve winding of one converter transformer feeding a six-pulse bridge is wye connected. The valve winding of the converter transformer feeding the series-connected, second, six-pulse bridge is delta connected. The 30-degree phase difference between the wye and delta connections displaces the voltage waveforms of the second six-pulse group from the first six-pulse group to provide 12-pulse operation.

Converter stations built in 1969, and earlier, utilized mercury-arc valves in the conversion equipment. After 1969, solid-state technology had progressed to the point where silicon-controlled rectifiers, also called thyristors, could be used for power transmission. The Eel River converter station, which is an asynchronous tie installation near Dalhousie, New Brunswick, was the first application of all solid-state converters for power transmission. The solid-state converters proved to be far superior than their mercury-arc predecessors, and today all new HVDC power converters are solid-state. The mercury-arc converters are subject to arc-backs and require a great deal of maintenance to keep them operating. Solid-state converters have proven to be exceptionally reliable and require very little maintenance.

Solid-state converters utilize low-voltage silicon-controlled rectifiers in a matrix of series and parallel connections with associated circuit elements to control the current and voltage environment at the individual cells. A picture of a semiconductor assembly is shown in Figure 7.15.

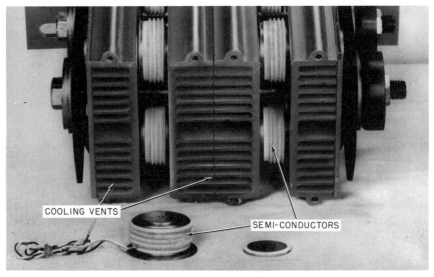

Figure 7.15. Semiconductor assembly.

DIRECT-CURRENT TRANSMISSION

Several semiconductor assemblies are mounted on a power module. Various power modules are then mounted in an insulated housing to make up the solid-state valve. The power modules can be easily replaced in a valve so that maintenance is simplified and down time reduced. Figure 7.16 shows the valve housings for one terminal.

Solid-state conversion equipment offers great flexibility in voltage and current rating because of the series and parallel arrangement. Larger silicon-controlled rectifiers are being developed that result in increased voltage and current ratings of the individual cells. This makes possible substantially increased power capabilities of HVDC terminals.

HVDC converter units generate harmonic currents, and harmonic filters are used to minimize the harmonic currents that flow into the ac system. The filters usually consist of tuned filters for the lower-frequency harmonics and high-pass filters for the higher-frequency harmonics.

The dc reactor smooths the current ripple in the dc line. The reactor also limits the fault current and isolates the valve from the dc transmission line or cable from impulse voltages on the line.

Figure 7.16. Typical valve hall. (Courtesy of Minnesota Power and Light Company.)

HVDC System Characteristics

The dc system in its present state of development has unique operating characteristics that differ from those of the ac system, including:

1. *Circuit configuration.* The dc systems planned or in operation involve point-to-point transfer of power. Control of more complex configurations with intermediate stations is possible, allowing future modification of systems that are initially point-to-point.

2. *Asynchronous ties.* A dc transmission line is inherently an asynchronous tie that is insensitive to the power angles and frequencies of the interconnected ac systems. In cases where parallel ac ties are utilized, the dc system behavior is determined by the ac voltages at the converter terminals, and by the control provided for the dc system. Examples of asynchronous ties between divergent ac systems are the Eel River Power Station in New Brunswick and the Stegall Station, Nebraska. In Japan, the Sakuma Station is a dc interconnection between 50-Hz and 60-Hz systems.

3. *Control of power flow.* The amount and direction of power flow may be easily controlled by adjusting the dc voltage at the two terminals. The terminal equipment is readily designed to operate as either a rectifier or an inverter.

4. *Speed of response.* The dc system is capable of rapid response to changes in control setting, with dc power transfer changes occurring faster than in ac systems. This rapid response offers possibilities in the stabilization of the ac networks at the converter terminals or in parallel ac ties.

5. *Economical for long distances.* HVDC transmission is extremely economical for long distances where generation sites are remote from load centers because of lower transmission line costs. Two examples are (1) the Nelson River Project which transmits remote northern hydro power to Winnipeg, Manitoba, and (2) the Square Butte Project which transmits power from the lignite fields of North Dakota to the iron mining region of Duluth, Minnesota.

6. *Damping oscillations.* The transmission capability of ac lines is sometimes limited by undamped oscillations under "steady-state" conditions. The dc transmission power flow can be modulated to dampen the oscillations and increase their power-carrying capability. Modulation of up to ±50 MW on the Pacific Intertie in the western United

States has been used to stabilize the parallel 500-kV ac lines, increasing their transmission capacity by 400 MW.

This ability to dampen oscillations is also significant because ac lines may be limited by transient stability considerations when the sudden loss of a generator or a second transmission line causes a power swing on a tie line. If the oscillations cannot be damped out, the two systems will become so far out of phase that the tie will be lost. The dc interconnection has the ability to respond rapidly to generator speed signals and increase or decrease the power flow as needed to dampen the transient oscillations. For example the Square Butte interconnection can modulate the dc power flow by ±20% in 5% increments.

7. *Reliability factor.* The use of thyristor valves has increased the reliability of HVDC converters to impressive levels. Since its installation in 1972, the Eel River Station has attained full availability of almost 99% with an annual load factor of over 100 percent.

8. *Right-of-way consideration.* HVDC lines can transmit greater amounts of power for a given right-of-way than ac lines. This benefit can be very important to utilities since obtaining rights-of-way for transmission lines is very difficult, especially in heavily populated areas.

9. *Environmental considerations.* Experience and research on corona effects from HVDC lines show considerable advantage over ac lines particularly with respect to radio interference and audible noise. Residual ionic effects do require attention, but they are entirely manageable with proper line design.

10. *Reactive power.* Rectifiers and inverters, as presently developed for HVDC transmission, represent reactive power (kvar) loads to the connected ac systems, and each terminal consumes reactive power of approximately 50–60% of the active power flow. As reactive power cannot be transmitted over dc lines, each ac system must supply the reactive power for the connected dc terminal. The ac harmonic filters furnish part of the required kvar, and where the system is unable to supply the rest of the reactive power, shunt capacitors may be utilized. The proper operation of an inverter requires an ac system short-circuit capacity in the order of three to five times the transferred power, and where short-circuits are lower than this level, the addition of synchronous condensers may be necessary.

Figure 7.17. Typical 500-kV ac tower structure under construction. (Courtesy of Bonneville Power Administration.)

11. *Fault protection.* High-speed protection for faults, either in the dc line or the dc terminal equipment, is provided by control of the valves. In the event of faults on the dc line, the control oper-

ates to limit the fault current to less than twice rated current in about one cycle, and the time required for deionization of the arc path is less than for faults on an ac line. The reduction in deionization time results from lower short-circuit currents, shorter fault duration, and gradual restoration of voltage to the line. It should be noted that the dc transmission system does not contribute to the fault current flowing in the ac system during the ac faults.

12. *Harmonics.* The converter units generate harmonics which flow in the ac and dc lines. The six-pulse bridge unit generates current harmonics in the ac line of 5th, 7th, 11th, 13th, etc., and voltage harmonics in the dc line of 6th, 12th, 18th, etc. By means of Y and Δ connections in the converter transformers, a 30-degree phase shift between the six-pulse units results in twelve-pulse operation. As a result, the current harmonics on the ac side are 11th, 13th, 23rd, 25th, etc., and the voltage harmonics in the dc side are 12th, 24th, etc. Harmonic filters for these harmonics are generally provided on the ac and dc sides of the converter terminals.

13. *Earth return.* The earth provides a low-resistance conducting path for direct current. In experimental work, it is shown that direct currents flow between widely separated earth electrodes by going directly into the earth and returning only to the earth's surface at the remote electrode. The use of earth return offers attractive possibilities in the event of conductor failure on the line or contingencies in a converter terminal, permitting operation of the remaining half of the system. This substantially reduces the impact of contingencies.

Metallic return switching can be provided to utilize the dc line of an unused pole as the return conductor. This circuit arrangement will eliminate the ground current, but the dc line losses will be doubled.

14. *Tower structure.* Dc needs only two conductors per line, one positive and one negative, as opposed to the three per line used for three-phase ac transmission. Earth is used as a return conductor in case either aerial conductor is out of service. For the same amount of delivered power, less insulation is needed on a dc line. Therefore, tower construction is lighter and simpler, and the rights-of-way are narrower. Figures 7.17 and 7.18, pictures of typical 500-kV ac and ±400-kV dc tower structures, illustrate this difference in tower construction and design.

Figure 7.18. Typical ±400-kV dc tower structure. (Courtesy of Bonneville Power Administration.)

Economics

The nature of the dc system is such that overhead lines and cables are less costly, and terminal station more costly than ac alternatives. A minimum break-even distance is necessary for the savings in the dc line or cable to equal the extra cost in terminal equipment. While the break-even distance varies widely with different costs of all equipment, it is generally over 400 miles (640 km) for overhead lines and over 20 miles (32 km) for underground cables in metropolitan systems and for underwater cables.

While these figures provide general guides, the evaluation of dc for a specific case must recognize actual conditions, including reliability requirements, need for supplying future intermediate loads, and so on.

As overhead ac transmission distances increase, it is necessary to modify system design to maintain adequate transmission line loadings. These modifications include series capacitors to reduce the through-transmission-line impedance, shunt reactors to compensate the line capacitance, and switching stations to sectionalize the transmission lines to minimize the impact of transmission line loss. These considerations tend to increase the cost of ac transmission at a rate that is considerably greater than a linear increase with distance. With dc transmission, there are no inductance and capacitance effects on steady-state power transfer, and there is merely increased resistance with distance; with no stability problems in the dc link itself, the cost of dc transmission becomes a linear function with distance.

In high-voltage ac cable systems, capacitance effects impose severe limitations when the distances are long. To minimize these effects, shunt reactors are required, with resulting complexity and cost. With direct current, the steady-state capacitance effects do not result in a reactive component of current, thus permitting increase in current capability. Higher direct voltages than alternating voltages may be applied to cables, resulting in power transfer capability for direct current of approximately three times that possible with alternating current for a given cable.

The main areas of application for dc transmission include cases where savings in the line construction offset increased terminal cost, long cable transmission (underwater or underground), or cases where a nonsynchronous tie is required.

SUMMARY

The term "transmission" usually denotes the highest voltage circuits on a given system. They may be as low as 69 kV. Originally, transmission voltages were used principally to transmit power long distances, as from distant hydroelectric plants to the load. Now the high-voltage system also connects the various generating stations to the primary substation, and to each other, becoming the backbone that holds the system together.

Transmission circuits use their own rights-of-way. They are supported on steel towers by means of suspension insulators. Conductors are usually made of aluminum with a steel core. The line must be designed to minimize the effects of weather conditions—e.g., wind, rain, ice, dust, lightning, and extremes of temperature. Switching-surge performance is an important part of line insulation design.

Consideration must be given to the possibility of corona and its probable effects such as audible noise, radio and TV reception, and occasionally to its power loss. If the line is long, the effect of its shunt capacitance must also be considered.

Ground-level electric fields must be kept to acceptable levels by proper line design and right-of-way width.

If the transmission distance exceeds 400 miles (640 km) (20 miles [32 km] for cable), the economics of HVDC transmission should be investigated. The characteristics of dc transmission differ from ac transmission and may dictate its use as is the case for the asynchronous tie in Nebraska between the eastern interconnected system and the western system.

REFERENCES

1. Anderson, J. G., and Barthold, L. O., "METIFOR-A Statistical Method for Insulation Design of EHV Lines" *IEEE Transactions,* **PAS-83,** 271–280 (March 1964).
2. Anderson, J. G., et al., *Transmission Line Reference Book -345 KV and Above,* first edition, Electric Power Research Institute, Palo Alto, California, 1975.
3. Anderson, R. B., and Eriksson, A. J., "Lightning Parameters for Engineering Applications," Report ELEK 170, CSIR, Pretoria, South Africa, June 1979.
4. Deno, D. W., and Zaffanella, L. E., "Electrostatic and Electromagnetic Effects of Ultrahigh-Voltage Transmission Lines," EPRI Research Project 566-1, Final Report, EL-802, Electric Power Research Institute, Palo Alto, California, June 1978.
5. Dunlop, R. D., Gutman, R., and Marchenko, P. P., "Analytical Development of Loadability Characteristics for EHV and UHV Transmission Lines," *IEEE Transactions,* **PAS-98,** 606–617 (March/April 1979).
6. IEEE Working Group, "Electrostatic Effects of Overhead Transmission Lines, Pt. I—Hazards and Effects, IEEE Transactions, **PAS-91,** 422–426 (March/April 1972).

REFERENCES

7. IEEE Working Group, "Electrostatic Effects of Overhead Transmission Lines, pt. II—Methods of Calculation," *IEEE Transactions,* **PAS-91,** 426–444 (March/April 1972).
8. IEEE Working Group on Insulator Contamination, "Application of Insulators in a Contaminated Environment," *IEEE Transactions,* **PAS-98,** 1676–1690 (Sept./Oct. 1979).
9. LaForest, J. J., ed., *Transmission Line Reference Book-345 KV and Above,* second edition, Electric Power Research Institute, Palo Alto, California, 1982.
10. Peek, Jr., F. W., "The Law of Corona and the Dielectric Strength of Air," *AIEE Transactions,* **PAS-30,** (pt. III), 1889–1965 (1911).
11. *The Aluminum Electrical Conductor Handbook,* first edition, September, 1971, The Aluminum Association, 819 Connecticut Ave., N. W., Washington, D.C.

8
TRANSFORMERS

DESCRIPTION AND USE

A transformer is a static device for transferring electric energy from one circuit to another magnetically, that is, by induction instead of by conduction. Its usual function is to transfer energy between circuits of different voltage. A transformer has a magnetic core on which there are two or more windings. These windings are insulated from each other and from ground. In autotransformers, however, the windings are connected together. The assembly of core and coils is normally insulated and cooled by immersion in mineral oil or other suitable liquid within an enclosing tank. Connection to the windings is by means of insulating bushings, usually through the cover.

Transformers are found in all substations except the comparatively few that are switching stations only. In a modern utility system, the energy may undergo four or five transformations between generator and ultimate user. As a result, a given system is likely to have about five times more kilovolt-amperes of transformers than of generators.

RATIO

The "ratio of transformation" is determined by the relative number of turns in each of the windings. This is known as the "turn ratio," and it is the ratio of the no-load voltages. When the unit is carrying load, the ratio of the actual voltages is slightly different because of the drop caused by the flow of load current through the impedance of the transformer windings. At rated load, this drop is known as the "voltage regulation." The amount of voltage drop varies with the power factor of the output, even when the kilovolt-amperes remain constant.

Whether the voltage is increased or decreased in going through a transformer, the opposite happens to the current, in more or less the same degree. Suppose we assume that the effects of exciting current and voltage regulation are small enough to be neglected. If the ratio of output (secondary) turns to input (primary) turns is N, then secondary voltage is N times primary voltage, and secondary current is $1/N$ times primary current. The product of current and number of turns in each winding is the same; that is, "ampere-turns must balance." This is a useful relationship to remember.

TAPS

Practically all power transformers and many distribution transformers have taps in one or more windings for changing the turn ratio. Changing the ratio is desirable for two reasons (1) to compensate for varying voltage drop in the system and (2) to assure that the transformer operates as nearly as possible at the correct core density. For the latter purpose, the taps should be in the winding subject to the voltage variation.

De-energized tap changers are used when it is expected that the ratio will need to be changed only infrequently, because of load growth or some seasonal change. The desired tap is selected by means of a ratio adjuster (no load taps).

Tap changing under load (LTC) is used when changes in ratio may be frequent or when it is undesirable to de-energize the transformer to change a tap. A large number of units are now being built with load tap-changing equipment. It is used on transformers and autotransformers for transmission tie, for bulk distribution units, and at other points of load service.

It seldom makes much difference to the user which winding or windings are tapped; therefore, the choice is usually made by the designer on the basis of cost and good design. Both winding current and voltage must be considered when applying LTC equipment. High-voltage and high-current applications require special considerations to arrive at an optimum location for the LTC equipment. Step-down units usually have LTC in the low-voltage winding and de-energized taps in the high-voltage winding.

SINGLE-PHASE VERSUS THREE-PHASE UNITS

The bulk of all transformers, except for large extra-high voltage (EHV) and distribution units are three-phase units. In the early days of the industry, it was almost universal practice in the United States to use three single-phase units connected in a three-phase bank. Insulation clearances and shipping limitations for certain large EHV units now require this design. The distribution systems serve mainly single-phase loads in residential areas and are served from single-phase transformers.

AUTOTRANSFORMERS

When energy is to be transferred between two circuits of nearly the same voltage, the use of autotransformers affords cost savings over two-winding units. The nearer the voltages are to each other, the smaller will be the autotransformer per kilovolt-ampere of output, and the greater the savings. The simplicity of phasing out systems has increased its use.

Most autotransformers are Y-connected, and it has been a standard American practice to add a low-capacity, delta winding. This is frequently referred to as a "delta tertiary." Its primary purpose has been to provide an internal path for third harmonic currents (required for excitation), thus reducing those currents on the power system. It also helps to stabilize the neutral and to ground the system better. In recent years, the use of shielded telephone cables has reduced the requirements for the delta tertiary.

Because an autotransformer does not afford electrical separation between the two circuits, disturbances originating on one circuit can be communicated to the other. This difficulty is minimized by solidly grounding the neutral of the autotransformer. Solidly grounding the neutral, however, causes (among other things) current of short-circuit magnitude to flow through the delta-connected tertiary winding during ground faults on either system. Autotransformers are not inherently self-protecting and, therefore, all windings must be examined for mechanical strength as applied to the system where they will be used. Tertiaries are normally 35% of the physical size of the largest winding of the autotransformer, unless otherwise specified by the user.

Autotransformers are used for most ties between moderate-voltage and EHV transmission circuits. Modern practice is to provide EHV-reactive compensation, in part, through the use of switched reactors on a tertiary bus. Some installations also switch capacitors on the tertiary bus or either capacitors or reactors, as system conditions require. LTC is also used to provide "fine" voltage control for the low-voltage bus.

PHASE-ANGLE TRANSFORMERS

Sometimes it becomes desirable to make an electrical connection that closes a loop through one or more power systems. Before the loop is closed, there may be a phase-angle difference in voltage between the points to be connected together, caused by the flow of currents in parts of the circuits forming the loop. Since this voltage would be short-circuited by the new connection, undesirable currents may flow after the loop is closed. It may appear necessary to insert a phase-angle transformer to correct the condition. Voltage transformation may or may not also be required.

Sometimes a permanent shift in phase angle is adequate. More often it appears desirable to vary the amount of shift to suit changing load conditions. This requires the use of load-tap changing.

CORES

Transformer cores are made from thin laminations of grain-oriented silicon steel to keep core loss to a low value. Cold-rolling during the manufacturing process establishes a preferred path-grain orientation in the rolling direction. It is essential that the steel be applied so that the flux is parallel to the grain direction to obtain minimal loss. This is achieved in single-phase distribution transformers by making wound cores, constructed by winding the steel strip into a coil. Large single-phase and three-phase "plate" cores are constructed by stacking consecutive layers of laminations which have the straight sections connected at the corners by 45° mitered joints.

As energy costs have increased in recent years, the economic value of the transformer core loss has also increased. Evaluations as high as $10 per watt have been placed on core loss, based on lifetime owning costs. This has led to the following developments:

1. Thinner laminations, down to .009 inch (.23 cm) thick, to reduce eddy loss in the steel.
2. Better chemical formulations of the steel and improved processing to achieve lower loss at high flux densities.
3. Improved joint configurations to minimize the flux disturbance (and loss) in the corner region of plate cores.

The audible sound level produced by core excitation also constitutes a major application problem. The source of the audible sound is mechanical vibration of the core, generated by a steel characteristic known as magnetostriction. Since the magnetostrictive motion increases with increased flux density, lowered flux density is one means of controlling sound amplitude. Other means that may be required for noise-sensitive areas include close-fitting enclosures or barrier walls in the substation.

WINDINGS

There are three basic winding designs used in core-type transformers: concentric layer, helical, and disc. The helical winding, the basic low-voltage winding for power transformers, is used for high-current, low-voltage applications. It will usually be applied to low-voltage windings with ratings up to 15 or 23 kV. Layer and disc windings are more suited for medium- to high-voltage applications, having less conductor cross-sectional area for the lower currents involved. For each power transformer application, the winding design that meets the electrical and size requirements at the lowest operating cost and loss levels can be selected.

LOSSES

Total losses are made up of no-load loss and load loss. No-load loss is predominantly hysteresis and eddy-current loss in the core. It is independent of load and exists whenever the transformer is excited. Load losses—consisting of I^2R loss in the windings, and stray losses in windings, clamps, and fittings—are caused by load current, and vary as the square of the load. On large units, the total losses at rated output amount to approximately 0.3–0.6% of the rated kilovolt-amperes. Within limits, the designer can vary the ratio of no-load to load losses to give the lowest overall cost for a given method of cooling, assigned loss evaluations, and assumed load curve.

COOLING

Small transformers have enough tank surface to radiate all the heat caused by their losses without exceeding the permissible temperature rise. As size increases, the losses increase faster than the tank surface which soon becomes inadequate. Various methods have been developed to get the heat out of the tank more effectively.

As larger and larger transformers were built, the losses outgrew any means of "self-cooling" then available, so water-cooling was introduced. This was accomplished by placing a coil of metal tubing in the top oil, around the inside of the tank. Water was circulated through this cooling coil to remove heat from the losses. A variation on the water-cooled transformer consisted of pumping the hot oil through an external oil-to-water heat exchanger; this was known as forced-oil-to-water cooling (FOW). It differed from modern forced-oil cooling in that there was no attempt to direct the flow of cool oil inside the tank. For many years this method was more popular in Europe than it was in the U.S. Modern methods use external heat exchangers. Pumps are utilized to increase the cooling efficiency.

Water was not always available at locations where it was desirable to reduce transformer size by

COOLING

supplemental cooling. This led to the practice of forced ventilation of the radiators. The equipment for this purpose first took the form of a blower with duct-work to convey the air to the various radiators. That soon gave way to an individual fan on each radiator, or a group of fans on a radiator bank. Originally the fans merely provided a "supplemental" rating above the self-cooled rating. It was intended that the unit would carry loads in excess of the self-cooled rating only during a temporary emergency. Present practice is to assign fan-cooled ratings that produce no higher internal temperatures than those caused by operation at the self-cooled rating without fans.

Prior to the outbreak of World War II, development work was in progress looking toward still more effective cooling methods. It seemed to some engineers that if the utility industry was going to demand larger and larger transformers, it would be desirable to learn how to get more rating out of less material. Inevitably, that led to the necessity for better cooling. Progress in this direction was greatly speeded during World War II by the hard-hearted attitude of the War Production Board. This board consistently refused to be generous in the allocation of critical copper and steel for transformers of "conservative" design.

Out of this search for better cooling came the "directed flow," forced-oil unit. In this design, the oil is pumped both through the external cooling devices (air or water heat-exchangers) and through internal channels that are located nearest the points where the heat is generated. Thus, the transfer of heat to and from the oil is far more effective than in the plain self-cooled or fan-cooled unit where the oil is allowed to circulate by convection.

The largest units being built today are commonly of the forced-oil-cooled, directed-flow type. This gives the highest rating for a unit of given weight and dimensions; and therefore, for units of high rating, the minimum amount of disassembly for shipment.

The simplest of these forced-oil-cooled transformers uses an external bank of oil-to-air heat exchangers through which the oil is continuously pumped. It is designated as type FOA. A second type, called FOW, is identical except that it utilizes oil-to-water heat exchangers. Pumps and heat exchangers are operated whenever the transformers are energized, and the transformers have essentially no load-carrying ability without the cooling equipment in operation.

Figure 8.1. Type OA self-cooled transformer.

Where a self-cooled rating is desired, a third type of cooling is employed, designated as triple-rated. Such units carry up to about 60% rating by natural circulation of the oil. At the temperature corresponding to approximately 60% of maximum nameplate rating, the first stage of auxiliary cooling is automatically started, and at 80% the final stage is put into operation. Such units have been made with both stages of auxiliary cooling consisting of fans on the radiators, and designated as OA/FA/FA; with both stages' forced circulation pumps, in addition to the fans and radiators, designated as OA/FOA/FOA; and with various other combinations of fans and pumps to accomplish the desired ratings. Current practice automatically employs fans for the first stage and pumps for the second, and is designated as type OA/FA/FOA.

Several different methods of cooling, as applied to transformers in current production, are shown in Figures 8.1 through 8.4.

Figure 8.2. Type OA FA/FOA self-cooled/fan-cooled/forced-oil, air-cooled transformer. (Courtesy of Rochester Gas & Electric Corporation.)

OIL PRESERVATION

A great many transformers, including all the large ones and all the high-voltage ones, are immersed in mineral oil, which serves the double purpose of cooling and insulating the windings. The insulating value of oil is substantially reduced by even a small amount of moisture. It is important, therefore, to exclude all moisture.

Mineral oil is a mixture of a great many organic compounds, many of which are susceptible to oxidation, at rates which increase with temperature. The products of oxidation are other organic compounds, some of which are solids, or "sludges." Water may also be formed, as well as organic acids which may attack the insulation on the windings.

Because air contains both oxygen and moisture, it is important to keep it from direct contact with the oil, especially with the hot oil. This would be easy if the volume of the oil did not change. However, in going through a temperature change of 100°C, which is not unusual, oil expands by about 8%. This change in volume must be factored into any method for preserving the oil from deterioration.

Conservator

This device (Figure 8.5), which is basically an oil expansion tank, was one of the earliest attempts at keeping moisture away from the oil. Although it did not separate oil from air, it represented progress. The effectiveness of the conservator was greatly improved by the later adoption of the gooseneck, or U-tube, in the pipe between the main tank and the conservator. This stopped the thermal circulation of oil between the two tanks and reduced the rate of transfer of moisture from the conservator to the main tank. It also kept the oil in the conservator cool, and greatly reduced the rate of oxidation.

The conservator is a simple method of oil preservation and requires the least volume of main tank plus auxiliary tank. This is still the most common method used throughout the world, except in the United States.

Atmoseal® Constant-Pressure System

An improved type of atmospheric-pressure oil preservation system, designated as the Atmoseal System (Figure 8.6), was introduced in 1957, and is now used on nearly all of the large power transformers made by General Electric Company. Similar to the conservator in operating principle, it employs a vapor-tight air cell in the expansion tank to prevent atmospheric contamination of the oil. The air cell floats on the oil and vents air to the atmosphere via a weathertight breather. Damage to the cell will be indicated by a liquid-level gauge, while the unit can continue to operate like a conservator. Other manufacturers have now adopted similar devices.

Gas Seal

The gas seal (Figure 8.7) uses a bottle of dry nitrogen to keep the space above the oil at a positive pressure as the oil cools and contracts. When the oil expands, excessive pressure is prevented by venting some nitrogen to the atmosphere. These functions are performed automatically by control and relief valves. Some operators prefer the gas seal because it assures a positive pressure inside the tank to exclude air and moisture. Other operators feel that the possibility of faulty operation of the gas control equipment, the additional maintenance, and the cost of periodic replacement of nitrogen bottles, are objections that outweigh the advantages of this method. In this system, the regulators must hold the

Figure 8.3. Type FOA forced-oil, air-cooled transformer. (Courtesy of Niagara Mohawk Power Corporation.)

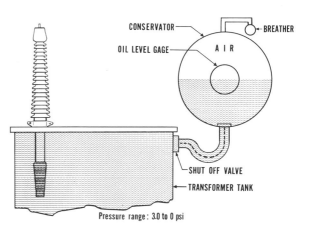

Figure 8.5. Conservator system.

Figure 8.4. Type FOW forced-oil, water-cooled transformer during manufacturing test.

151

pressure to rather narrow limits to prevent the oil from reaching a supersaturated condition and releasing nitrogen bubbles owing to the temperature variations.

Sealed Tank

The sealed tank (Figure 8.8) is the simplest way to preserve the oil. It is now used on practically all distribution transformers, and sometimes on power transformers up to about 50,000 kVA or even higher. It is seldom used on larger units because it necessitates the largest tank of any method of oil preservation. The popularity of the sealed-tank method is increasing, and it has been applied on units as large as 100 MVA.

Methods of Attaching Cover to Tank

The former practice of attaching cover to tank by means of a bolted, gasketed joint has given way to welding. Most operators feel that the latter method gives greater assurance than gaskets of a tight joint. Also, when they have to remove the cover, they claim that they can chip off a weld more quickly than they can take out cover bolts and reweld the joint more quickly than they can bolt it up again.

It should be understood that welding on the cover is not a method of oil preservation. The cover of a sealed tank may be either welded or bolted; the cover of any transformer may be welded on, regardless of the method of oil preservation.

REDUCTION OF FIRE HAZARD

It is frequently desirable to place a transformer indoors or otherwise near the load where the flammability of the insulating oil would be unacceptable. For these applications, transformers with fire-resistant or fire-proof insulating liquids or transformers with no insulating liquid at all are available.

Silicone-Filled Units

Transformers with silicone (polydimethylsiloxane) insulating fluid are accepted by the National Electric Code as high fire-point units since the liquid has a minimum fire point of 300°C. They may be placed near a load to reduce the length of the relatively

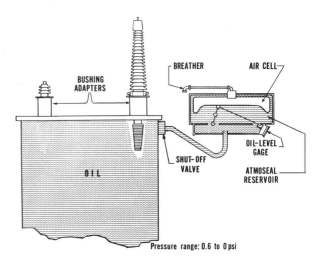

Figure 8.6. ATMOSEAL constant-pressure system.

heavy secondary bus, often allowing considerable savings in the total system cost. The fluid is relatively inert, has suitable dielectric and heat transfer characteristics, compatible with other transformer materials, and does not pose a threat to the environment.

VaporTran® Units

A new concept in transformers was introduced in 1978, utilizing vaporization-cooling and a vaporizable dielectric liquid known as R-113, or trichloro-

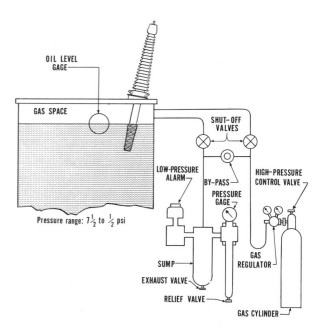

Figure 8.7. Automatic gas seal system.

trifluoroethane. It is totally nonflammable, and the superior heat-transfer characteristics allow for a cooler transformer and a significant increase in fan-cooled loadability. The electrical characteristics of the unit are essentially the same as for the silicone-filled transformer. The physical appearance is quite different since the cooling or condenser tubes are on the top of the main tank; but it can be connected to switchgear to form an integrated secondary substation in a manner similar to other types of transformers. The liquid will vaporize readily in case of an unintentional escape from the transformer and it is not considered harmful to the environment in this application.

Dry-Type Units

For applications where any liquid, even a nonflammable one, is objectionable, the dry-type transformer is available. The ventilated dry-type unit is cooled by a continuous natural draft of air and consequently is not suitable for locations where the air is wet or dirty. For these locations a completely enclosed unit, the sealed dry-type, is available, having a core-and-coil in a tank that is sealed and filled with an insulating gas. Dry-type transformers are completely nonflammable, using inorganic insulating materials such as glass, porcelain, and asbestos compounds.

EFFECT OF TEMPERATURE ON ORGANIC INSULATION

The transformer is the most likely to be overloaded of any major piece of equipment on the utility system. Generators are required to carry only as much load as their prime movers can deliver; however, there may be excess current caused by low power-factor operation, or low system voltage. The load on a motor cannot exceed the needs of whatever it is driving, which usually is within the safe limit of the motor. But a transformer has no such built-in load limiter. Sometimes the temptation to overload is very great, and the apparent reward for doing so is very high.

Overloading a transformer increases the losses, however, and therefore the temperature rise. It is known that organic insulations deteriorate with time—more rapidly as the temperature is raised. The question then is, How much effect does the higher temperature associated with a moderate

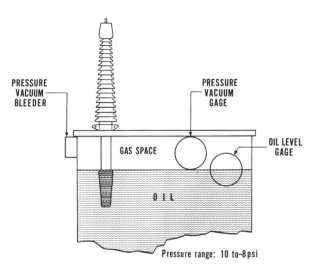

Figure 8.8. Sealed-tank system.

overload on a transformer have on the life of the insulation?

Few problems have received as much attention from the electrical industry over the years as the problem of finding some relationship between time, temperature, and useful life of organic insulation. Many years ago the utilities and the electrical manufacturers raised the question of permissible overloads on pole-type distribution transformers. At that time the domestic load served by these transformers was predominantly lighting; therefore, it lasted at most two or three evening hours in the summer, when the ambient temperature was high, and in the winter not over six or seven hours in the evening and perhaps an hour or so in the morning. The rest of the time the load was practically nothing so that the average temperature rise was low; furthermore, the highest rise due to the longest load duration occurred in the winter when the outside temperature was lowest. The operators, therefore, asked if these circumstances did not justify carrying a more than rated load on their pole-type transformers, and if so, how much.

In response to these questions, extensive tests were made to try to find an answer. Strips of varnished cambric insulation were immersed in mineral oil at different temperatures for varying periods, and then tested. The first tests demonstrated that insulation does not essentially deteriorate electrically until it has lost all mechanical strength, and has become charred and sufficiently weakened and brittle to crumble. Generally, the dielectric strength *increases* until the material cracks, so that it is hopeless to try to determine the rate of deterioration of insulation by its electrical strength. This leaves only the mechanical strength for a criterion.

Further tests sought a correlation of the time and temperature experience of insulation samples, with a reduction in their tensile strength. The results of these tests were plotted on semi-log paper. There was much scattering, as would be expected, but a crude average of the points could be represented by a straight line, the slope of which indicated that the rate of reduction of tensile strength doubled for every 8°C increase in temperature. This is the famous "eight-degree" rule announced in a 1930 AIEE paper by V. M. Montsinger of General Electric Company's Power Transformer Engineering Department. It has been said that the enthusiasm with which this rule was acclaimed was more of a testimonial to its simplicity than to its correctness.

The eight-degree rule merely states that if the temperature is raised 8°C, the rate of loss of tensile strength doubles; it refrains from stating either a normal rate of loss or a normal life expectancy of insulation. Furthermore, even after the strength reaches zero, nothing happens until some force causes fracture. Such a force might be due to a short circuit or to moving the unit. Even when the insulation is broken, nothing happens unless the break occurs at a point of voltage stress.

Later investigations under more closely controlled conditions disclosed that variations in the acidity or moisture content of the oil in which the samples were heated, or in the temperature, or in the prior history of the sample, could change the increase in temperature at which the aging rate doubled from 8°C to anything between 2°C and 12°C or 15°C. Despite this inexactitude and the limitations stated in the preceding paragraph, the eight-degree rule or modifications of it have received universal acceptance; it forms the sole foundation upon which rest the ANSI guides for operations (i.e., overloading) of transformers.

Many other tests have been made, including some on fractional-horsepower motors, seeking a solution to the same problem—the effect of temperature on the life of the insulation. The phenomenon is so complex, however, and there are so many variable factors, that the subject is still more art than science; therefore, about the most positive and precise statement that can be made is to repeat that "organic insulations deteriorate with time—the more rapidly as the temperature is raised."

In recent years, several types of insulation have been developed to permit higher temperature operation of insulation. Most of these systems use improved enamels and thermally upgraded paper, and are designed for 65°C average temperature rise instead of the standard 55°C rise. The 65°C transformer is widely accepted, and almost all new power transformers are specified with this rating. ANSI Standard C57.92 is a guide for loading oil-immersed power transformers.

IMPULSE VOLTAGE

Nonuniform Voltage Distribution

One or more of the windings of many transformers are connected to overhead feeders or transmission lines, exposing them to lightning voltages. As already illustrated, lightning voltages are of short duration and high rate of rise.

The maximum voltage to ground at the transformer terminal is usually limited by a surge arrester. When a lightning voltage is impressed on the terminal of a transformer, the distribution of voltage across the winding is not uniform. This is because the transformer is a complicated network of resistances, inductances, and capacitances between turns, between coils, between windings, and from each turn and coil to ground. In this network, an impulse voltage tends to appear initially on a few turns next to the terminal, as shown in Figure 8.9. Subsequent oscillations of voltage within the winding may cause parts of the winding to swing to a voltage even higher than that impressed on the terminal—as shown in the figure. Even more serious is the increase in voltage between parts of the winding. This not only raises the requirement for major insulation, it calls for sufficient insulation to withstand these transient voltages between turns and coils.

Shielding

The distribution of impulse voltage throughout the winding can be improved and the resultant oscillation reduced by proper electrostatic shielding of the winding. In principle, this shielding consists of various means of increasing the series capacitance within the winding. It may take the form of separate, noncurrent-carrying electrodes, or it may in-

DIELECTRIC TESTS

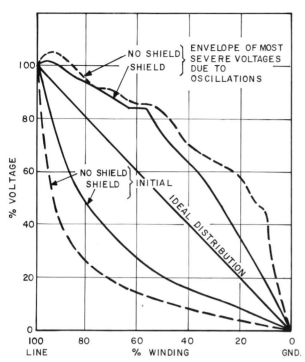

Figure 8.9. Impulse-voltage distribution in 138-kV, grounded-wye winding.

volve special "interleaving" of the winding turns. By these means, transformers have been built with impulse-voltage distributions approaching a straight line. In the moderate voltage range—46 kV through 138 kV—less complete shielding is used. The designer can control the degree of shielding to obtain the optimum over-all performance characteristics.

The voltage distribution in the transformer can also be materially improved by connecting disks of nonlinear resistance material between critical points in the winding. This method of damping out transient voltage oscillations is coming into increased use as a supplement to shielding for tap sections and series windings.

Neutral Grounding

Most high-voltage transformers are Y-connected on the high side. Whether or not the neutral point is connected to ground, and if so, how, has a very decided effect both on the transformer itself and on the system of which it is a part.

If the transformer neutral is directly grounded, it can be protected with a lower-rated surge arrester which will hold the voltage to a lower value than if the neutral is not solidly grounded. The industry trend is to neutral grounding of EHV units.

DIELECTRIC TESTS

The purpose of dielectric tests in the factory is to check the material, workmanship, and thoroughness of processing of the insulation system.

Standard dielectric tests consist of low-frequency, impulse, and switching surge tests. Low-frequency tests, and certain quality control impulse tests on distribution transformers, are usually routine. Complete impulse and switching surge tests are becoming routine on EHV transformers but are normally made only when specified by the purchaser. The low-frequency test may be either an applied-potential test (in which the test potential is applied between the entire winding and ground), an induced-potential test (in which voltage is induced between turns), or both. If the transformer has reduced insulation at the neutral, it is, of course, impossible to make an applied test at full potential. In this case, an induced potential test is used. All EHV transformers are tested with an induced potential test.

A trial-use standard for dielectric testing for power transformers operating on effectively grounded systems at 345 kV and above was instituted in 1977. At this time, it has not become a permanent standard but will be referred to as IEEE Std 262B-1977. Although several changes from previous standards are included, a major one is an induced test at 150% of operating voltage for one hour with corona levels not exceeding 150 μV.

Corona Testing

The service life of a transformer depends largely on its insulation life. If corona is present in a transformer, it is an indication that some portion of the insulation system is overstressed, resulting in erosion of solid insulation, chemical decomposition of oil, and possible early transformer failure. Conversely, the absence of corona is one assurance of long transformer service life.

Corona generates two signals—one as sonic or ultrasonic waves and the other as an electrical signal. In testing a transformer for corona, full induced voltage is applied, and both ultrasonic and RIV (radio interference voltage) measurements are utilized. A RIV level of 500 μV is the maximum allowable level at full induced voltage whereas 150 μV is the limit during the 1-hour induced test. If any sonic signal is detected, triangulation methods are employed to locate the cororna source so that corrective measures can be taken.

Standard impulse tests for transformers consist of one application of a full wave at reduced voltage, two applications of a chopped wave, and then one application of a full wave at full-rated voltage. A nominal 1.2 × 50 μsec voltage wave is the standard impulse wave for transformers. Waves of negative polarity for oil-immersed transformers and of positive polarity for dry-type or compound-filled transformers are recommended and are used, unless otherwise specified.

The ANSI rules recognize that it is impractical to require exactly a 1.2 × 50 wave for commercial tests. These tests require only that the time on the front of the wave, from virtual time zero to actual crest, shall not exceed 2.5 μsec for test voltages up to and including 650 kV, and 3 μsec for test voltages above this value. The time on the tail of the wave to half-crest value is to be not less than 40 μsec.

The oscillogram of applied voltage may be hard to decipher, especially at the start. If it is, virtual time zero may be determined by locating points on the front of the wave at which the voltage is, respectively, 30% and 90% of the crest value and then drawing a straight line through these points. The intersection of this line with the time axis is taken as virtual time zero. Also, for convenience, the time to crest may be considered as 1.67 times the actual time between the points on the front of the wave at 30% and 90% of the crest value.

The first impulse applied to a transformer during the impulse test is at reduced voltage, having a crest value somewhere between 50% and 70% of the crest value of the full-wave test specified in the standard. This test is made to obtain a cathode-ray oscillographic record of both voltage and current for comparison with the record of the final full-wave test. This comparison determines whether or not there was a failure. The first test is made at a reduced voltage level to assure that failure does not occur during this test. If the two waves are not identical in shape, then something is wrong. An investigation must be made to ascertain whether or not the difference is due to the test circuit itself or to an actual failure within the transformer.

The chopped-wave tests follow the first full-wave test at reduced voltage. The wave is chopped by flash-over of an external gap at a time not less than that specified in the standard. This time is 1.5 μsec for transformer equipment in the 1.2-kV insulation class, and increases to 3 μsec for the 25-kV and all higher insulation classes. The magnitude of the chopped-wave test is approximately 15% higher than the crest value of the full-wave test (but 10% higher for units covered by IEEE Std 262b-1977).

Another test specified by some customers is called the "front-of-wave" test. This was originated by one of the large management corporations which was disinclined to use surge arresters. The test is similar to the chopped-wave test, but the voltage is cut off by an external gap, set at a specified spacing, on the rising front of the wave rather than shortly after the crest of the wave. The rate of voltage rise for the test is 1000 kV/μsec for all insulation classes 69 kV and higher, and the time to flash-over in microseconds is 0.5 for all insulation classes 46 kV and below. The front-of-wave test is not included in the ANSI Transformer Standards.

The final full-wave test at full voltage serves two purposes. It is a test on the transformer itself, and it provides an oscillographic record from which, by comparison with the record of the first test, it can be determined whether or not the transformer has failed.

Surge-arrester characteristics have continually improved so that a large protective margin exists in the lightning or short time-protective coordination area. The emphasis on coordination has now shifted to the switching surge region which characteristically has slower rising fronts of from 100 μsec to 2000 or 3000 μsec to crest. The switching surge strength of a transformer is approximately 83% of the BIL (basic insulation level) of the winding. The normal arrester application allows a 15% margin between its maximum switching surge spark-over and the switching surge strength of the transformer.

The standard for the transformer switching surge test specifies that the applied switching surge must be above 90% of the transformer switching surge strength for at least 200 μsec and have a total duration of not less than 1000 μsec.

SUMMARY

Energy is transferred between circuits of different voltage by means of transformers. In nearly all distribution and power transformers, the core and windings are immersed in oil or other liquid within a weatherproof tank. Nearly all transformers are installed outdoors.

EHV transformers are predominantly generator step-up units and transmission-tie autotransformers. The large bulk distribution substation transformer is presently utilized at voltages 230 kV and below.

The tanks of very small distribution transformers have enough radiating surface to dissipate the heat caused by losses, without excessive temperature rise. Heat is removed from large units by one of these methods:

1. Providing radiators for cooling the oil by thermal circulation (fans may be used to increase the effectiveness of the radiators).
2. Pumping the oil through ducts in the core and coils and then through external air- or water-cooled heat exchangers.

Several methods are used to keep oxygen and moisture away from the oil: (1) sealed tank, (2) Atmoseal system, (3) conservator, and (4) gas seal. In locations where there is a fire hazard, nonflammable liquid replaces the oil. For users who prefer not to have any liquid at all, dry-type units are available at moderate voltages.

The rate of insulation deterioration increases with temperature. The ANSI Transformer Loading guide provides data for loading both with and without moderate loss of life.

One or more of the windings of many transformers are connected to overhead lines exposed to lightning. To prevent the concentration of steep-front lightning voltage across the few turns nearest the line terminal, the windings are protected by shielding or by bridging them with nonlinear resistors.

Transformers are designed to withstand specified impulse tests, although not all units are impulse tested. Switching-surge strength often represents the limiting condition for the determination of the required level.

REFERENCES

1. Blume, L. F., Boyajian, A., Camilla, G., Lennox, T. C., Minneci, S., and Montsinger, V. M., *Transformer Engineering*, Wiley, New York, 1951.
2. Degeneff, R. C., "A General Method for Determining Resonances in Transformer Windings," *IEEE Transactions*, **PAS-96**, 423–430 (March/April 1977).
3. Degeneff, R. C., McNutt, W. J., Neugebauer, N., Panek, J., McCallum, M. E., and Honey, C. C., "Transformer Response to System Switching Voltages," *IEEE Transactions*, **PAS-101**, 1457–1470 (June 1982).
4. Kaufman, R. B., and Meador, J. R., "Dielectric Tests for EHV Transformers, *IEEE Transactions*, **PAS-87**, 135–145 (January 1968).
5. McNutt, W. J., McMillen, C. J., Nelson, P. Q., and Dind, J. E., "Transformer Short-Circuit Strength and Standards—A State-of-the-Art Paper," *IEEE Transactions*, **PAS-94**, 432–443 (March/April 1975).
6. McNutt, W. J., Kaufmann, G. H., Vitols, A. P., and MacDonald, J. D., "Short-Time Failure Mode Considerations Associated With Power Transformer Overloading," *IEEE Transactions*, **PAS-99**, 1186–1197 (May/June 1980).
7. McNutt, W. J., Blalock, T. J., and Hinton, R. A., "Response of Transformer Windings to System Transient Voltages," *IEEE Transactions*, **PAS-93**, 457–467 (March/April 1974).
8. MIT Staff, *Magnetic Circuits and Transformers*, MIT Press, Cambridge, Massachusetts, 1943.
9. Montsinger, V. M., "Loading Transformers by Temperature," *AIEE Transactions*, **49**, 776 (April 1930).
10. Patel, M. R., "Dynamic Response of Power Transformers Under Axial Short Circuit Forces—Part I—Winding and Clamp as Individual Components," *IEEE Transactions*, **PAS-92**, 1558–1566 (Sept./Oct. 1973).
11. Patel, M. R., "Dynamic Response of Power Transformers Under Axial Short Circuit Forces—Part II—Windings and Clamps as a Combined System," *IEEE Transactions*, **PAS-92**, 1567–1576 (Sept./Oct. 1973).
12. Say, M. G., *Alternating Current Machines*, Halsted, New York, 1976.

9
SWITCHGEAR

DEFINITION

Switchgear is a general term covering switching and interrupting devices, also assemblies of those devices with control, metering, protective, and regulatory equipment with the associated interconnections and supporting structures.

FUNCTION

Switchgear is the vehicle for performing two distinctly different functions. Under normal conditions, it is the means of carrying out a multitude of routine switching operations, e.g., disconnecting and isolating any piece of apparatus for maintenance or replacement; disconnecting a generator from the system when it is no longer required to serve the load; sectionalizing a line for inspection, maintenance, or construction purposes; transferring loads; isolating regulators; by-passing circuit breakers; and performing the reverse of these various operations. Under abnormal conditions, switchgear provides the means for automatically disconnecting the part of the system in trouble to prevent excessive damage and to confine the trouble to the smallest possible part of the system. Under these conditions the switchgear equipment is performing a protective function.

COMPONENTS

Some of the devices we call switchgear are connected directly in or to the circuit, or are closely related to it. In this category are circuit breakers and disconnecting switches or disconnecting devices; fuses, instrument transformers, buses and connections between all these components, and supporting insulators, structures, or housings.

Other switchgear devices may be located at some distance from the circuit with which they are associated. These are the indicators such as instruments, annunciators, and signal lamps; meters; protective and control relays; control switches; generator-voltage regulators; and the panels on which these various devices are mounted.

SHORT CIRCUITS

The "abnormal conditions" mentioned above usually mean short circuits, or the system disturbances that they cause. Short circuits interfere with giving good service. They may be dangerous to life and property. When the fault is accompanied by arcing, as it usually is, it produces a very high temperature. This excessive temperature may start a fire and melt a cable sheath or the core of a generator or motor. The heat of the arc causes the violently rapid expansion of the nearby air and the gases generated by the arc. Almost immediately, the arc may spread to other phases or circuits, or to ground. When the short circuit occurs in a confined space, as in the slot of a rotating machine, or under the oil of a transformer, the pressure generated can cause a destructive explosion.

The amount of heat generated and the amount of damage caused by a short circuit are a function of the amount of current and of its duration. During the short circuit, the flow of greater-than-normal current causes high voltage drops which often result in widespread reductions in system voltage. Under-voltage release may cause load interruption during

CALCULATION OF FAULT CURRENT

periods of low voltage. The system generators may tend to drift apart (system stability). Those switchgear components that are in series with a circuit (circuit breakers, disconnecting switches, buses, connections, etc.) should be able to withstand the electromagnetic forces and the thermal effects caused by short-circuit currents of specified duration.

Short circuits come uninvited and seldom leave voluntarily. To protect equipment from damage and to maintain system operation, the short circuit should be removed as quickly as possible. Interrupting (opening) the circuit on both sides of the short circuit will isolate and deenergize it. The devices used to disconnect circuits or equipment are fuses and circuit breakers. Fuses are generally used in the lower voltage part of the system where a permanent outage can be tolerated. Circuit breakers are capable of reestablishing the connection very quickly. If the fault (short circuit) is temporary, as most are, the circuits and/or equipment may be reconnected quickly by circuit breakers, but not by fuses.

CALCULATION OF FAULT CURRENT

Switchgear devices must be able to carry the maximum short circuit current experienced in that location. This means it must be physically strong enough to withstand the mechanical forces associated with the high current and not be destroyed by heating (I^2R). Also interrupting devices must be able to interrupt the current flowing during interruption. Note that this is not necessarily the maximum short-circuit current at that location.

The magnitude of the short-circuit current depends on the impedance or impedances between the source or sources of voltage and the point of fault. The current from a generator to a short circuit at its terminals is limited only by the internal impedance of that generator. If the fault is elsewhere, the current is limited by the internal impedance of the machine, plus the impedance of the intervening circuit, including any transformers. In the typical system more than one source of power exists, and there are normally several paths through which the current can flow to a fault. Multiple paths reduce the impedance and increase the fault current.

Unless the system fault occurs at the instant of normal current zero of the short-circuit current wave, the current will not be symmetrical about the zero line. Its axis will be displaced, or offset. In power circuit-breaker applications, it must be assumed that a short circuit on any ac system can produce the maximum offset (dc component) of the current wave. The resulting asymmetrical or total current gradually decays (see Figure 9.1) to a symmetrical current, with the rate of decay of the dc component determined by the X/R ratio of the system supplying the fault current. Power circuit-breaker application multiplying factors must recognize this current offset and decay.

In 1964, new standards were published and adapted as supplements to existing standards for power circuit breakers. [These standards are presently published by the American National Standards Incorporated (ANSI)]. The ANSI standards introduced considerable change from their previous rating structure, the test code, and other related requirements. A significant new requirement is the change in interrupting current rating from a total current basis to a symmetrical current basis of rating. All new breaker designs have been transferred to this new rating basis.

Intended for use with the new breaker rating structure is application guide, ANSI C37.010, which presents two methods of short-circuit current determination for breaker selection. The first, an "E/X Simplified Method," provides an application procedure that will give the required accuracy in the majority of cases. Where application closer to the breaker rating is required, a second method is presented which provides knowledge of the short-circuit characteristics of electrical equipment and systems. Common to both methods are the following two steps of procedure:

1. Calculate an rms symmetrical value of fault current, using the equations of Figure 9.2. For three-phase faults the symmetrical current value is E/X_1. For single line-to-ground faults, $3E/(2X_1 + X_0)$ can be used to determine the symmetrical current value since X_2, as seen from the fault location, is generally equal to X_1. This is not true near a generator since the generator X_2 is less than X_1. (Appropriate reactance values for these calculations are defined in the application guide.)

2. Multiply the result by a factor whose value depends on the application.

For step 2, the simplified method employs generalized application multipliers and the method can be used without knowledge of the system X/R ratio. In

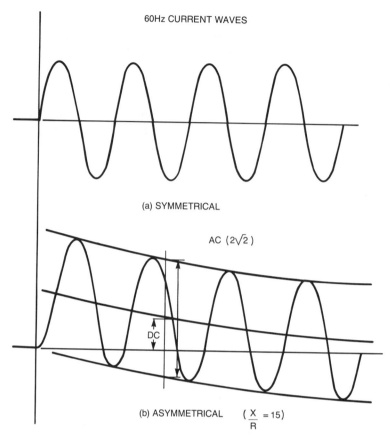

Figure 9.1. Ac current offset and decay during fault.

the more precise method of calculation, application multipliers are selected from a family of curves in the C37.010 guide based on system fault location, breaker-contact parting time, and system X/R ratio as viewed from the fault point. This latter method, which accurately accounts for the decay of both the ac and dc components of fault current, requires an X/R calculation.

In the determination of system X/R ratio (X_1/R_1) for three-phase faults, or $(2X_1 + X_0)/(2R_1 + R_0)$ for single line-to-ground faults, the recommended procedure is that of reducing system resistances to a single value with complete disregard for resistance, and then reducing system resistances to a single value with complete disregard for reactance.

Both methods of short-circuit calculation for power circuit-breaker application contained in C37.010 can be adapted to analog and digital computers.

INTERRUPTION OF FAULT CURRENT

The primary function of interrupting devices such as circuit breakers and fuses is to stop the flow of current. During the process of interruption, an arc is created that must be extinguished. In an ac circuit interrupter, this involves taking advantage of a current zero and then injecting dielectric strength between the contacts faster than the recovery voltage builds up. The phenomenon is substantially the same regardless of the medium in which the arc is drawn. (With the exception of air or gas where the step for decomposition of a liquid is absent.) An understanding of the mechanism of extinguishing an ac arc may be gained by following the phenomenon

TYPE OF FAULT		FAULT-CURRENT AMPERES	
THREE-PHASE	(symbol)	$I_{3-\phi} = \dfrac{E}{X_1}$	E — X_1
SINGLE-PHASE LINE-TO-GROUND	(symbol)	$I_{L-G} = \dfrac{3E}{X_1 + X_2 + X_0}$	$3E$ — $X_1 + X_2 + X_0$

Figure 9.2. Calculation of fault current.

INTERRUPTION OF FAULT CURRENT

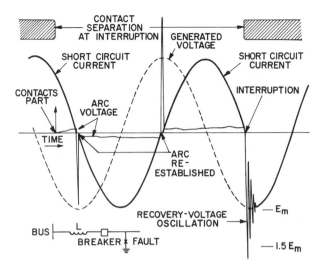

Figure 9.3. Interruption of short-circuit current.

in a plain-break circuit breaker. (Reference to Figure 9.3 will be helpful.)

As the breaker starts to open, the current flowing prior to when the moving contact leaves the stationary contact will continue to flow and draw an arc. The electric arc acts as an energy source to the insulating medium (oil or gas), and the temperature of the medium within the arc may approach 30,000°K. When a current zero occurs, a transition must be made from the high-temperature arc to a cool insulating medium before the voltage across the opening contacts builds up sufficiently to cause an interruption failure. If the voltage builds up in a very short time, 20 μsec or less after the current zero, ionized gas is still present in the arc chamber and a thermal failure may occur with no perceptible cessation of current. Should cooling be adequate to prevent a flow of current immediately after current zero, a dielectric breakdown is still possible hundreds of microseconds later, if the voltage across the opening contacts exceeds the dielectric withstand of the insulating medium.

In an ac circuit, this is exactly what happens every half cycle as the current goes through zero. Under these circumstances the question that immediately arises is whether a plain-break oil breaker interrupts the circuit at the first current zero after the contacts part. The answer is that it could, but in most cases, arcing continues beyond the first current zero.

During a short circuit, the current lags the system voltage by upwards to 90°, depending on the relative values of R and X. At the instant the fault current goes through zero, the system voltage may be nearly at maximum. It is at this instant that the arc goes out temporarily. As long as current flows through the breaker, the voltage across it is only arc drop, which usually is low. But when the arc goes out and the current stops, the voltage across the breaker suddenly shoots from this low value toward full system voltage. The inductance and capacitance in the circuit, however, cause the voltage to overshoot to a value which, neglecting damping, is as far above the system voltage as it was below it at the start. When this occurs, the transient voltage across the breaker may be nearly double the steady-state system voltage. In addition, if the sudden increase in arc resistance (as the current approaches zero) should cause the current to reach zero prematurely, that would cause another transient and generate more voltage because of the greater rate-of-change of current in an inductive circuit.

The result of these voltages, known as recovery voltage, appears across the breaker. Since it is caused by transients of higher-than-system frequency, it reaches maximum value much more quickly than a 60-Hz voltage wave. The rate at which it builds up is known as the "rate of rise of recovery voltage." The recovery voltage appears between the breaker contacts and is applied to the insulating medium and hot gas between them, which an instant before was carrying current. If this mixture has insufficient dielectric strength to withstand the recovery voltage, it breaks down and current flows again. Should this phenomenon occur during the first quarter cycle following interruption, it is known as a reignition; after one-quarter cycle, it becomes a restrike.

In a plain-break circuit breaker, the above cycle may be repeated several times. At each current zero, the moving contact is a little further away from the stationary contact. As a result, the arc becomes longer and comes in contact with more of the insulating medium. At the same time, its temperature and conductivity go down, current decreases, arc drop goes up, and the gap (to be broken down by the recovery voltage until it can withstand this voltage) is lengthening.

Assuming that the breaker is successful in opening the circuit before the end of its stroke, there comes a current zero at which the recovery voltage is no longer able to break down the gap between the contacts. At this point, arc extinction is permanent and circuit interruption is complete.

As illustrated, the problem of circuit interruption is the problem of extinguishing an arc, and this problem involves taking advantage of a current zero and injecting dielectric strength between the contacts faster than the recovery voltage builds up. Many ingenious ways of efficiently accomplishing this are embodied in the various types of modern-day breakers described in this chapter.

INTERRUPTION OF LEADING CURRENT

A special problem in arc extinction is encountered in interrupting the leading current of a capacitor or an unloaded transmission line. The explanation of this phenomenon will be more easily understood by reference to Figure 9.4.

At the first current zero of a leading-current arc, interruption occurs quite easily. It does so because the capacitor retains the instantaneous voltage of the source at the instant of current zero, and for several hundred microseconds thereafter the voltage across the breaker is low. For this reason, interruption seems to be complete at the first current zero of arc even though contact separation is extremely small. A quarter cycle later, however, the system voltage reverses, and a quarter cycle after that approximately double voltage appears across the breaker. If the breaker withstands this voltage without restriking, as it may when opening at low current, then the interruption is complete. At high currents when the arc generates an increased amount of hotter gas, the short arc gap cannot withstand double voltage. At this time, there may be enough voltage across the contacts to cause a restrike during the half cycle immediately after the first interruption. If the restrike occurs just one-half cycle after the first interruption, the capacitor has a charge whose polarity is opposite to that of the system voltage, and there follows a flow of current that is limited by the inductance of the circuit. Because of the capacitance, the voltage overshoots its mark by the initial difference. In this manner, the voltage can reach minus $3E$ at the exact instant that the transient current passes through zero.

From this point two possibilities of major interest arise:

1. Interruption of the high-frequency current at its first current zero (case 1, Figure 9.4).
2. Continuation of the high-frequency current, which slowly dies away, and the reestablishment of the normal frequency capacitor current (case 2, Figure 9.4).

If the latter occurs, then another interruption will take place at the next current zero of the normal frequency current. This is just one cycle after the first attempted interruption. By this time the gap will be longer, its insulation will have increased, subsequent restriking is usually avoided, and final interruption is probably achieved. In this case, the voltage to ground may rise to only three times normal crest on a single-phase basis.

On the other hand, if the breaker interrupts the high-frequency current at its first current zero following restrike, then the capacitor remains charged to as high as three times normal voltage crest. One-half cycle later, when the system voltage again reverses, a maximum of four times normal voltage is impressed on the gap insulation. Subsequent restriking can usually be expected under these very severe conditions. In this manner, voltages up to the breakdown values of circuit or equipment insulation can be generated.

This rough analysis of the interruption of leading current, although applied specifically to a capacitor, is equally valid for a long open-circuited high-voltage line or cable having distributed capacitance. The voltage oscillations are more square-topped because of the traveling wave characteristics of the line. If restriking occurs at the peak of the source

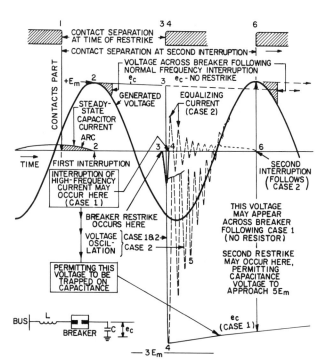

Figure 9.4. Capacitive circuit interruption.

voltage reversal (following the first interruption), a voltage wave equal to the difference, or twice normal voltage crest, travels along the line and established a charge equal and opposite to the initial trapped charge. The current through the breaker, caused by this voltage wave, is equal to the voltage divided by the surge impedance of the line. When these waves reach the far end of the open line, the voltage wave reflects the double voltage without change of sign, and returns to the breaker, charging the line to approximately three times normal voltage crest. The accompanying current wave, however, reflects with change of sign to cancel the forward-wave current. When this occurs, current zero literally travels back to the breaker, at which point the breaker arc is usually extinguished. This leaves a charge on the line that raises it to a voltage of nearly three times normal crest. On a 200-mile (320 km) line, the complete round trip time of such a wave from restrike to current zero at the breaker is about 2000 μsec or about one-eighth of a cycle (on a 60-Hz basis). On a 10-mile cable, the round trip time is about 200 μsec.

The preceding discussion illustrates why it is more difficult to switch off capacitors and open-circuited transmission lines than to open the same amount of lagging current, and why doing so may generate high voltages. Again, modern breakers incorporate design features that aid in the interruption of leading current.

METHODS OF ARC INTERRUPTION

Three broad classifications of interrupting equipments are: (1) high-voltage disconnect switch, (2) load-break switch, and (3) the circuit breaker.

About the simplest, yet crudest, way of opening a circuit is to use a horizontal-type disconnecting switch. Breaking current in this kind of device forms an arc in air which, on a still day, may rise many feet and last several seconds before it is stretched out long enough to be chilled at some current zero. During this time, the blade is moved so far from the stationary contact that the recovery voltage, when the arc finally is extinguished, does not cause a restrike.

The disconnect switch is the simplest switch on the basis of function, operating only in the absence of appreciable current. This switch cannot open normal load current, and its function is to disconnect or connect transformers, circuit breakers, other pieces of equipment, and short length of high-voltage conductors, only after current through them has been interrupted by opening a circuit breaker or load-break switch. It may, however, open minute "charging" currents to these unloaded equipments being disconnected. When opening, the switch blade is swung upward roughly 90°, creating a long, simple gap in air.

A load-break switch will switch normal load or power but will not interrupt short circuit currents. A wall switch in a home fits this classification.

Circuit breakers will perform the switching functions of the other two classes, but will, if applied within rating, interrupt all short circuits that may occur on the system.

Circuit switches are available that have limited interrupting capability but are primarily used for line, transformers, capacitor, and reactor switching.

The most effective method of current interruption depends on the circuit voltage, current to be interrupted, and circuit in which the interrupting device is to be applied. Different kinds of interrupting devices use different methods. Over the years new discoveries and developments have led to improvements in circuit breakers.

Air

Air at standard pressure of 14.7 pounds/in^2 is a very poor interrupting medium. As indicated in Figure 9.5, a 1-inch-long (2.54 cm) electric arc carrying more than about 1 ampere or less, convention air currents will remove heat energy at a sufficiently high rate to extinguish the arc.

Examples of practical switches that operate simply by the opening of contacts in free air include simple wall switches in homes, the points in an automobile ignition system, and simple relays. Typical ratings of such devices are 20 amperes at 240 volts, or 0.005 MVA, and 30 amperes at 600 volts, or 0.02 MVA. In a breaker with arcing characteristics similar to those in Figure 9.5, the rating would be 1 ampere times 1000 volts, divided by 1,000,000, or 0.001 MVA.

It would appear that the larger the gap between contacts, the greater the interrupting capacity. However, in the case of a 230,000-volt disconnect switch, the open gap measures about 11 feet (3.4 meters)—yet even a 2-ampere arc does not always extinguish unless there is a wind blowing across the

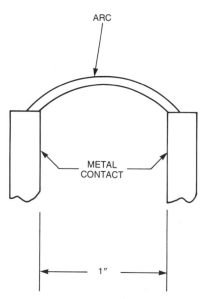

Figure 9.5. Free arc in air at room pressure.

arc. Instead, the arc bows upward in a curved path longer than the 11 feet from the stationary to the moving contact. If this switch did interrupt, its rating would be about 0.3 MVA. This is far from the 20,000-MVA capacity of the top 230,000-volt oil or air circuit breaker. The 0.3-MVA capacity of such disconnect switches is so small and so uncertain that there is no guarantee of interruption. High-voltage disconnect switches take many seconds to open and close, and are operated by hand or by a motor through gears. An example of such a device is shown in Figure 9.6.

Just as one can extinguish a weak flame by quickly inserting cold metal or insulating material, electric arcs can be extinguished by moving them against such materials. If a series of metal or insulating fins is placed as shown in Figure 9.7, the interrupting capacity is greatly increased.

The cooling action is primarily due to injection into the arc of relatively cool vapors from the surface of the metal or insulating material over which the arc plays. In addition, arc chutes with insulating fins elongate the arc in a tortuous path as it rises. The longer arc may have such a high electrical resistance that the magnitude of the current is reduced. The combination of cooling and increased resistance to conductivity acts to extinguish the arc.

A further, almost separate interrupting effect, is achieved when the arc is made to enter a stack of metal fins in such a manner that the single long arc is replaced by a series of short arcs between successive pairs of metal fins, similar to Figure 9.7.

Many circuit protective devices and switches are based on the arc chute principle. Some use insulating arc chutes of the chamber or slot type. Others use a simple metal fin shaped to form a chamber. In some molded case breakers, the arc chutes look very much like Figure 9.7, and are capable of interrupting as much as 25,000 amperes at 600 volts (15 MVA).

The magne-blast circuit breaker is used in metal-clad equipment of power switchgear assemblies. It includes an arc chute, generally with insulating fins. Rather than depending on simple convection to move the arc and elongate it, the magne-blast circuit breaker has a magnetic field. This field acts on

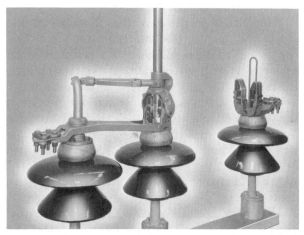

Figure 9.6. Disconnect switches interrupt current by rapidly elongating the arc in air.

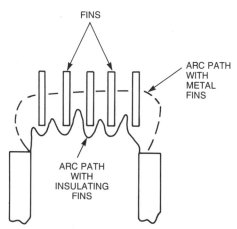

Figure 9.7. Insulating fin or metal plate arc chute.

METHODS OF ARC INTERRUPTION

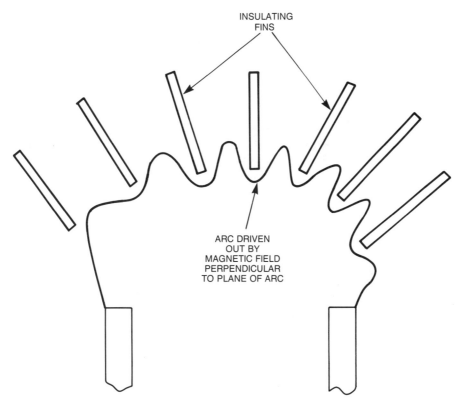

Figure 9.8. Arc interruption by magnetic blowout.

the arc just as the magnetic field in a motor acts to push the wires carrying current. This principle is illustrated in Figure 9.8.

The result is that the arc is rapidly elongated and forced into the tortuous path through the arc chute. The arc resistance is increased by elongating the arc and by vaporization of insulating material from the arc chutes as the arc is blown against it.

The combination of the arc chute, the magnetic blow-out, and the much larger size raises the interrupting capacity to as high as 42,000 amperes at 8000 volts, or 340 MVA per break. A three-phase breaker will have three times this rating, or 1000 MVA at 13,800 volts. A schematic of the breaker is shown in Figure 9.9.

In recent years, improved technology of air-blast circuit interruption has resulted in new circuit-breaker designs with higher interrupting capacities.

Used primarily in indoor power stations at 14,400–34,500 volts is a family of high-current air-blast circuit breakers. In this type of rather limited usage air-blast breaker, an arc chute with one "splitter" or insulating fin is used. This splitter is called an arc barrier. The arc is drawn between the upper and lower gap electrodes. During interruption, a blast of air is directed across the arc, pushing the arc against the barrier or splitter. At current zero, the arc string is broken and carried down-

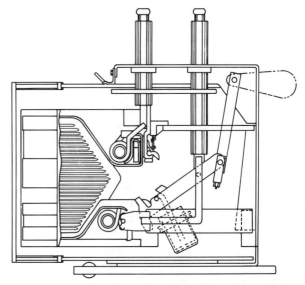

Figure 9.9. Schematic of air magnetic breaker.

Figure 9.10. 550-kV air-blast breaker.

stream. The contacts are driven both to closed and open positions by compressed air pistons.

Another type of circuit breaker utilizing the axial air-blast principle at transmission voltages can interrupt 63 kA at 800 kV and 80 kA at 242 kV. In operation, high-pressure compressed air (up to 800 pounds/in^2) is confined to one side of the interrupter head by means of a vertical plate. During interruption, the opposite side of the interrupter head is open to the atmosphere, so its air pressure is low. When an arc is formed between the upstream arcing electrode and the downstream arcing electrode, the compressed air blasts forth from the pressure chamber through the interrupting orifice. The air stream flows along the axis of the arc giving the interrupter its name "axial blast."

At the instant the current is at its maximum, the arc is so thick that is approximately fills the orifice. By the time current zero is reached, the arc is reduced to a thin "string." At current zero, the arc "string" is broken and is carried downstream faster than the speed of sound, completing the extinction of the arc and the interruption of the current.

Just as with some of the large oil breakers, the axial air-blast breaker uses a power resistor. This resistor remains connected across the main interrupter gap until after the main power current is interrupted. Then the cross-arm of the resistor switch moves upward, interrupting the remaining current through the resistor. The main purpose of the power resistor is to control transient voltages that would otherwise appear when an arc occurs a short distance down a transmission line.

Each interrupter tank usually has two breaks with capabilities of up to 140 kV (three-phase) per break. A 550-kV breaker would use two interrupters or four breaks in series, as illustrated in Figure 9.10.

Oil

If contacts are separated in oil rather than air, both the voltage and current that can be interrupted are increased greatly in comparison with air at room pressure (excluding air blast). There are two basic reasons for this. First, the dielectric strength (ability to withstand the electric pressure without breakdown and initiation of a spark) for oil is many times greater than for air. Second, the arc generates hydrogen gas from the oil, and this gas is superior to air as a cooling medium.

While almost all forms of oil circuit breakers have two or more arcs per pole in series, one is shown in Figure 9.11. In this example, the arc is hot enough to first decompose the oil so that gas

METHODS OF ARC INTERRUPTION

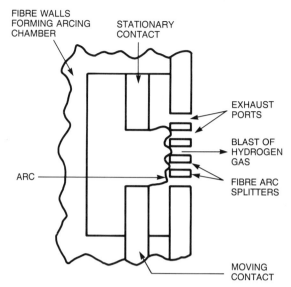

Figure 9.11. Oil-blast interruption using arcing chamber.

evolves and, second, to ionize the gas so that it is conducting. The heat generated in the arc is absorbed, first, in decomposing the oil and, second, in ionizing the resultant gas. It also melts and gasifies some of the contact material while ionizing it. The presence of the ionized contact material may or may not be of importance in the problem of arc extinction, but it is the reason why, after many operations, the contacts partially disappear and have to be replaced. The ionized material in the arc stream forms a high-conductivity path between the contacts.

Since the gas is not ionized when it is evolved from the decomposing oil, it is nonconducting, and the ionized gas in the arc stream is continually being diluted with fresh nonionized gas. This dilution lowers the temperature and the conductivity of the gas in the arc stream.

The bubble of gas between the contacts is forced upward by its buoyancy, and it is replaced by oil. At low-contact separation and high current, however, the arc can maintain itself indefinitely by decomposing the oil as it flows in to replace the rising gas bubble, and then ionizing the fresh gas. If the current is gradually reduced, there comes a time when the arc no longer generates sufficient heat to both decompose the oil and ionize the evolved gas. At this point, there is a sudden reduction in the conductivity of the gas, the amount of current, and the temperature. The result is that the arc disappears and the circuit is open.

Even though immersing the contacts in a bath of oil increases interrupting capacity greatly with respect to air, the resulting capacity is very small compared with the 8000 MVA capacity per break needed for the highest ratings.

Enclosing the contact and the arc in a small fiber arcing chamber with exhaust ports on one side (Figure 9.11) can provide the needed increase in capacity. This is achieved for two reasons: (1) the hydrogen gas released by the arc in the confined chamber builds up pressures that increase the dielectric strength of the gas bubbles in the oil, and (2) the blast of hydrogen passing through the arc is extremely powerful in cooling and extinguishing the arc. The addition of these insulating chambers to oil circuit breakers has multiplied the interrupting capacity of a break of similar dimensions by about 500 with reference to bulk oil and by 1,000,000 with respect to open air.

In Figure 9.11, the current and arcing path has been illustrated as it probably exists just before current zero. At this stage, the blast of hydrogen gas has scavenged the inside of the interrupting chamber of most of the gas that would have been hot enough to remain as an electric conductor. Most of the volume inside the arcing chamber is occupied by a relatively cool gas bubble stretching from one contact to the other, over to and including the exhaust port region. The arc then looks like the thin string arc that is pushed up against the series of exhaust ports and splitters in Figure 9.11. (These are the pieces of fiber that form the exhaust ports and function somewhat like the insulating fins in an arc chute.)

Just after current zero, the arc is split by the arc splitters into several sections in series, with each small section being carried rapidly out through the exhaust ports. Thus, the flow of current is interrupted and the arc is extinguished.

There are two forms of oil breakers. The first is the bulk oil type in which the interrupters are submerged in oil in a metal tank. This type is also referred to as a dead-tank design. The previous discussion has used this design as a vehicle for describing the interruption process. The second form is the minimum oil or live-tank design. This design uses a small portion of the oil found in the bulk oil type. The principles of interruption are very similar. In the United States, the bulk oil type has been the predominate choice due to the explosive hazard of the minimum oil type. The bulk oil has the metal enclosure as the ground. The interrupter of

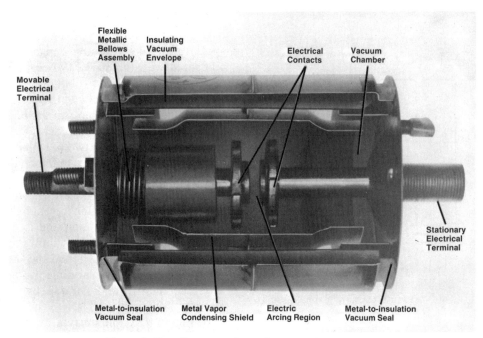

Figure 9.12. Cutaway view of vacuum interrupter.

the minimum oil type is mounted above ground on an insulator.

Vacuum

Perhaps the most significant recent advance in the art of circuit interruption is the design of circuit breakers that utilize a high vacuum as the interrupting medium.

The vacuum interrupter employs a moving contact and a stationary contact enclosed within an evacuated, hermetically-sealed envelope, as shown in Figure 9.12. The moving contact rod passes to the outside through a high-strength metal bellows. The contacts are parted by means of an external mechanism affixed to the moving contact.

To achieve interruption, the contacts need to be parted only a short distance since the power arc that results will be extinguished at the first current zero. The exceptionally high dielectric strength of vacuum prevents reestablishment of the arc.

Several major user benefits result from the following inherent characteristics of vacuum interruption:

1. High dielectric strength at small contact-gap distance results in superior capacitor switching ability without restrikes.

2. A high rate of recovery of the dielectric means that the arcing time in the vacuum is shorter, and total interrupting time is faster. Interrupting time is not affected by the magnitude of current values within the interrupter rating. Vacuum interrupters are insensitive to evolving faults, and are self-protecting during multiple lightning surges.

3. The contact erosion rate is low, providing unusually long contact life with maintenance reduced to a minimum.

4. Interruption occurs in a nontoxic, nonflammable atmosphere, permitting consideration of application in semihazardous areas.

5. Interruption does not produce ground shock or reaction forces. As a result, foundations can be simpler and units can be readily mounted in structures or poles.

Since their introduction in late 1961, the vacuum breakers and reclosers have been applied widely to utility and industrial substation circuits up to 34.5 kV.

Sulfur Hexafluoride

The use of sulfur hexafluoride, SF_6, as an arc-interruption medium has found increased acceptance in

METHODS OF ARC INTERRUPTION

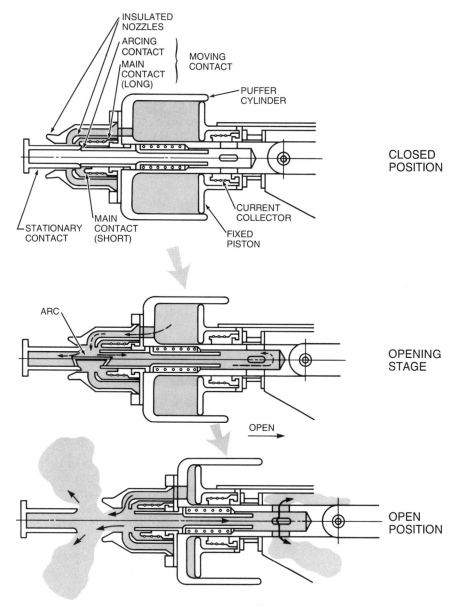

Figure 9.13. SF_6 gas-puffer interrupter.

the world because of its excellent dielectric properties. Sulfur hexafluoride, an electronegative gas, has a particularly high electron attachment energy and an ability to bind electrons to itself for considerable time as an SF_6 negative ion. This limits the number of free electrons that are available with sufficient energy levels to initiate and maintain an arc in the SF_6 gas. The voltage-withstand capability is three times that of air.

Two types of SF_6 circuit breakers are in common usage, a two-pressure system that operates similar to an axial air blast design and a single-pressure puffer system. The two-pressure system has high-pressure SF_6 gas surrounding the interrupter and acts as the interrupting medium when the breaker opens. A low-pressure region is maintained for exhausting the high-pressure gas during interruption for later recycling.

Interruption is initiated by a gas-driven piston that initiates movement of a moving contact. When this contact parts with the contact fingers, an arc is formed and is immediately driven inward by the high-velocity flow of gas.

Two-pressure SF_6 gas circuit breakers are generally of the "live tank" type similar to the air blast breaker shown in Figure 9.10. Some two-pressure designs have been built in a grounded enclosure ("dead tank") configuration for use in gas-insulated substations.

Two-pressure SF_6 designs suffer from a need for heaters to prevent liquefaction of the stored high-pressure gas at low ambient temperature. To overcome this deficiency, a new SF_6 puffer circuit-breaker design is finding wider use. The SF_6 gas puffer circuit breaker is a simple device consisting fundamentally of a piston and cylinder assembly integral to the moving contact, fully insulated by low-pressure SF_6 gas within a cylindrical enclosure as shown in Figure 9.13. With a single motion, the operating mechanism opens the contacts, compresses SF_6 to interrupting pressure in the puffer cylinder, and applies the high-pressure burst of arc-extinguishing gas to the contact gaps. Since the gas is stored at low pressure (75 psig), liquefaction does not occur at temperatures above $-30°C$. The design simplicity has increased reliability and made it ideal for "dead tank" installation in gas-insulated substations.

Attractive features of SF_6 breakers are:

1. Contact separation kept to a minimum because of dielectric strength provided by the high-pressure SF_6.
2. Arc reignition is minimized because of the chemical properties of SF_6.
3. Low arcing time because of the very high rate of recovery of the dielectric.

ARC INTERRUPTION IN FUSES

Power fuses consist essentially of a fusible element and an arc-extinguishing means. In some fuses, the fusible element is made of silver, but usually it is tin, copper, aluminum, or some alloy. This element is of proper shape and cross-section to carry rated current continuously and to melt in accordance with a specific time–current characteristic on heavy overload or short circuit. There are four generally used methods of extinguishing the arc when the fuse melts.

The first is to place the fuse element in a fiber tube. When the fuse melts, the arc decomposes the tube wall and generates gas. The fresh, nonionized gas cools the arc, and the high pressure expels what is left of the fusible element from the bottom of the tube and blows out the arc.

In the second type of fuse, the fusible element is surrounded with boric acid powder. Under the action of the arc, the powder turns to liquid and then to steam, which cools the arc. The resulting blast action extinguishes the arc.

In the third type of fuse, the fusible element is immersed in a nonflammable liquid, such as carbon tetrachloride, which cools and puts out the arc.

The fourth type, a silver–sand fuse, has the characteristics of current limiting devices discussed in the next section.

In all fuses, except the silver–sand type, the fusible element is attached to a spring to increase the speed with which the ends of the element separate after melting. The spring accelerates that increase in arc resistance and electrode separation upon which extinction depends.

CURRENT-LIMITING DEVICES

Modern power systems inherently grow by increasing short circuit capacity. Operating the system with the existing circuit breakers may require a split system to avoid exceeding in-service breaker ratings, but this would complicate system operation. System splitting can also be avoided by replacing in-service breakers with new breakers having higher short-circuit ratings. This is a costly, time-consuming, repetitive requirement as system capacity increases. In addition to circuit breakers, the effect of higher short-circuit currents on substation transformers, buswork, and current transformers also needs to be considered.

An attractive alternative is to use a current-limiting device which prevents excessive short-circuit capacity and attendant problems. Figure 9.14 illustrates how current limiting can restrict system fault current.

The simplest type of current-limiting device used on distribution systems is the silver–sand fuse. This fuse is "current limiting" since it opens the circuit before the current can reach the crest of the first half cycle of short circuit. It cannot wait for a natural current zero; instead it forces an earlier-than-normal current zero. The fusible element consists of silver wires or ribbons embedded in sand. When the wires melt, the arc is immediately in intimate contact with the sand that cools it. Since the sand itself does not give off ionizable gas and, by its presence,

EHV CIRCUIT BREAKERS

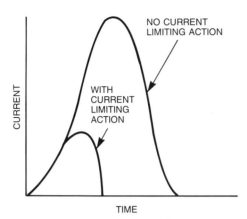

Figure 9.14. Current-limiting action.

excludes all but a very little ionizable air, the arc dies of starvation as well as of chilling. Death is almost instantaneous. In fact, special design features have to be incorporated to prevent the current from dropping to zero so suddenly as to generate dangerous overvoltage.

Other types of current-limiting devices under development or investigation for distribution and transmission systems rely on commutating the high-fault current into an inductor or resistor within 2–4 msec to limit the crest during the first half cycle of fault current. These include use of cryogenic or superconducting conductors that can be switched, mechanical separation of conductors by explosive charges, a cross-field tube to interrupt and commutate current, a high-arc voltage in an air-magnetic or vacuum device, a charged capacitor to force a premature current zero, a saturable reactor, and parallel-series resonant circuits.

HIGH-SPEED OPERATION

We have already seen that it is desirable to get rid of a short circuit as promptly as possible. About 30 years ago, the industry began to demand higher-speed breakers than were then available. Since that time, breaker-interrupting ratings have been progressively reduced to eight, then five, then three, and finally to two cycles (on a 60-Hz basis). These are the total times from energizing the trip coil until the interruption of current on the primary arcing contacts. Shorter arc durations ultimately mean less contact burning, resulting in less maintenance and a better breaker.

Faster interrupting ratings of circuit breakers, available in the transmission voltage, offer significant advantages to the transmission system. Faster switching time is advantageous not only in terms of system stability, but also from the standpoint of minimizing possible equipment damage during the fault period, and allowing faster reclosing times.

The term "immediate reclosure" means the immediate return to service of a circuit that has been tripped automatically. In this case, the breaker is reclosed without intentional delay. To be of much help to stability, it is essential that the breaker be reclosed just as quickly as possible. This requirement has brought about the development of breaker operating mechanisms that are not only high-speed opening, but are also high-speed reclosing. On high-voltage transmission circuits, the limiting item is fault arc deionization time.

EHV CIRCUIT BREAKERS

EHV substation breakers, utilizing different interrupting media, are available. Above 345 kV, however, oil is not used as an insulating medium. Most of these breakers have closing resistors which reduce the switching-surge voltage.

Some of the characteristics of the different types of EHV breakers are noted below:

1. Bulk oil breaker characteristics
   a. Old, established technology.
   b. Low cost.
   c. Easy to install.
   d. Low maintenance required.
   e. Closing resistors not applicable.
   f. Possible oil fire hazard.
   g. Slower interrupting time.
   h. Dead tank design.
2. Air blast breaker characteristics
   a. Substantial experience.
   b. Live tank design.
   c. High interrupting current capability.
   d. Extinguishes arc with high pressure, dry air.
   e. Fast interrupting time of one to two cycles.
   f. Relatively expensive.
   g. Requires high-pressure air storage.
   h. Considerable audible noise.
   i. Requires opening resistor.
3. SF_6 breaker—Two-pressure-type characteristics

a. Sulfur hexafluoride, SF_6, an electronegative gas with excellent dielectric properties, is used to extinguish arcs.
b. Requires SF_6 tank heaters at low temperatures.
c. Live or dead tank design.
d. Substantial experience with live tank design.
e. Requires high-pressure SF_6 storage.
f. High SF_6 leakage.
g. High interrupting current capability.
h. No opening resistors required.

4. SF_6 breaker—Puffer-type characteristics
   a. Simpler design than two-pressure type.
   b. SF_6 is used to extinguish arcs and provide insulation.
   c. SF_6 stored at low pressure (75 psig).
   d. No SF_6 tank heaters required down to $-30°C$.
   e. Limited operational experience in the United States up to 1983.
   f. Very low SF_6 leakage.
   g. Moderate interrupting levels without massive operating mechanism.
   h. No opening resistors required.

Modern EHV breakers have an average span of about two cycles from the time the relays energize the trip coil to complete interruption of the fault. Clearing times of this order are necessary in many instances to maintain stability when a fault occurs on the system. Faults that last for nine or more cycles generally produce instability.

Interrupting ratings are now up to 63 kA at 500 kV and 80 kA at 230 kV. Steady-state continuous current ratings up to 4000 amperes are available with some special designs built for 8000 amperes.

ADDITIONAL REQUIREMENTS OF BREAKERS AND SWITCHES

Circuit breakers and disconnects should not be blown open or otherwise damaged by short-circuit currents within their short-time ratings. The circuit breakers and disconnecting switches should be designed or protected to withstand normal operating voltages across the device in the open position.

Circuit-Breaker Standards

The importance of the circuit breaker cannot be over emphasized. If a circuit breaker should fail to perform its task, the results could be catastrophic. For this reason, there are numerous standards defining the capabilities of ac high-voltage breakers, circuit-breaker tests, preferred circuit-breaker ratings, switches, fuses, relays, assemblies, etc. Several of these are listed here. They are all ANSI publications.

C37.04 Standard Rating Structure for AC High Voltage Circuit Breakers Rated on a Symmetrical Current Basis.
C37.06 Schedule of Preferred Ratings for AC High Voltage Circuit Breakers Rated on a Symmetrical Current Basis.
C37.07 Interrupter Capability Factors for Reclosing Service.
C37.09 Test Procedures for AC High-Voltage Circuit Breakers.
C37.20 Switchgear Assemblies Including Metal-Enclosed Bus.
C37.32 Schedule of Preferred Ratings, Manufacturing Specifications and Application Guide for High Voltage Air Switches, Bus Supports and Switch Accessories.
C37.46 Specification for Power Fuses and Fuse Disconnect Switches.
C37.100 Definitions for Power Switchgear.

Application of circuit breakers within the guidelines defined in the standards will generally avoid breaker misapplication.

Types of Circuit Breakers

The appearances of several types of circuit breakers are illustrated in Figures 9.15–9.20.

1. Drawout air circuit breakers are used for the 220- and 440-volt circuits. They are used to feed small station auxiliary motors and lighting in substations and generating stations. They are furnished in metal-enclosed equipments and are shown in Figure 9.15.
2. For higher-voltage circuits of moderate interrupting requirement, there are magne-blast air and vacuum circuit breakers. These are available up

ADDITIONAL REQUIREMENTS OF BREAKERS AND SWITCHES

Figure 9.15. Low-voltage, drawout air circuit breaker.

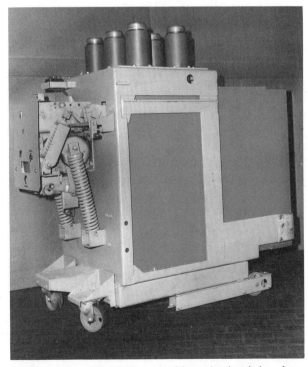

Figure 9.16. 1000-KVA magneblast air circuit breaker.

Figure 9.17. Vacuum metal-clad switchgear.

Figure 9.18. Metal-clad switchgear line-up.

to 1,000,000 kVA at 13.8 kV. This corresponds to an interrupting rating of 41.4 kA

$$\text{kVA} = \sqrt{3}\, V_{l-l} I_\phi$$

where V_{l-l} is system line-to-line voltage in KV and I_ϕ is system phase current in amperes. They are furnished in complete metal-clad equipments. Figure 9.16 illustrates an air magnetic breaker, and Figure 9.17 a vacuum breaker, with associated metal-clad equipment in Figure 9.18.

3. A typical tank-type, outdoor oil circuit breaker, used at distribution and subtransmission voltages, is shown in Figure 9.19.

4. The "dead tank" design, SF_6 puffer breaker is used for protection of transmission voltage systems of 115 kV and above. A 362 kV breaker is shown in Figure 9.20.

OUTDOOR GEAR

The early practice of housing all switchgear in a building has given way to the present practice of placing most of it outdoors. Even components that are basically suitable only for indoor use can be made weatherproof by a metal enclosure. When the gear is placed outdoors, building space is saved. This is an important item at the higher voltages. Also, the fire hazard is reduced. With circuit breakers of modern design, this is a less important consideration than it was some years ago when an automatic operation might blow the top off an oil breaker and shower the surrounding landscape with oil.

Obviously, there is no point in having the control board of an attended station outdoors. All other components are either built for outdoor service or can be suitably enclosed. The higher the voltage of a device, the more certain it is to be out in the weather.

Figure 9.19. Outdoor oil circuit breaker, 345 kV.

CONTROL POINT

Figure 9.20. 362-kV SF$_6$ puffer breaker.

CONTROL POINT

Supervision of an attended station or substation is exercised by an operator in the control room. Here will be found the control panels on which are mounted various devices that tell the operator what is happening. Included are indicating instruments, signal lamps, annunciator, etc.; also, the control switches and wiring that the operator uses to open and close the breakers. Panel boards have been built in many forms and shapes. Figure 9.21 shows a duplex board of modern design.

Figure 9.21. Duplex control board.

SUMMARY

The function of switchgear is to control and protect other electrical equipment and circuits. Switchgear consists of circuit breakers, disconnecting devices, instruments, relays, control switches, buses, panels, and assemblies of these devices. The largest single item is the circuit breaker.

Circuit breakers are the usual means of interrupting the flow of current in a circuit. Frequently the current to be interrupted is abnormally high because of a short circuit. A successful breaker must be so designed that there is enough dielectric strength between the stationary and moving contact to withstand the recovery voltage at current zero soon after the contacts part. Prompt interruption of the current minimizes contact burning. It also minimizes the undesirable effects on the system of long-duration short circuits.

Circuit breakers, disconnecting switches, current transformers, and other series devices must withstand the maximum current that flows in the circuit, must carry rated current within specified temperature rise, and must withstand the circuit voltages.

There are many types of circuit breakers; the exact form depends on the rated current, voltage, and interrupting capacity, and whether it is designed for indoor or outdoor service. The interrupting medium is either oil, vacuum, air, or arc-quenching gases such as sulphur hexafluoride.

The switchgear of large attended stations is controlled from a centralized location known as the control room where the control switches and relays, instruments, and indicating lights are mounted on panels.

REFERENCES

1. Flurscheim, C. H., editor, *Power Circuit Breaker Theory and Design*, IEE Monograph Series #17, Peter Peregrinus Ltd., Stevenage, Herts., England, 1975.
2. Lee, T. H., *Physics and Engineering of High Power Switching Devices*, MIT Press, Cambridge, Massachusetts, 1975.
3. Rieder, W., "Circuit Breakers—Physical and Engineering Problems I—Fundamentals," *IEEE Spectrum*, **7**, 35–43 (July 1970).
4. Rieder, W., "Circuit Breakers—Physical and Engineering Problems II—Design Considerations," *IEEE Spectrum*, **7**, 90–94 (August 1970).
5. Rieder, W., "Circuit Breakers—Physical and Engineering Problems III—Arc Medium Considerations," *IEEE Spectrum*, **7**, 80–84 (September 1970).

10
SUBSTATIONS

DEFINITION

"A substation is an assemblage of equipment for the purpose of switching and/or changing or regulating the voltage of electricity."* Wherever transformers or switchgear are located in a power system, there will probably be a substation. The substation can be large or small, and it can be tailor-made or factory fabricated. Control can be automatic or manual, and manual control may take place from a distance or by a local attendant. In addition, the substation can contain equipment to control the voltage at its location by supplying the proper amount of reactive power into the system. Figure 10.1 shows an example of a substation.

The purpose of a substation is to perform one or more of the following functions:

1. Switching. Connecting or disconnecting parts of the system from each other. This is accomplished using breakers and/or switches.
2. Voltage transformation. Power transformers are used to raise or lower the voltage.
3. Reactive power compensation. Shunt reactors, shunt capacitors, static var systems, and synchronous condensers are used to control voltage. Series capacitors are used to reduce line impedance.

In addition to fulfilling these functions, the substation may be designed to accommodate overvoltage and overcurrent protection, reliability, shielding, equipment-grounding procedures, and substation control.

* Glossary of Electric Utility Terms—Edison Electric Institute.

BUS ARRANGEMENTS

There are five basic methods for connecting buses, breakers, and circuits; and each method may be varied to suit particular requirements. The choice of arrangement depends largely on the cost, the application, and the required degree of service continuity and reliability.

Single Bus

The single bus arrangement is shown in Figure 10.2. It is simple and lowest in cost since only one circuit breaker per circuit is utilized. No transfer breakers and few disconnect switches are required. Relaying is relatively simple since the only requirements are relays on each of the circuits plus a single bus relay. Maintenance is rather difficult since most of the work must be done with the station completely or partially out of service. The single-bus reliability is the lowest of any substation arrangements, since a bus fault or breaker failure can shut down the entire station for an extended outage. Considerable improvement and flexibility can be obtained by adding bus tie breakers in the bus between transformers. This modified single-bus arrangement permits faulted bus sections to be isolated rapidly from the remainder of the station.

Main and Transfer Bus

The main and transfer bus, illustrated by Figure 10.3, is next highest in cost and in reliability. Some engineers prefer to call this arrangement a main and auxiliary bus; they restrict the designation "transfer" to a slightly different arrangement. The ar-

177

Figure 10.1. Outdoor substation—steel work composed of rolled-shape members. (Courtesy of City of Seattle, Department of Lighting.)

rangement shown in Figure 10.3 is widely used and is preferable to the single-bus arrangement since it permits insulators to be cleaned, breakers serviced, etc., without an interruption to service and with only temporary compromise of complete breaker protection. Breaker maintenance requires interruption of service to the circuit supplied by that breaker. While the transfer bus arrangement does not prevent a complete outage in case of a fault on the bus, it does provide means of readily restoring service.

In large or very important stations where the maximum practical reliability is required, the schemes that are used are the ring bus, the breaker-and-a-half, and the double bus-double breaker arrangement.

Ring Bus

The ring bus (Figure 10.4) is about the same in cost as the single-bus arrangement, but has the advantage of a double feed to all circuits. It uses the same number of breakers, fewer switches, and somewhat more steel. In addition to preventing a complete outage due to a bus fault, it permits maintenance of the circuit breakers without a circuit outage or a sacrifice in protection. The arrangement requires, however, that the ring be opened when a breaker is out for maintenance, thus increasing the load current through the remaining breakers. During this time, an automatic operation of an adjacent breaker would cause an unnecessary interruption to one circuit. More than one circuit might be interrupted if the number of circuits supplied by the bus greatly exceeded the number of circuits feeding the bus. For this reason, the use of a ring bus is questionable if the ratio of outgoing to incoming circuits exceeds two or at the most three.

If future expansion of a substation is planned, the initial construction may be a ring bus with adequate space provided for development of a breaker-and-a-half scheme as additional circuits are needed.

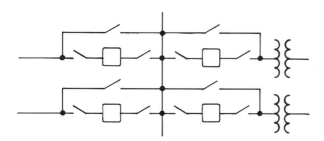

Figure 10.2. Single bus.

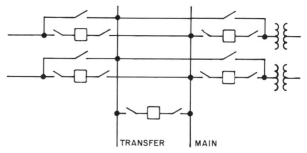

Figure 10.3. Main and transfer bus.

BUS ARRANGEMENTS

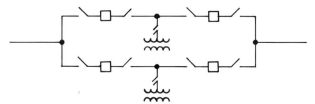

Figure 10.4. Ring bus.

Breaker-And-A-Half

The breaker-and-a-half scheme (Figure 10.5) has advantages for maintenance and reliability. All transfer switching is by means of circuit breakers. Each circuit has breaker protection, even with one breaker out for maintenance. A bus fault or breaker fault will have little or no affect on the reliability of the station since the fault can be isolated. With one breaker out, however, an automatic operation of an adjacent breaker may cause an unnecessary circuit outage, as in the ring bus arrangement. With a breaker out for maintenance, the increase in loading on the remaining breakers is slight. Two buses are available so one can be taken out of service for insulator cleaning or maintenance.

This scheme cannot logically be used on substations having only four circuits, since the result would be a ring bus with two extra breakers. It is also at a disadvantage for substations that have an odd number of circuits, although this is of small importance in large stations.

Using the breaker-and-a-half scheme, substation expansion is permitted without interruption of service. Primary relaying is more complicated than for a single-bus arrangement since it is necessary to open two breakers in the event of a circuit fault. However, bus potential is permissible for relaying if a potential through-over scheme is used when a fault occurs.

The breaker-and-a-half scheme is probably the most commonly used arrangement for large transmission and subtransmission substations which require good reliability at reasonable cost.

Double Bus-Double Breaker

The double bus-double breaker arrangement (Figure 10.6) is used when the utmost in service reliability, continuity, and ease of switching is required. It provides full protection for each circuit with any breaker out of service. No breaker operation can affect more than one circuit. During periods when one breaker is unavailable, the remaining breaker does not carry the load of more than one circuit. Also, the scheme uses two buses, assuring that one is available for maintenance and cleaning without disturbing service. This scheme is more expensive than any of the others and is used only at locations that are considered very important.

Some operators believe that the claimed increase in reliability with the more complex bus arrangement is not all realized. They contend that the loss of simplicity tends to increase the chance of an operating error.

EHV Bus Design

The massive transfer capabilities of EHV systems demand a level of reliability that must be flexible. Typical EHV substations will probably begin with a ring bus arrangement. This provides a flexible and dependable bus system without incurring excessive

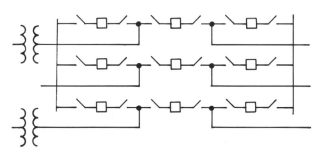

Figure 10.5. Breaker-and-a-half.

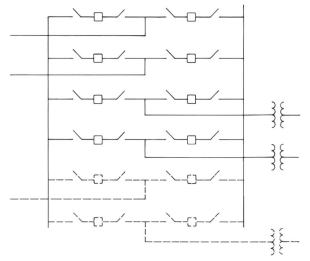

Figure 10.6. Double bus-double breaker.

initial cost of breakers. These stations are invariably designed to be converted to a breaker-and-a-half scheme at a later date as the number of terminal positions increases. Recently, there has been a growing trend to move directly into a breaker-and-a-half scheme.

CIRCUIT BREAKERS

In selecting the proper circuit breaker, three basic facts must be considered:

1. The continuous load current that the breaker must carry under normal or emergency conditions.
2. The short-circuit current that the breaker must interrupt.
3. The speed of short-circuit interruption.

The continuous load current can be obtained from substation loading data or from system load-flow studies. It is possible to overload the circuit breakers under certain operating conditions, but, generally, the loads are limited to nameplate rating. Short circuit data commonly determines the selection of the breakers. It is based on nominal three-phase megavolt ampere duty, or on the basis of rated short-circuit current.

As additional circuits are added to a substation each circuit breaker in the substation experiences the extra short-circuit duty. Consequently, it is necessary to base the selection of circuit breakers on future rather than present requirements. This selection procedure has resulted in a demand for symmetrical short-circuit ratings as high as 80 kiloamperes.

Generally speaking, the faster the breaker interrupts a system fault, the better it is for the system. Fast interruptions reduce the possibility of extensive damage and, in the case of transformers, reduce the possibility of an oil fire. A primary reason for faster circuit breakers on transmission systems is improved transient stability. This has led some users to request independent pole operation, which provides that even during a breaker malfunction, a three-phase fault will be limited to a single-phase fault until backup clearing occurs. In a breaker without independent pole operation, all three poles operate as one. The failure of any one pole prevents all poles from clearing the fault. A three-phase fault remains a three-phase fault until back-up clearing occurs. The consequence is reduced stability margin.

The final selection of breakers is usually based on economics, the type of breakers available in the voltage class being considered, and the advantages and disadvantages of competing interruption methods which are primarily oil, vacuum, air, and SF_6. Each of these are discussed in more detail in Chapter 9.

TRANSFORMERS

The selection of the proper transformer can have a major impact on the cost of a substation, since the transformer represents the major cost item. Nameplate rating is only a guide to transformer application, and should only be used as a first step in the selection process. The transformer is available as a self-cooled unit, or it can be purchased with additional steps of forced cooling that use fan or fans and oil pumps. Transformer ratings can be increased from 25% to 66% by the addition of fans and pumps. The nameplate rating is based on a continuous load producing a 55°C to 65°C conductor temperature rise over ambient. Since many transformers do not carry continuous loads, advantage can be gained from the thermal time lag to carry higher peak loads without exceeding the temperature limits.

Transformer ratings are based on the assumption that only an extremely slow deterioration of insulation will take place with normal operation. A substantial increase in rating can be achieved by accelerating the loss of insulation life. This increase in rating might approach 200% for an hour or two, and approximately 20% for 24 hours.

For substations that are designed to carry full load under the outage of any one transformer, a high emergency rating for a 24-hour period (e.g., until the failed unit can be replaced) could mean the selection of a smaller transformer and a substantial saving in station cost. A Power Transformer Loading Guide C57.92-198 (1 or 2) will cover 100 MVA ($33\frac{1}{3}$ MVA single-phase) transformers and under 100 MVA. The guide for over 100 MVA should be available by 1985.

The selection of the transformer should involve a careful evaluation of a number of other factors:

1. Impedances should be selected considering their effect on short-circuit duties and low side-

breaker ratings, both for initial and future station developments. In addition, impedance is important to achieve a proper load division in the parallel operation of transformers.

2. No load tap ranges should be selected to provide an adequate low-side bus voltage.
3. If the high-side or low-side voltages vary over a wide range during the load cycle, it may be necessary to provide bus regulation. The actual regulation can be calculated using the system and load characteristics. If regulating equipment is needed, it may be desirable to provide it in the transformer by using load tap-changing (LTC) equipment. If the need for bus regulation is not presently evident, but may be required in the future, it may be economical to leave space in the station for future regulators, and buy transformers without LTC equipment.

Autotransformers are used almost universally in EHV substations, even for transformations from EHV directly to subtransmission voltage levels. For this application, low impedance continues to be desirable. The autotransformer is self-protecting in the prescribed environment (see ANSI C57.00-1980).

REACTIVE POWER COMPENSATION

The flow of alternating current through an inductive electric circuit causes a voltage drop that can either increase or decrease the voltage magnitude, depending upon whether the current is leading or lagging the voltage. The variation in system load, from peak to minimum, may be two to one or even more. This means that the voltage drop from generator to consumer tends to vary in the same ratio. Good service requires that the customer's voltage be kept within prescribed limits. If customer voltage is maintained at a fairly constant value, despite a two-to-one variation in load, generator voltage must vary, or some intermediate device must produce a voltage correction. Variations in generator voltage are not acceptable in a network, since many load voltages would be affected by a change in generator voltage.

There are two basically different methods of circumventing the variations in voltage drop caused by variations in load. One is by changing the ratio of voltage transformation by means of the taps on a transformer, or by means of a feeder voltage regulator. The other method is by eliminating the principal cause of the voltage drop, the flow of reactive power through the circuit impedance.

Reactive power compensation equipment is often located in substations to control the power frequency voltage, to increase power transfer and to improve the efficiency of the adjacent portion of the power system. For example, either the series inductive reactance of transmission lines or the shunt capacitive reactance may be compensated. In addition, the vars taken by a load may also be "compensated." There are a number of types of compensation equipment available. The choice of equipment depends upon consideration of the particular application needs, the performance required and costs of the equipment.

Shunt Reactors

A shunt reactor is an inductor that is connected line to ground. It consists of one winding insulated for the proper operating voltage. This winding may or may not have a magnetic core. At lower voltages, the reactor might be "dry," while at higher voltage levels it will be oil immersed. In general, low-voltage reactors may have an air core, while higher-voltage units will have magnetic cores.

The application of shunt reactors in the power system has grown as a result of the advent and growth of EHV transmission lines. Shunt reactors are the primary form of compensation used on EHV systems. These reactors are applied to compensate for the undesirable voltage effects caused by the capacitance of the line. The degree of compensation provided by a reactor is usually quantified by the percentage of the line capacitance that is compensated. The percent shunt compensation of EHV lines in service ranges from 0% to 90% with the reactors located in the substations at one or both ends of the line.

The amount of shunt compensation required on a transmission line depends on the characteristics of the line, the anticipated line loading, system operating philosophy and other factors. The amount of compensation is determined by means of load flow and stability studies. The capacitive reactive power associated with a transmission line increases directly as the square of the voltage and is proportional to line capacitance.

The capacitance of a line has two related voltage effects. One is the rise in voltage along the line resulting from the capacitive current of the line flowing through the line inductance. The second effect is the rise in voltage resulting from the capacitive current of the line flowing through the source impedances and the generator field magnetization. These effects are corrected by the generator voltage regulators. If the line delivers too much charging current, the generator field excitation will become very low which reduces the stability limit and is unacceptable. These voltages can be reduced by the application of shunt reactors.

Reactors may be either directly connected to the transmission line or connected on transformer tertiary windings, each location offering specific system benefits. Tertiary reactors can be economically switched as system reactive power requirements and voltage profiles vary. Permanently connected line reactors offer the advantage that they cannot be separated from the compensated line section during line-switching operations. Line reactors almost always reduce line-switching surge magnitudes.

EHV shunt reactors have been furnished at 345 kV, 500 kV, and 765 kV. Typical single-phase sizes have been 50 Mvar at 500 kV, and 100 Mvar at 765 kV. Three-phase shunt reactors have been furnished in sizes up to 150 Mvar at 345 kV and 70 Mvar at 500 kV. Tertiary reactors are generally three-phase with a typical size being 25 Mvar at 13.8 kV.

Shunt Capacitors

In planning system kvar supply, many utilities today are focusing their attention on bulk kvar supply at the substation to meet system needs for emergency conditions as well as for normal load situations. Over the past 60 years, installations of shunt capacitor banks have had rates of growth significantly greater than the rates of growth of active power generation.

An estimated 60% of the industry-installed kilovar supply is on distribution feeders in the form of pole-mounted capacitor banks, and about 30% is in relatively small distribution substation banks. The remaining 10% of the installed capacity is in larger transmission and subtransmission substation equipments, and the current trend is toward increasing use of these larger higher-voltage banks. Shunt capacitors are only occasionally applied on an EHV system.

Substation capacitor bank design economics dictate the use of individual units assembled in appropriate series and parallel connected groups to obtain the desired bank voltage and capacity rating.

The industry has undergone many stages of improvement since the first application of capacitors for power-factor correction in 1914. The introduction of thin Kraft paper to replace linen about 1930 made possible individual unit ratings up to 10 kvar. A few years later, a 15 kvar with askarel instead of oil was standard. By the early 1960s, the 100 kvar rating was introduced, evolving from comparatively small, very costly refinements in the basic paper/askarel dielectric. In 1965, a 150-kvar unit using a paper/polypropylene film/askarel dielectric system was introduced. Further refinements made possible individual unit ratings up to 600 kvar. Today, however, the most economical unit rating is the 200-kvar size. Recent replacement of askarel with non-PCB fluids has not had a significant effect on unit sizes or ratings. The voltage rating of individual units range from 2.4 to 21.6 kV rms.

Distribution voltage-class capacitor banks can be connected in grounded wye, ungrounded wye, or delta. Transmission banks are generally connected grounded wye to lessen the recovery voltage requirements on the capacitor switch or breaker.

The basic purpose of a shunt capacitor bank is to increase the local circuit voltage and/or improve the load power factor as seen by the circuit carrying this load. Many large capacitor banks are switched on and off as the system need for reactive kilovolt amperes changes. System requirements govern whether a certain bank should or should not be switched. If the voltage at the capacitor would otherwise be too high during light load, some or all of the capacitors are switched off. Very large banks are usually switched in steps. This procedure has the disadvantage of requiring more switches and thus increasing the total equipment cost per kilovar. It does, however, provide a means of keeping the voltage change per step within permissible limits. Care should be taken in selecting switches and breakers for capacitor switching. Automatic equipment that responds to time, voltage, kilovars, or combinations of these quantities is available to control the switching. Control by time of day is the most common.

It is general practice to fuse each unit in a bank. When a unit develops a short circuit, its fuse interrupts the fault current, taking only that unit out of service. This is done quickly enough to prevent rupture of the unit which could cause serious damage.

REACTIVE POWER COMPENSATION

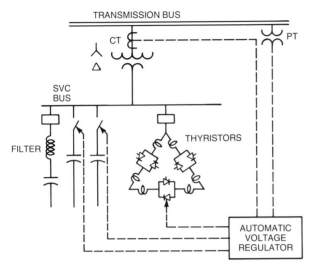

Figure 10.7. Static var control power and control circuit.

The bank remains in service and the faulted unit can be replaced at the convenience of the station servicepeople.

Pole-top capacitor units do not have individual unit fuses, since this is not justified economically. Each phase has parallel capacitors with one terminal of each unit connected to ground. The other terminals are all connected together, and are connected to a phase conductor through one fuse.

For personnel safety, each capacitor has an internal resistor that will discharge an unenergized unit.

Static Var Control

Static var control (SVC) equipment utilizes reactors and capacitors in combination with high-power semiconductors to provide a source of controlled reactive power.

The circuit for one SVC is shown in Figure 10.7. This design facilitates voltage control of a transmission bus. A transformer interface is used to allow the SVC to be designed for practical voltage levels such as 13.8 kV or 34.5 kV. Solid-state power switches or thyristors are connected in series with reactors and the thyristors are phase controlled to allow for continuous adjustment of the inductive current. The capacitor can be fixed or varied by switching discrete capacitor segments in and out as dictated by system conditions. The automatic voltage regulator (AVR) is programmed to regulate the transmission bus voltage within a preselected tolerance, despite varying transmission conditions. The AVR performs this function by controlling the duration of the current through the inductor on each half cycle via the thyristors. If a number of capacitor segments are employed, the AVR would also orchestrate their switching as required. In addition, Figure 10.7 shows filters that consist of series inductor-capacitor branches. These may be required where harmonic currents caused by the thyristor phase control of the reactor current must be filtered to prevent unacceptable transmission bus voltage distortion.

A SVC controls the voltage by varying the reactive current drawn by the combination of capacitors and reactors. An important advantage of this equipment is speed. The SVC equipment can respond to a system change in a few cycles. As a result, this equipment has application in reducing voltage flicker resulting from load variation; it also provides dynamic voltage control for the transmission system during major disturbances. The latter function can promote system stability.

The first application of an SVC to provide flicker control for an arc furnace was in 1973. Since then, numerous units have been installed to solve large industrial load problems. This equipment is also being applied to solve transmission system problems.

Synchronous Condensers

Synchronous condensers have played a major role in system voltage and reactive power supply for more than 40 years. In recent years, there has been increased interest in the application of synchronous condensers to solve transmission problems. This has led to the installation of units up to 250 Mvar.

A synchronous condenser is basically a synchronous motor running at synchronous speed with no mechanical load. The condenser has a control circuit that controls the field excitation to provide voltage control.

When the system voltage starts to fall below the desired values, the control circuit will automatically increase the field excitation which causes the synchronous condenser to supply vars to the system. This will increase the system voltage at that point. A reduction of the system voltage can be accomplished by decreasing the condenser field current. In the "overexcited" mode, the synchronous condenser acts electrically like a capacitor. If the sys-

Figure 10.8. Installation of a 167-MVA hydrogen-cooled synchronous condenser, including cooling towers, control house, and transformer. (Courtesy of Philadelphia Electric Company.)

tem voltage starts to rise above the desired value, the control circuit will automatically decrease the field excitation and reduce this high voltage by absorbing reactive power. In the "underexcited" mode, the condenser acts electrically like an inductor on the system. Changes in the output of the condenser are smooth and automatic.

The rating of a synchronous condenser is the magnitude of reactive power that the unit can supply continuously at rated voltage. In addition, condensers typically have the capability to supply above rated reactive power for short periods of time such as 1.5 times rating for one minute. The ability of the condenser to absorb reactive power is generally about 35–40% of its rating. The speed of response of the synchronous condenser is greatly influenced by the effective time constant of the field, which is typically 2–3 seconds.

The rated voltage of a synchronous condenser is in the order of 13.8 kV. Therefore, for transmission applications, a transformer is used. A typical installation is shown in Figure 10.8, together with its major auxiliary equipment components. Condensers are generally salient pole machines and modern units are designed for operation at a speed of 720 or 900 r/min.

Most modern condensers are totally enclosed, hydrogen-cooled units, containing integral hydrogen-water heat exchangers (coolers) suitable for outdoor installation. They have direct connected main exciters (no pilot exciter) operating in the hydrogen gas. An advantage of the hydrogen-cooling is lower operating losses.

Practical starting methods for large condensers include reduced-voltage starting motor and static starter. Full-voltage or across-the-line starting is not recommended for large condensers, both from the standpoint of system voltage dips and the severity of condenser duty. Generally, starting is infrequent, since the units are shut down only at planned maintenance intervals. Likewise, starting time is generally not critical, with 15–20 minutes being acceptable.

Series Capacitors

All shunt compensation methods are effectively connected line to ground; these methods influence the power system by absorbing or supplying reactive power. Series capacitors are unique in that they are connected in series with the line to reduce the total effective series impedance of the line. Series capacitors are very effective in increasing power transfer and reducing load voltage variations since they directly cancel some of the series reactance of the line.

Series capacitors were originally applied to distribution circuits to reduce voltage regulation; however, this application is not common today. Instead, their application on EHV transmission has grown, particularly in the western United States, because series capacitors permit increased power transfer at a modest cost. On occasion, this EHV application has resulted in a new power system problem referred to as subsynchronous resonance. Subsynchronous resonance will occur in some highly compensated transmission circuits; this problem can be defined as an oscillation occurring at a frequency of less than 60 Hz. Its most serious side-effect is to damage steam turbine-generator units, usually by flexing, even breaking, the shaft.

Series capacitors are installed on platforms insulated from ground for the system line-to-ground voltage. The rating of a bank is related to the maximum continuous current anticipated for the bank. The ohmic value of a series capacitor bank is generally between 30–70% of the total series reactance of the line in which it is located.

A system fault can result in excessively high voltages across the series capacitors. To prevent damage from overvoltages, series capacitor installations have overvoltage protective devices. These

OVERVOLTAGE PROTECTION

devices consist of special gaps, switches, and, recently, nonlinear resistors all mounted on the platform with the capacitors and connected across the capacitors, i.e., in parallel.

PROTECTIVE RELAYING

It is common practice to provide protective relaying for all of the major equipment in a substation. The protection provided will always include primary and back-up relaying for the detection of short circuits and, in some instances, relaying for the detection of abnormal operating conditions. The protective systems are designed to detect abnormal conditions quickly (commensurate with system requirements) and to open a minimum of circuit breakers to isolate the defective equipment. (See Chapter 12.)

OVERVOLTAGE PROTECTION

Surge Arresters

Surge arresters are devices that are capable of conducting high-surge currents (such as the current from a lightning stroke) harmlessly to ground without producing an excessive voltage across the arrester. In this manner, they prevent excessive voltage from appearing on equipment insulation. The surge arrester is connected from each phase conductor(s) to ground. It may consist of gaps in series with silicon carbide nonlinear resistors or it may be one of the newer arresters utilizing zinc oxide nonlinear resistors without series gaps.

The gap silicon carbide type of high-voltage arrester is made of a number of similar series units. Each unit contains a single nonlinear resistor and multiple gaps in series. The gaps spark over when the voltage approaches the protective level of the arrester and current begins to flow to ground. The arrester protective level is the voltage across the arrester, which the manufacturer guarantees will never be exceeded either before or after arrester spark-over. The arrester is designed to spark over at less than the protective level voltage for any surge at the arrester terminals. In addition, the voltage across the arrester following spark over is approximately the same magnitude as the spark over and is dependent upon current magnitude. When the current through the arrester drops to a sufficiently low value, the gaps reseal (i.e., open) stopping further flow of current through the arrester.

The series gaps in this type of arrester may be either current limiting or noncurrent limiting. The arrester, using a noncurrent-limiting-type gap, limits current following spark-over for the nonlinear resistance element only. As the current goes through zero, the gap tends to prevent further current flow, unless the prior current through the gap was very high, leaving excess ionization in the gap. Surge arresters with current-limiting type gaps utilize the arc resistance build-up capability of the gap in series. The nonlinear resistance limits the power-follow current that flows after arrester spark-over, and improves the arrester reseal capability.

The protective level of an arrester is directly related by design to arrester rating. The arrester rating is established in its application to the normal 60-Hz voltage and to the temporary overvoltages anticipated at the arrester location. A series gap arrester is designed to reseal after an operation if the system voltage is less than or equal to its rated voltage. If the system voltage is higher, the arrester may fail. Generally for this type of arrester, the rating is selected to be greater than the line-to-ground voltage during the temporary overvoltages resulting from system faults. In addition, arrester rating should be at least 1.25 times the maximum continuous line-to-ground system voltage. Arresters are not designed to withstand their rated voltage continuously.

The important characteristics of a surge arrester constructed with series gaps are the following:

1. Its spark-over on impulse or switching surge voltages.
2. The voltage across the arrester (*IR* drop) during passage of various values of impulse current.
3. The maximum 60-Hz voltage at which the gap will reseal.

The introduction of zinc oxide nonlinear resistors has led to the development of surge arresters that do not have series gaps. These resistors are significantly more nonlinear than are silicon carbide resistors. High-voltage arresters are constructed by stacking disks of this material in series until the desired volt ampere characteristic is achieved. Since there are no series gaps to spark over, this arrester will immediately begin to conduct more and more current to ground as the voltage across it approaches the protective level. As the surge is dissipated and the voltage across the arrester decreases, the arrester current decreases to a low leakage value. Significant arrester current flows only during overvoltages. The current magnitude is in accordance with the arrester volt-ampere characteristic. There is no large power-follow current that persists until a power frequency current zero is reached. The absence of the series gap eliminates the gap clearing problems when the arrester is exposed to temporary overvoltages in excess of its rating.

The important characteristics of zinc oxide arresters without series gaps are nearly the same as those of the series gap-type arresters, except that the spark-over and reseal considerations are absent. Instead of spark-over and gap reseal, consideration need only be given to the arrester discharge voltage (*IR* drop) for various values of current and thermal equilibrium of the arrester material. Depending on the arrester thermal design and the characteristics of the zinc oxide material, it is possible to operate this type of arrester for extended periods of time at voltages in excess of its rating. The allowable duration of temporary overvoltages can be determined from the published arrester characteristics. The application of these arresters must be consistent with their ability to withstand normal line-to-ground system voltage continuously.

Neutral Grounding

The method of grounding the neutrals of the transformers affects the degree of protection that can be given them and the amount of insulation they need. If the system neutral is not grounded at all, and a ground occurs on one conductor, the arresters on the other phases may have to operate and cease conduction with full line-to-line voltage. For this condition, a "100 percent arrester" is often selected. This means that the arrester rating is 100% of the system line-to-line voltage.

The voltage on the healthy phases, during a single line-to-ground fault, will be closer to normal if the neutrals are solidly and directly grounded everywhere, without the insertion in the ground connection of any intentional impedance. In such a case, the arrester rating could be substantially lower.

The coefficient of grounding is the ratio of the maximum sustained line-to-ground voltage during faults to the maximum operating line-to-line voltage. For many years, the accepted ratio was 0.8. A 115-kV system (121-kV maximum operating voltage) used 121-kV arresters if its neutrals were isolated, but only 97 kV (80% rating, i.e., $0.8 \times 121 = 97$) if its neutrals were permanently and solidly grounded. Most equipment rated 230 kV and above is designed to operate only on an effectively grounded system. An effectively grounded system is one in which the coefficient of grounding does not exceed 0.80. At higher voltage levels, insulation is more expensive and, therefore, more economy results from reducing the insulation. If the insulation is to be as well protected after it is reduced, the arrester rating must be correspondingly reduced. For the higher voltages, a ratio as low as 0.7–0.75 is common. Factors such as system-resonant conditions, as well as overvoltages during faults, affect this decision and must be considered on each individual system.

Insulation Coordination

Insulation coordination is the process of determining the proper impulse insulation and switching-surge strength required in various electrical equipment together with the proper surge arrester. This process is determined from the known characteristics of voltage surges and the characteristics of surge arresters.

To most engineers, insulation coordination now means (a) the selection of the minimum arrester rating applicable, based on system conditions and (b) the choice of equipment having an insulation level that can be protected by the arrester.

Standards have been established covering insulation levels for transformers, breakers, and related apparatus, corresponding to each standard 60-Hz voltage level. These are known as "basic impulse insulation levels" (BIL) and are recognized throughout the industry. Various groups are giving this whole subject continuing study.

The maximum impulse voltage that can appear at a point within a substation is usually the protective level that is held by the arrester. This voltage has some assumed value of impulse current through it, plus an additional voltage related to the distance between the arrester and that point.

The impulse insulation level of a piece of equipment is a measure of its ability to withstand impulse voltage. It is the crest value, in kilovolts, of the wave of impulse voltage that the equipment is to withstand. Usually figures are given for the 1.2×50 μsec full wave.

ADDITIONAL SUBSTATION DESIGN CONSIDERATIONS

The switching-surge insulation level is the dominant factor on extra-high-voltage (EHV) systems. This results from the fact that the switching surge strength of most external insulation is lower than the corresponding impulse strength and also because the switching surge strength does not increase linearly with increased air gap distance. Modern EHV arresters must be capable of discharging these high-energy switching surges. The maximum switching surge level used in the design of a substation is either the maximum surge that can occur on the system or the protective level of the arrester, which is the maximum switching surge the arrester will allow into the station.

ADDITIONAL SUBSTATION DESIGN CONSIDERATIONS

The design of a high-voltage substation includes basic station configuration and physical system layout, protective relaying systems, lightning protection, circuit-breaker selection, transformer selection, switches, bus designs, lightning shielding, and grounding.

The fundamental information required in determining the selection of a basic system configuration is as follows:

1. Consideration of initial and future loads to be served.
2. A study of the transmission facilities and operating voltages desired.
3. Examination of service reliability requirements.
4. Determination of space availability for station facilities and transmission line access.

Space and Reliability

Items (1) and (2) are usually known or can be estimated. Items (3) and (4) are interrelated and are the most difficult to determine because improved reliability generally requires extra equipment which, in turn, needs more space.

Figures 10.9, 10.10, and 10.11 show the relationship between reliability and station space requirements.

The bay size depends on the voltage, and varies from about 24 square feet at 34.5 kV to about 52 square feet at 138 kV.

More substation area will usually be required to raise the substation reliability. This also increases the substation cost. The determination of reliability requires a knowledge of the frequency of outages of each piece of equipment and the cost of the outage.

Failure rates for substations vary widely with substation size, degree of contamination and the definition of what is a failure. Transformer failures are currently in the range of 0.3–1.0% per service year.

Shielding

Vertical masts, lightning rods, and shield wires are commonly used to intercept direct lightning strokes before they reach the substation equipment.

Shielding incoming lines with overhead ground wires is also advised. The lines, connected to the substation, may have overhead ground wires

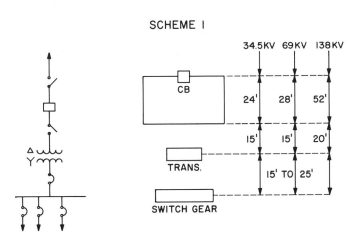

Figure 10.9. Single supply line and one transformer.

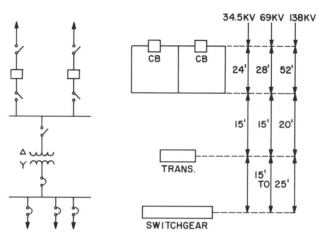

Figure 10.10. Two supply lines and one transformer.

throughout the entire line. In low-isokeraunic level areas, it is customary to use overhead ground wires on 0.5–1 mile (0.8–1.6 km) of line nearest the substation. In this case, the traveling waves caused by lightning strokes must travel under 0.5–1 mile (0.8–1.6 km) of ground wire to reach the substation. This assumes no shielding failure. When the wave travels under the ground wire, the front of the wave (the rate of rise) and the peak magnitude are both reduced. This makes the voltage surge less severe on the insulation in the station. The lightning currents generally assumed for surge arresters take this reduction into account. When no shielding is provided, higher lightning discharge currents must be assumed for the arresters. It is important that the arresters are located as close as possible to the transformer terminals.

Grounding

During a fault, the flow of current to earth will result in voltage gradients on the surface within and around a substation. Unless proper precautions are taken, the voltage differences along the ground may be great enough to endanger a person walking there. In addition, such voltage differences can sometimes exist between "grounded" structures or equipment frames and the nearby earth.

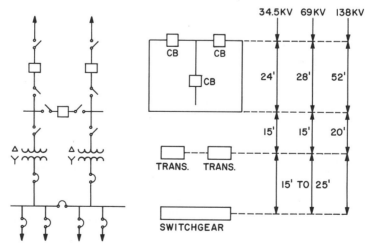

Figure 10.11. Two supply lines and two transformers.

OTHER TYPES OF SUBSTATIONS

As a result of these concerns, it is common practice for substations to have an electrical ground system consisting of a grid of horizontal, buried conductors. The wide use of the grid is due to the following advantages:

1. In systems where the maximum ground-fault current may be very high, it is seldom possible to obtain a ground resistance so low as to insure that the total rise of the grounding system voltage, during a fault, will not reach values unsafe for human contact, e.g., "step voltage." Therefore, the hazard can be corrected only by control of local potentials. A grid is usually the most practical way to do this.

2. In a substation of any size, no ordinary single ground electrode is adequate to provide needed conductivity and current-carrying capacity. However, when several of these are connected to each other, and to structures, equipment frames, and circuit neutrals that are to be grounded, the result is naturally a grid. If this grounding network is buried in soil of reasonably good conductivity, this network alone may provide an excellent grounding system.

Some power companies depend on the use of a grid alone. However, ground rods are relatively inexpensive and are useful if the resistivity of the upper layer of soil in which the grid is buried is of much higher resistivity than that of the soil beneath.

OTHER TYPES OF SUBSTATIONS

Primary Unit Substations

A unit substation is a factory-fabricated substation in which one or more step-down transformers are mechanically attached to metal-clad switchgear. These units offer some equipment cost advantages and reduction in assembly work at the site.

The unit substation is particularly applicable to small- and medium-sized step-down installations involving two voltage levels, at least one of which is not greater than 34.5 kV. This class includes the distribution substations that feed the primary distribution feeders.

Unit substations having a secondary voltage between 2400 and 15,000 are known as primary unit substations. They constitute the connecting link between subtransmission circuits and primary distribution circuits. Hence, the usual secondary voltages are the primary distribution levels of 2.4, 4.16, 7.2, 12.5, and 13.8. The supply voltages in common use are 12, 13.2, 14.4, 23, 34.5, 46, 69, 115, and 138 kV.

The transformer used in primary unit substations is three-phase and is usually oil insulated. Cooling is by means of tubes and fans. The fans enable the unit to continuously carry a load that is 25% over the self-cooled rating. For large transformers, both fan and oil pumps can provide continuous operation at 67% above the self-cooled rating. These ratings are available for peak load or emergency conditions; they provide added capacity without a proportional increase in cost.

When the incoming lines terminate in metal-enclosed switching equipment, the transformer busings are brought out the end of the tank to a metal-enclosed transition compartment or throat for direct connection to the bus in the switching equipment. When the power comes in through a cable that terminates in a junction box or switch on the transformer, the bushings are brought out the end of the tank into the junction box. Otherwise, cover bushings are used.

Unit-substation transformers are usually built with load-tap-changing equipment which is automatically controlled by a voltage-regulating static-control package. The standard equipment provides $\pm 10\%$ voltage change in 32 steps of $\frac{5}{8}$ percent each. A line-drop compensator is included to permit holding voltage at the load rather than at the substation. It is energized by current from a current transformer mounted under oil in the main transformer tank; the potential transformer is located in the metal-clad switchgear.

Mobile Substations

The mobile unit substation is a primary-unit substation mounted on a trailer or semi-trailer. It combines incoming switching and/or fuses, surge arresters, transformer, and low-voltage switchgear similar to the master substations (see Figure 10.12). A mobile substation usually includes several voltage combinations so as to be usable at a number of points on a customer's system. It is limited in size by the permissible weight allowed on the highway, as well as its length and width. The viability of the mobile substation as a reserve unit has played an important part in the acceptance of three-phase transformers

Figure 10.12. Mobile substation.

and in the use of radial unit substations. These mobile units have materially lessened the fear of transformer or bus failures and have greatly shortened such outages when they occur. Their greatest advantage is in serving temporary loads and substituting for fixed substations during planned outages for preventive maintenance or system cutovers.

Load-Center Unit Substations

The major problems of secondary distribution within factories and large buildings are voltage regulation, continuity of service, interrupting requirements, and cost of low-voltage cables. Voltage regulation is best and cable costs lowest when power is carried as near as possible to the center of the load area at high voltage, and there stepped down for distribution over short secondary cables. This philosophy leads to the use of small substations. Keeping these substations electrically separate on the secondary side holds down the interrupting duty, and, therefore, breaker costs are low. Normally open tie breakers between the low-voltage buses assure acceptable continuity of service. This describes what is known as the load-center unit substation, which has a secondary rating of 600 volts or less. It is used to supply 208-, 240-, and 480-volt loads.

Load-Center Unit Transformers

Transformers for use in load-center units may be liquid-filled or dry types. The latter may be either ventilated dry-type construction or sealed dry type. Oil-filled transformers are used in those few installations that were made outdoors or in suitable vaults indoors. Most load-center units are installed indoors. Silicon liquid-filled (Figure 10.13) Freon vaporization-cooled (Figure 10.14) or dry-type (Figure 10.15) transformers are commonly used to avoid the expense of fireproof vaults. A dry-type transformer may be used indoors without a vault only when no combustible material is within specified distances vertically and horizontally. Because this transformer has an insulation level roughly half that of a liquid-filled unit, it is not used where it is directly exposed to lightning. The silicon liquid-filled, the Freon-cooled, and the sealed dry-type units have a completely sealed tank and are preferred for application in dirty, moist, or corrosive atmospheres.

SF_6 Substations

Another type of substation applied to HV and EHV systems makes use of sulfur hexafluoride, SF_6, as

UNATTENDED SUBSTATION CONTROL

Figure 10.13. Load center unit substation with liquid-filled transformer.

the insulating medium instead of air for the substation bus. The SF_6 gas has superior insulating properties compared to air so that insulation distances can be considerably reduced. The bus generally takes the form of an isolated-phase coaxial conduit with an inside conductor line and an outside conductor ground. The space between the line and the ground is filled with SF_6 under pressure. The bus is equipped with special interfaces allowing direct connection to other equipment, such as breakers. Air-to-gas entrance bushings allow line connection.

The SF_6 substation is superior to the conventional air insulated substations because it requires substantially less land area and the high voltage is enclosed within a grounded enclosure.

UNATTENDED SUBSTATION CONTROL

Growth and Advantages

The development of a dependable automatic reclosing relay made the unattended substation practical. Automatic operation, without the continuous presence of an attendant (operator), was first tried on the synchronous converter substations supplying an interurban electric railway in 1914. In the early 1920s, the development of automatic reclosing relays, protective relays, and means for automatic control of voltage made it possible for ac substations to be completely automatic.

During the period when the substations had to be controlled manually, it was necessary that they handle enough power to justify the operator's wages. This dictated large stations, frequently serving large areas by means of many feeders. The feeders that served the more distant parts of an area had to go a long way before they picked up any load.

Furthermore, as the load on a large substation grows, transformer capacity has to be added. This increases the short-circuit current on the secondary side; sometimes the short circuit exceeds the interrupting capacity of the circuit breakers already installed. Appropriate remedies include bus sectionalizing, or the installation of reactors or some combination of the two. Such measures are frequently expensive and inconvenient and they constitute an undesirable limitation on operating flexibility.

Once the unattended substation became a reality, the substation could be made smaller, located nearer the load, and long feeders were no longer necessary. A new method of distribution had become available since load growth could now be accommodated by installing another substation that could carry parts of the area formerly carried by adjacent substations. Usually, there is no interconnection between the circuits energized by the various substations. Consequently, there is no increase in short-circuit severity. However, at the distribution level, tie switches are used to switch feeders or feeder sections between transformers in the same station or to adjacent substations for substation transformer load management.

Automatic control has been applied to larger and larger substations, and to many hydroelectric generating stations. It is also used on individual circuits and equipment in many attended stations. For example, even in many stations where an operator is on duty, outgoing overhead feeders are controlled by automatic reclosing relays.

Under some circumstances, it may be impractical to make a station completely automatic. These circumstances could exist because of technical, ec-

Figure 10.14. VapoTran® load center unit substation with Freon vaporization-cooled transformer.

Figure 10.15. Load center unit substation with sealed dry-type transformer.

onomic, or safety considerations, or because of reasons related to intercompany agreements or labor relations. A station need not be attended even though it is not fully automatic. It is possible to control a station from another location, such as a larger attended station or the system operating office. This can be done by means of supervisory control (a form of remote control).

Supervisory Control and Data Acquisition (SCADA) Systems

Supervisory Control and Data Acquisition (SCADA) systems are an indispensable tool in the operation of today's power systems. Through advancements in technology, most SCADA systems today operate in a scanning mode, providing multistation operation, continuous self-checking of both the communication path and remote station equipment, and high-speed data acquisition.

SCADA requires two-way communication between the master station, normally at a higher-level control center on the power system, and stations with a remote (SCADA equipment), which are often at the substation level.

SCADA comprises one or more of the following functions:

1. Alarm conditions, such as fire alarm, illegal entry, over-temperature, low battery voltage, and indication alarm of any uncommanded change-of-state.
2. Control and indication including:
   a. Control of, and indication from a device having only two positions. Examples are circuit breakers (tripped or closed) and motor starters (stopped or started).
   b. Control of, and indication from a device having three positions. An example is a valve (closed, midposition, and open).
   c. Indication without control of a device having two or three positions.
   d. Indication with memory capable of storing single or multiple changes of status that occur between scans.
   e. Control without indication of a device. Examples are tap-changing transformers (tap position is brought to the master by telemetering means) and raise–lower control of generator load.
   f. Set-point control to provide a set-point reference to a controller located at a remote.
3. Data acquisition including recording or displaying:
   a. Analog quantities. Examples are ac watts, vars, volts, and amperes.
   b. Kilowatt-hour data from digital pulses.
   c. Digital data such as levels that are digitalized by external devices and read digitally by the SCADA equipment.
4. Sequence-of-events that recognize the occurrence of predefined events, associated time of occurrence with each event, and presentation of the event data in the order of occurrence of the events.

By means of a SCADA system, an operator at a dispatch center can cause operations such as the opening and closing of breakers, the starting and stopping of condensers, and the changing of the taps on load-ratio-control transformers. The operator can receive an indication that the operation has been completed. All of this can be done over a voice-grade communication channel. It is the use of voice-grade, two-way communication channels that distinguishes supervisory control from direct-wire remote control; the latter requires one direct-wire circuit for each controlled device. The term supervisory control is seldom used by itself today. The usual reference is to Supervisory Control and Data Acquisition systems or SCADA systems.

SCADA—A Supplement to Automatic Control

SCADA equipment supplements automatic control. All of the protective relays and most of the control relays required for automatic control are also necessary when supervisory control is used. Only the initiating devices may be different or omitted. For example, a fully automatic synchronous condenser is started and stopped by an initiating device responsive to voltage or current or both. In addition, there are many protective and control relays. If the condenser is to be controlled by supervisory control, all the other devices are still necessary, and only the initiating devices can be omitted.

With the recent revolution in microelectronic performance and costs, digital automation systems are now being developed for commercial installation on utility systems. These automation systems provide an integrated systems concept for protection, control, and monitoring functions.

TABLE 10.1. Transmission Substation Automation Functions

Line protection
Synchronism check
Automatic reclosing
Transformer and bus protection
Shunt reactor protection
Automatic bus sectionalizing
Voltage and var control
Sequence of events
Fault distance and reporting
Substation transformer load monitoring
Breaker health monitoring
Data acquisition, monitoring, alarm and status, and logging

TABLE 10.2. Distribution Substation and Feeder Automation Functions

Feeder deployment switching and automatic sectionalizing
Substation transformer load balancing and load monitoring
Automatic bus sectionalizing
Integrated volt/var control
Time overcurrent and instantaneous overcurrent protection
Synchronism check
Underfrequency protection
Automatic reclosing
Breaker failure, bus fault, and transformer protection
Downed conductor protection
Data acquisition, monitoring, alarm and status, and logging

The approach utilizes distributed microprocessor-based modules located within the substations and out on the transmission and distribution system. This provides protection and control decisions out on the system where the data is located. Reporting to the operator is done automatically "by exception," but the operator retains the option of overriding or initiating action of his own.

By automatic transmission and distribution systems, major benefits will be obtained. These include greater utilization of existing facilities, reduction of communication requirements, deferral of major system investments, optimized operations and maintenance, improved system reliability, more efficient system operation through lower losses, rapid restoration of service after a fault, and more timely, significant data on utility system conditions. A totally integrated approach to automation is achieved.

The functions implemented by using an integrated systems approach represent an improved concept over present approaches, or entirely new functions that had not been feasible to implement previously. These functions may be classified as:

1. Transmission substation automation functions.
2. Distribution substation and feeder automation functions.

A tabulation of these functions is listed in Tables 10.1 and 10.2.

SCADA Operation

SCADA provides two-way communication between the master station and remote station. A representative SCADA systems is shown in Figure 10.16. Only one remote station is shown, but other remotes are similar. SCADA supplements rather than replaces fully automatic control or manual control of a substation.

SCADA is a system for the selective control and automatic indication of remotely located substations. It utilizes a digital code to transmit and receive information. This information can be commands to remote locations (TRIP-CLOSE, position, set points, etc.) or it can be the transmission of status indication, position, and analog or digital data from the remote to the master station.

Upon initiation at the master station, the SCADA equipment performs the desired operation according to the logic and security incorporated in the SCADA design. As shown in Figure 10.16, the master station provides capability for several operation functions and efficient means of display and logging. The remote station provides the interface for the points to be monitored or controlled. Communication is provided by means of a two-way modem (modulator/demodulator) suitable for operation over many different types of communication medium.

The digital code between the master station and remote uses a word structure containing synch bits, function address, address modifier, cyclic code-check bits, and stop bits. The communication bit rate between the master and remote may vary from 600 to 2400 bits per second. There are also various code forms and security measures incorpo-

UNATTENDED SUBSTATION CONTROL

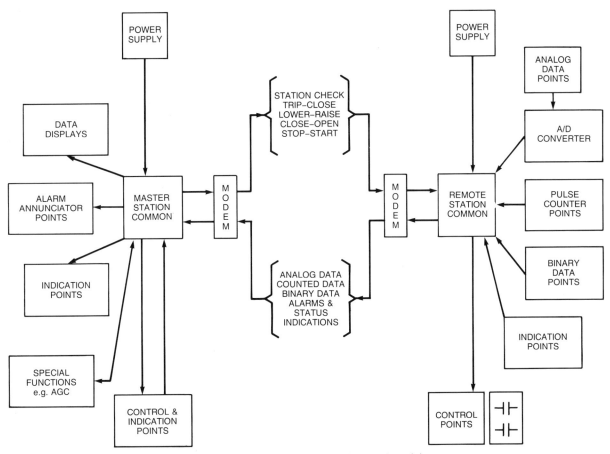

Figure 10.16. Supervisory Control And Data Acquisiton system.

rated in the design of SCADA systems. Security measures are used to minimize false operations or prevent erroneous data. They may vary from "select-before-operate," "check-back," to "double transmission," to various forms of built-in code message self-checking methods.

In "check-back" methods, a message is sent from the master station to the remote, received at the remote, and then sent back to the master. The response from the remote verifies at the master the initial signal or message before proceeding with the transmission of data or a command. If the received message matches the initial signal, the action desired may then proceed.

In double-transmission methods, the initiating message is sent more than once. At the remote, two successive transmissions must match at the received location before action can proceed.

Operating modes of the SCADA system permit the following:

Data scan, in which the master station continuously requests and receives data from one or more remotes.

Status scan, in which the master station continuously scans for status changes at each remote when data is not requested.

Control sequence operation, in which the scan of data and status is interrupted and the operator selects a point to be operated and the control function is performed. The select-before-operate and check-back sequence is completed before the control function is executed.

Master station SCADA equipment that utilizes a computer-based system is capable of communicating with hundreds of remotes through complex communication networks. The communications input/output function is handled by several microcomputers capable of operating with a number of communication lines. The system supports simultaneous processing of communication and application software. Some of the functions performed by the computer communications controller are:

Polling the remote stations and updating its own status registers.

Converting to engineering units.

Checking against limits and setting flags for violations.

Executing control sequences.

A number of master station command codes are used in the operation of a SCADA system. Examples of these codes are:

1. Alarm scan
2. Continuous scan
3. Full scan
4. Reset all points
5. Reset counters
6. Store counters
7. Station check
8. Acknowledge
9. Control selection
10. Trip selected device
11. Close selected device
12. Set point
13. Special scan of continuous data, selectable data, and indication data

Conventional master station man–machine interface systems consist of standard panels with mimic bus, indicating lights, escutcheons, control switches, alarm panels, and indicating meters. Control and status indication of all power equipment in the substation is provided for local, automatic, or remote operation. Access to all control points for manned and unmanned operation is available.

Today, computer master station equipment and CRT displays with limited graphics—color or black and white—give the system operator the information he needs to control or to observe in the operation of a remote. Three basic input/output equipments are provided for the operator: (1) CRT display, (2) operator keyboard and push buttons, and (3) typer for hard-copy record.

The CRT provides a one-line diagram that dynamically displays the status of the substation and also alphanumeric displays; this saves space and wiring. The keyboard and special-function push buttons are used by the operator for the display function, data and events logging, limit violation, and other operator actions. Logging by the typer device provides a permanent hard-copy record of all things that have occurred at a remote.

Operator control allows the following functions to be initiated: control action, CRT display selection, alarm acknowledgement, data entry, demand logging, limit changes, and points IN or OUT of service in the substation. Means for PRINT and DISPLAY, data entry, control, and point selection and alarm functions are also provided.

The CRT "Page" display function provides the operator with the ability to obtain:

1. One-line diagrams that dynamically update the status of the substation.
2. Tabular substation displays for indication and status points, giving the substation identification, page number, data, and time (dynamically updated while the display is on the screen), point identification, points that are controllable, and point description and status indication, a display to indicate when a point is SELECTED and when it is out of service.
3. Off-normal summary displays.
4. Event displays that give the day of the month, hour, minutes, and seconds; point description; and status.

A PRINT function button allows the printing of any point in any substation as it appears on the CRT display. A LOG function button initiates the print-out of a periodic log of data points. A display key initiates the print-out on the I/O typer of the CRT information displayed on the screen. This is most useful in the event the CRT is out of service, since the operator can cause the typer to produce the page of the substation display. This allows the operator to perform any selection, control function, alarm acknowledgement, or any other function that could be performed from the display itself by looking at the print-out on the I/O typer; means for cancelling the printing or display are also provided. Flexibility for each expansion of the points in the substation, point modification, and deletion of points are provided by means of the I/O typer.

Channels for SCADA Systems

A two-way bidirectional channel is required between the master station and each outlying remote station so that signals can be sent both ways. Communication between the master and remote equipments utilizes an orderly communications protocol. This two-way channel provides a means for the remote continuously to update indication and telemetering to the master station, regardless of the action taken by the master station. The channel also permits the operator at the master station to perform a control operation at the remote. The normal state of operation is repetitive communication with the remote. With most SCADA systems, more than one outlying station may be on the same channel.

Generally, voice-grade channels are used for SCADA systems as follows:

1. Leased-wire line from the local telephone company.
2. Voice channel on a single sideband power line carrier system.
3. Microwave voice channel.

Programmable Remote

If a station containing a remote has microprocessors, it is possible for the substation remote to operate as a programmable remote. The microprocessor-based remote supports the very basic functions of status monitoring, analog retrieval, and control. However, new sophistication and message protocol permit a number of expanded SCADA functions including:

1. Status memory systems to stack status changes in a memory queue to permit the remote to track fully a breaker-recloser operation.
2. Multimode analog systems to provide not only high-speed analog retrieval but also slow-speed analog data. This permits the remote to be polled by exception, which allows reporting a new analog value whenever a predefined change is exceeded.
3. Multiple control systems for different classes of control in either a simultaneous or sequenced manner. For example, standard control, jog control, and multiple simultaneous generation control may be provided.
4. Multiple-mode digital accumulation for a number of classes of accumulators.
5. Integrated sequence-of-events with a ±1 msec resolution.
6. On-line real-time diagnostics that can continuously monitor the performance of the remote and transfer this information to the master station.

A programmable remote relieves the computer master with portions of its computational load and reduces the communication channel requirements. In the latter case, a programmable remote can achieve significant communication economies in the transmission of messages between the computer master and a number of smaller, lower-priority remotes in its vicinity. The smaller, lower-priority remotes can be coupled to the programmable remote via relatively slow communication links.

A programmable remote can act to transmit its own data plus that of the connected lower priority remotes to the computer master at a high-bit rate. It relays commands to the lower priority remotes from the master station. The computer master is able to communicate with an interconnected group of remote stations through a single high-speed communications channel which can save in communication channel costs.

The programmable remote also provides a flexible master station system operating mode, flexible data addressing, efficient data acquisition according to varying update cycles and degrees of importance, and security with "select-before-operate" and "check-back" for control functions. Transmission rates up to 2400 bits per second are feasible with high-density bit-per-point information storage and transmission. The interrogation cycle does not require transmission of status change unless it has occurred, or data transmission unless it has exceeded a prescribed limit. Flexibility is the key to programmable remotes. Ability to talk in the language of the computer master enhances the role of the programmable remote.

SUMMARY

In general, a substation includes power equipment for switching and/or changing system voltage level. It may also contain equipment for regulating the voltage.

There are many different design choices and equipment selections that go into the makeup of a substation. There are five basic arrangements for connecting buses, breakers and circuits in the most suitable way. There are circuit breakers of different ratings and types of interrupting media to accomplish the switching.

Transformers provide the system voltage-level change and must have the ratings appropriate for the power transfer. The control of the substation voltage can be accomplished by transformer load-tap changers and/or by reactive power equipment, i.e., shunt reactors, shunt capacitors, static var control, and synchronous condensers.

The overall control of a substation can be accomplished by an operator at the site or remotely from a system operating office. With SCADA systems, complete remote control and monitoring can be achieved.

REFERENCES

1. ANSI C62.2-1969, Guide for Application of Valve Type Lightning Arresters for Alternating Current Systems.
2. ANSI/IEEE Standard C37.1-1979, Definition, Specification and Analysis of Manual, Automatic and Supervisory Station Control and Data Acquisition.
3. Benko, I. S., Gold, S. H., Rothenbuhler, W. N., Bock, L. E., Johnson, I. B., and Stevenson, J. R., "Internal Overvoltages in EHV Compensated Systems—Series Capacitors and Shunt Reactors," CIGRE 33-05, 1976.
4. Brown, P. G., Otte, G. W., Saline, L. E., and Talley, V. C., "Economics of Switched Shunt Capacitors and Synchronous Condenser Kilovar Supply for Transmission Systems," *AIEE Transactions,* **73** (part III-B), 1553–1559 (1954).
5. Hauspurg, A., and Scherer, H. N., "Design of High Voltage Substations," *IEEE Transactions,* **IGA-2** (4), 286–296 (July/August 1966).
6. Dillow, N. E., Johnson, I. B., Schultz, N. R., and Were, A. E., *AIEE Transactions,* **71** (part III), 188–199 (1952).
7. Fundamentals of Supervisory Control Systems, IEEE Power Engineering Society Course Text, 81 EHO 188-3-PWR.
8. IEEE Std 80-1976, IEEE Guide for Safety in Substation Grounding.
9. IEEE Working Group of the Lightning Protective Devices Subcommittee," Lightning Protection in Multi-line Stations," *IEEE Transactions,* **PAS-87,** 1514–1521 (June 1968).
10. Sakshaug, E. C., Kresge, J. S., and Miske, Jr. S. A., "A New Concept In Station Arrester Design," *IEEE Transactions,* **PAS-96** (2), 647 (March/April 1977).

11
DISTRIBUTION

INTRODUCTION

At the end of 1980, the electric utility industry in the United States served almost 92 million customers with approximately 99% of them supplied power by a distribution system.

The elements of good service to a customer include continuity of service, proper voltage, and correct frequency. Because the frequency is determined by generation, only the first two elements apply to the distribution system. Good service continuity requires a system where interruption of service is rare and is limited to the smallest practical number of consumers. Proper voltage requires that (1) the voltage remains within a range established by an industry standard, and (2) sudden, momentary changes in load do not cause objectionable light flicker.

SUBTRANSMISSION SYSTEM

Subtransmission is a term used loosely to designate the circuits through which energy is delivered to the distribution substations (i.e., those substations that feed the primary distribution system). Usually the subtransmission system is fed from bulk power substations, but it may still be called subtransmission, even when fed directly by one or more generating stations. In some systems, there are two or more voltage levels in the subtransmission system.

Many subtransmission systems were originally transmission lines. As the load increased and it became necessary to bring more and and more power into the area, perhaps from a greater distance, the transmission voltage proved to be too low. The new transmission was at a higher voltage and became the transmission system, while the former transmission circuits became subtransmission. As a result, voltages from 230 kV down to 35 kV are found on subtransmission systems.

Some utility companies regard subtransmission as part of distribution. Others include only circuits of 35 kV or less within their concept of distribution. To them, subtransmission is part of the bulk power supply system.

DISTRIBUTION SYSTEM SCOPE

The distribution system includes the primary circuits and the distribution substations that supply them, the distribution transformers, the secondary circuits, including the services as far as the entrance to the customers' premises, and appropriate protective and control devices.

The primary system is three-phase, four-wire, and is operated in the 12–14 kV range. Current trends indicate an increased utilization of 25 or 35 kV levels in primary circuits. Technological advances have made higher primary voltages attractive to utilities from both the installation and operating standpoints. The secondary system serves most of the customers at levels of 120/240 volts, single-phase, three-wire; 208Y/120 volts, three-phase, four-wire; or 480Y/277 volts, three-phase, four-wire. The American National Standards Institute standard C84.1-1982 tabulates system voltages in use and outlines the preferred voltage levels to be used for new installations and uprating of existing systems.

CIRCUITS BY TYPE OF LOAD

The loads supplied by a utility may be classified as follows: (1) industrial, (2) commercial, or (3) residential, suburban, and rural. These are shown in Figure 11.1. Each kind of load differs from the others in service requirements and in the facilities needed to supply it.

In addition to the large industrial and commercial customers, there are approximately 91 million others that must be reached by the distribution system. Most of these are residential customers. The load and income per residential customer is small, and individual attention to each customer is economically out of the question. Short-cut statistical methods must be used in checking adequacy of service and in making plans for system strengthening or rebuilding, as required by load growth. This is an application where computers have been a most effective tool. Data indicating levels of power consumption, such as meter readings, is stored in the computer and is used to indicate growth rates, transformer load levels, and areas where uprating is needed. This capability has improved system planning efforts.

SERVICE TO INDUSTRIAL LOADS

Industrial loads include mills, factories, processing plants, and possibly other utilities and electric transportation. When these loads are large, they are usually served from "customer" substations which are fed from the bulk power supply system. A very large factory load may require several substations and may be fed from the transmission system. Such customer substations require special consideration as to transformer size, connections, location, provision for spare capacity, emergency operating requirements, service continuity, provision for load growth, and system-neutral grounding. All of this must properly coordinate with the industrial plant's requirements. It is also necessary to plan the utility relaying so that the industrial user can coordinate relaying with it. These substations can best be planned through a joint effort by the utility and industrial plant engineers. Small industrial loads are frequently located in or near residential areas and are supplied by the primary feeders in those areas.

SERVICE TO COMMERCIAL LOADS

The customers in commercial areas are stores, hotels, restaurants, theaters, office buildings, and possibly apartment houses. There are many customers, both large and small. The load density is high and these customers require high service reliability. The most reliable and flexible system for serving these commercial loads is the secondary network.

Secondary Network

A secondary network is a distribution system in which the secondaries of the distribution transformers are connected to a common network for supplying light and power directly to consumers' services. The circuits of the network are operated at 208Y/120 volts or 480Y/277 volts, three-phase, four-wire.

There are two basic types of secondary network—the grid network and the spot network. Grid networks are used for downtown commercial loads spread over a large area. Spot networks are used for commerical buildings, hospitals, shopping centers, and similar loads.

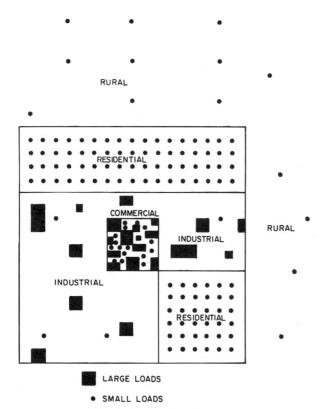

Figure 11.1. Convenient distinctions to make in load-area studies for distribution systems.

SERVICE TO COMMERCIAL LOADS

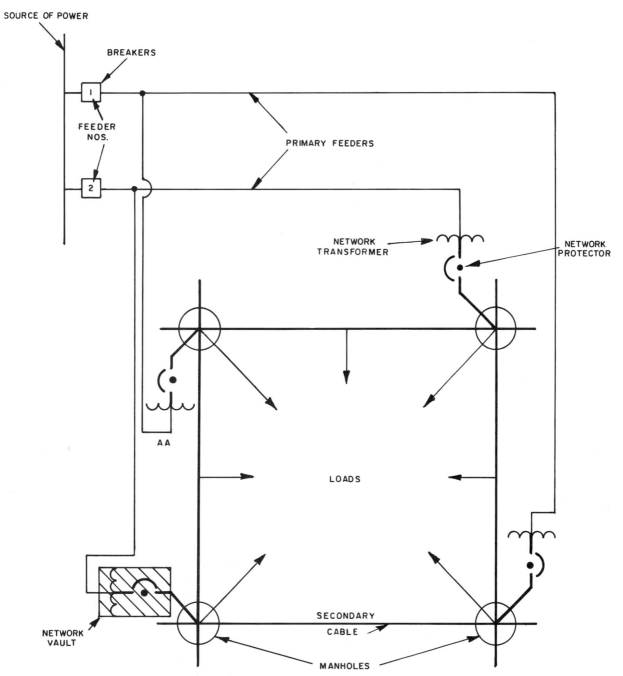

Figure 11.2. One-line diagram of secondary grid network and primary feeders.

Grid Network. The essential components of the grid network are illustrated in Figure 11.2. Secondary cables are installed along all streets in the commercial area, and all conductors of each phase are connected together at each street intersection to form a secondary cable grid. Customer connections are taken from these cables. The cable grid is energized by means of a number of network units (Figure 11.3), each of which consists of a network transformer and a network protector. At least two and frequently five or more primary feeders are used to supply the network units. These feeders are usually in a voltage range of 13 kV through 35 kV. Each feeder supplies several network units, and adjacent

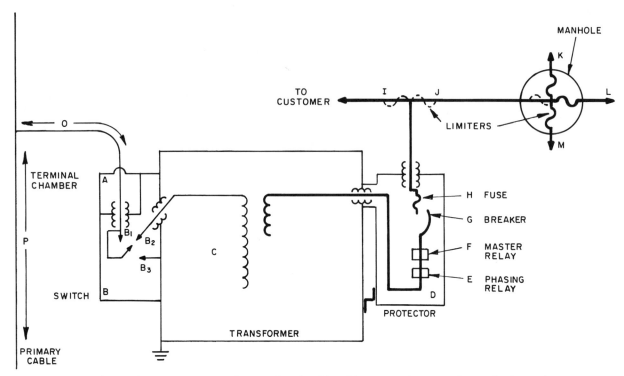

Figure 11.3. Arrangement of network unit with cable connections to primary feeder and secondary network.

network units are supplied from different primary feeders. The overall design of the grid network system provides voltage within prescribed limits to all customers with any major system component, i.e., primary feeder or network unit, removed from service. Some utilities use a double-contingency criterion for their networks, such that service will be maintained to all customers with any two major components out of service.

All parts of the network are usually underground: the network itself, service connections to customers, network units, and primary cables. Both primary and secondary network cables are in ducts under the street. Network units are in vaults under the street or sidewalk or in a building vault. At street intersections, there are manholes large enough to hold the cable bus work necessary for the formulation of the network and for the workers to pull and splice cables. At points where services to individual customers take off from the network, handholes provides access for doing necessary work from the street level.

Spot Network. It is usually not practical to serve new commercial buildings in the downtown area from the 208Y grid network due to the magnitude of the load. In addition, most new buildings are designed for a service voltage of 480Y/277 volts, with 480 volts three-phase used for motor loads and 277 volts single-phase used for fluorescent lighting. Many utilities use spot networks (Figure 11.4) to serve these loads. Spot networks may also be used to supply loads outside the downtown area that require high service reliability.

The design philosophy and the equipments used in the spot network are essentially the same as in the grid network, with the following differences:

1. The network units are at one location, frequently in the same vault, rather than dispersed over a number of city blocks.
2. The network units are connected to a common service bus rather than to a large cable grid.
3. Customer connections are taken from the service bus.
4. The service voltage is 480Y/277 volts rather than 208Y/120 volts.

Network Transformers

Network transformers (see Figure 11.5) are of more rugged construction and have lower losses than

SERVICE TO COMMERCIAL LOADS

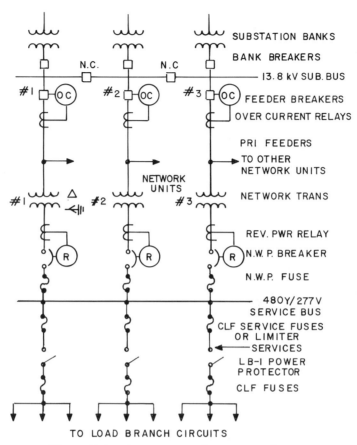

Figure 11.4. Arrangement of spot network.

conventional three-phase transformers. Ratings range from 300 kVA through 2500 kVA, with the smaller ratings used in 208Y/120-volt grid networks and the larger ratings used in 480Y/277-volt spot networks. Most network transformers are liquid-filled, either oil or non-PCB insulating fluid, with sealed-tank construction, and are self-cooled. A few dry-type units have been built. A three-pole, three-position, manually operated, liquid-filled switch is mounted on the primary side of the transformer. By means of this switch, the primary cable can be (1) connected to the transformer, (2) grounded, or (3) left open. An electrical interlock on this switch prevents opening it, or grounding the cable, while the transformer is energized.

Network transformer tanks, fittings, and accessories are designed for occasional or continuous submersion in underground vault applications. Secondary bushings are brought out through a throat for connection to the network protector.

Network Protectors

The network transformer is connected to the network through a network protector (see Figure 11.6). The network protector is an automatic air circuit breaker with relays and auxiliary devices and

Figure 11.5. Vault-type network transformer.

Figure 11.6. Network protector installation.

backup fuses, all enclosed in a metal case, which commonly is mounted on the secondary throat of the transformer. The network protector may be either of submersible construction for underground vault applications or dustproof construction for dry-vault applications. Pad-mounted network units are also available, and are generally used outside the downtown area.

The function of the network relays is two-fold: (1) to open the breaker on power-flow reversal, or in case of a fault in the transformer or primary feeder, (2) to reclose the breaker when the voltage of the primary feeder is of the correct magnitude and phase angle with respect to the network voltage, so that, when the breaker is reclosed, power will flow from the feeder into the network.

Control of Voltage

The voltage supplied to secondary network customers is controlled by holding correct voltage on the primary feeders. When the feeders came directly from generating stations, this was sometimes done by generator field control. As an increasing part of the energy is generated outside of the load area, however, more and more of the primary feeders originate in substations. In this case, voltage control is by transformer load-tap changing or step-voltage regulators.

It is preferable that all network primary feeders be fed from only one bus. This prevents exchanging kilowatts and kilovars between buses through the network.

Continuity

The almost complete freedom from service interruptions to secondary network customers is attributable to:

1. The design philosophy of the network system, which provides service to all customers with any primary feeder or any network unit out of service.
2. The protective devices (fuses, relays) used in the network system, which isolate faulted system components and keep as many network units in service as possible.
3. The 208Y cable grid system can experience short circuits. Since breakers or fuses would interrupt service to customers, another philosophy is used. When a cable experiences a short circuit, the cable will burn until a sufficient gap exists between it and the short circuit. The "sufficient gap" will actually extinguish the arc and permit the system to operate normally. The customers connected to this cable will still have service since the cable is energized from both ends.

SERVICE TO RESIDENTIAL, SUBURBAN, AND RURAL LOADS

Residential, rural, and suburban loads are similar in circuitry, equipments, and operating philosophy, differing primarily in load densities and, to some extent, in types of loads.

Residential loads in 1980 used 721 billion kWh of energy, roughly one-third of the total sold. Residential customers are more than 89% of the total number of customers and live in areas that are comprised mostly of residences, with relatively few shopping areas, service stations, etc. For this reason, the load is composed of many small and similar loads located rather close together over relatively large areas. The load density typically varies from 20 kVA to 300 kVA per 1000 front feet.

SERVICE TO RESIDENTIAL, SUBURBAN, AND RURAL LOADS

Service to residential loads is one of the fastest changing components of the electric utility industry. Historically, new residences have been served by overhead distribution feeders and laterals. In recent years, the trend has shifted toward underground distribution circuits. This trend has introduced new technologies, equipments, and practices along with new problems.

Typically, residential loads are served by three-phase, four-wire primary mains with single-phase laterals for area coverage. A few three-phase laterals may be required to serve other load types.

Rural loads differ from residential by having the lowest load density on the system, different load components, and a different shape to the load area. They are served from distribution lines (mostly single-phase) which are often 20 miles long or more. These lines extend along the highways and may serve about three or four customers per mile. The low load density means low revenue density too, and that requires careful engineering and planning to provide economical operation.

The rural loads are typically served by 7200-volt single-phase feeders from 12470Y/7200-volt sources. Some with very high load densities are served by 14400-volt single-phase feeders from 24900Y/14400-volt sources. Ordinarily, in rural areas, there is only one customer per distribution transformer.

Suburban loads are intermediate in load density and configuration from the residential and rural load types. They have several forms; one form being a small community located at the intersection of improved highways and ranging from a single service station to several hundred residences plus light commercial loads. Another type consists of residences built along highways in the more densely populated parts of the country and near large cities. This load is similar to the rural type in being spread out along a highway but differs in having a higher load density and includes both residential and light commercial such as stores, eating places, and other commercial enterprises. As the load grows with time, some sections will have side streets and loads added so that the load area acquires some breadth as well as length. Often, new real estate developments are located just off main highways and fall into this category.

These suburban loads are served using the same broad principles as outlined for residential and rural loads. The specifics will differ in order to accommodate the differences in size and shape of the load area, the total load and other similar factors.

Primary Circuits

Primary distribution circuits are predominantly of overhead construction, although in recent years, there has been a rapid and continuing trend towards underground construction. At present, the miles of three-phase overhead construction being added each year are about three times that of underground. For single-phase construction, the annual circuit additions are about the same for overhead and underground. This reflects the rapid acceptance of underground circuits, first, in single-phase and, later, in three-phase circuits.

Primary distribution circuits are radial in the sense that there is one source of power supplying a given circuit. A three-phase, four-wire main portion of the circuit extends from the substation into the load area with a number of lateral branches to cover the load area, as shown in Figure 11.7. Many of these lateral circuits may be single-phase, consisting of one-phase conductor plus the multigrounded neutral conductor. Most distribution transformers are fed from the lateral circuits, but some are also fed from the main circuit.

In underground distribution, the open-loop concept illustrated in Figure 11.8 is widely used to avoid excessive outage durations inherent in underground cable circuits. Two radial circuits are brought together with a normally open tie point. Switching facilities are provided so that any cable section can be isolated, with service maintained to all transformers on the two circuits.

One trend in primary distribution is the increasing use of higher primary voltages, i.e., voltages above the 15-kV class. This trend has an impact on substation and subtransmission practices because higher primary voltages almost axiomatically lead to larger distribution substations and higher subtransmission voltages.

To achieve economy, the higher primary voltages also require heavier feeder loadings, which could imply reduced service reliability since more customers are affected by primary faults. Greater use of automatic switching and protective equipment can do much toward preserving the level of reliability to which we have become accustomed. This is another reason most observers believe that an increased amount of automation is inevitable in our distribution systems.

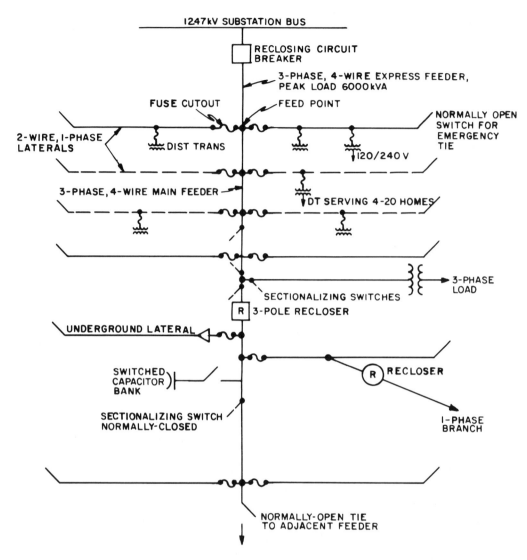

Figure 11.7. One-line, symbolic sketch of typical primary distribution feeder.

Underground Distribution

The use of underground construction for residential distribution systems has grown amazingly since its inception in the early 1960s. It is estimated today that more than 70% of the new residences built each year are served underground. Underground residential distribution (URD) is the standard construction for many utilities. A number of states have ordinances requiring URD for new developments of four to five homes or greater.

URD is more expensive than overhead construction (OH); however, the cost of URD relative to OH has declined due to volume of usage and to the evolution of materials, equipment, and practices. Initially, the cost ratio of URD/OH was perhaps five or greater. Recent data indicates that the median value of this ratio is now less than 1.5, with some utilities reporting that URD costs a little less than OH construction.

Two key developments that have made URD technically and economically feasible are solid di-

SERVICE TO RESIDENTIAL, SUBURBAN, AND RURAL LOADS

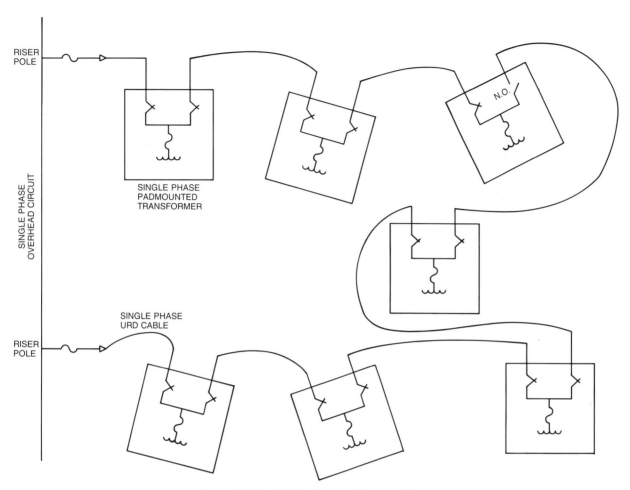

Figure 11.8. Single-phase underground loop.

electric concentric neutral primary cable and separable insulated connectors, commonly referred to as load-break elbows. A single concentric neutral cable is all that is required for a single-phase primary circuit. This cable is inexpensive and easy to install and to terminate. The cable can be directly buried in a trench and in some areas can be plowed into place. The central conductor is aluminum, the insulation is typically cross-linked polyethylene (XLPE) with an extruded semiconducting shield, and the bare concentric neutral consists of a number of strands of tinned copper.

Separable insulated connectors (Figure 11.9) have two major advantages. All external surfaces are shielded and grounded, which permits reduced clearances and therefore smaller equipment, as well as safer equipment. Second, the connectors provide simple, inexpensive and reliable load break switching at every transformer location. The connectors are rated for at least 10 switching operations at

Figure 11.9. Separable connector.

Figure 11.10. Single-phase pad-mount distribution transformer.

200 A, and in addition can be closed safely on a fault current of at least 10,000 A. These load-break elbows are easily installed on solid dielectric concentric neutral cable.

Concentric neutral primary cable and separable insulated connectors are in service on 15-kV, 25-kV, and 35-kV URD systems.

A typical URD system consists of single-phase, pad-mounted distribution transformers (Figure 11.10) with dead-front primary construction using load-break elbows and with 15-kV concentric neutral primary cable. Each transformer serves 8–12 homes, and is located on the front of a lot near the street. Primary and secondary cables are aluminum with solid dielectric insulation, and are directly buried in trenches adjacent to the street. The primary lateral circuit is of the open-loop configuration shown in Figure 11.8, with each half of the loop supplied from an overhead circuit by means of riser poles.

Until recently, underground distribution was predominantly URD, i.e., single-phase looped laterals supplied from overhead circuits. To an increasing degree, utilities are using underground construction of the three-phase main circuits also, so that the entire distribution circuit is underground. In some areas, overhead three-phase main circuits along arterial streets are being converted to underground. A typical UD primary circuit is shown in Figure 11.11. The open-loop configuration is used for the three-phase main circuits as well as for the lateral circuits. Pad-mounted switch and fuse equipment incorporates three-pole load-break switches to sectionalize the main circuits and single-pole fused switches for the lateral circuits. The lateral circuits may be single-phase or three-phase, and each is looped to a corresponding lateral from a different switch equipment.

Control of Voltage

The factors involved in voltage control are illustrated in Figure 11.12. The objective is to keep the voltage at the consumers' premises within an acceptable voltage range. An acceptable voltage range is one for which satisfactory operation is obtained for consumer equipment and also one which can be maintained by the electric utilities. The standard, ANSI C84.1, Voltage Ratings for Equipment Power Systems and Equipment, specifies these acceptable voltage ranges for both consumer and utility voltages. At the 120-volt level, this is 110–126 volts at the utilization point. It is customary for utilities to hold voltage at the meter location at 114–126 volts, which allows for a 4-volt drop to the utilization point.

Figure 11.12 shows that the locations of the voltage extremes are at the first and last customer locations. The variations from light to heavy load at these locations will establish the voltage range for the circuit.

As a first step in the control of voltage on such a circuit, most utilities will regulate the primary voltage at the substation. This takes care of variations in the voltage supplied to the substation transformer primary. The equipment used is either load-ratio control equipment in the substation transformer or step-voltage regulators. For most urban feeders, no other regulating equipment is needed, although shunt capacitor banks (Figure 11.13) are often installed to supply part of the kilovar portion of the load. On larger feeders, both step-voltage regulators and shunt-capacitor banks may be needed out

SERVICE TO RESIDENTIAL, SUBURBAN, AND RURAL LOADS

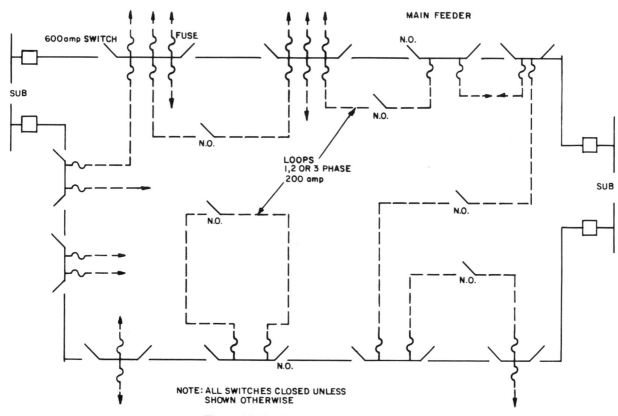

Figure 11.11. Underground feeder plan.

on the feeders to provide supplementary voltage control and kilovar supply.

In general, the control of voltage is more economical if both feeder-voltage regulators and shunt capacitors are applied than if either is applied alone. The capacitors are applied as an economic tool to reduce system losses by supplying kilovars locally. When applied this way, they also provide voltage profile benefits. The fine tuning of the voltage profile is then completed by applying feeder-voltage regulators.

Capacitors. Capacitors act as an economic tool in distribution engineering since they supply kilovars in the load area, eliminating the need for supplying these vars in the bulk power system. The load area kilovars reduce the current in all portions of the system. This releases capacity and reduces system losses. The reduction in flow of the reactive portion of current results in less voltage drop to the load area.

At light load, the capacitors installed for full load operation may cause too high a voltage at the distribution loads. Therefore, some capacitors may need to be switched off during these periods.

Step-Voltage Regulators. The equipment shown in Figure 11.14 is an autotransformer with automatic tap-changing under load. Automatic measuring and tap-changing equipment holds the output voltage within a predetermined bandwidth. By using the smallest practical bandwidth, more voltage drop can be allowed along the feeder, still keeping the consumer voltage within acceptable limits. Figure 11.12 shows that the substation voltage is higher at peak load than at light load. The means for achieving this are an integral part of the regulator controls called the line drop compensator.

When the voltage-regulating equipment, including capacitors, has been installed properly, it is often found that not all the allowable voltage range has been used. This permits a choice in the selection of the proper average voltage around which to center the voltage range. This decision is often made using the fact that a portion of the load is voltage sensitive. By reducing the average voltage

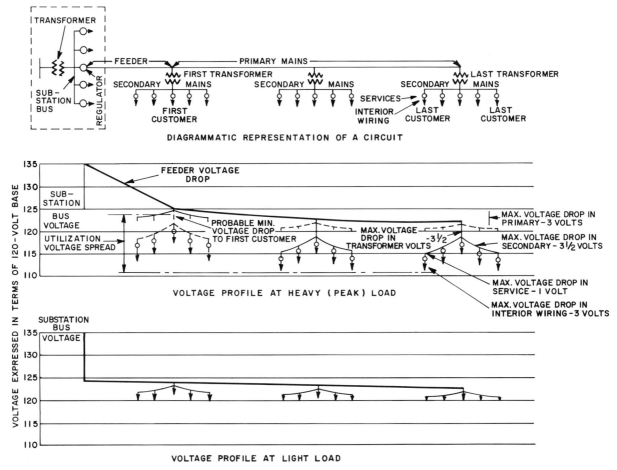

Figure 11.12. Representation of a distribution circuit with voltage profiles at heavy and light loads.

held by the regulator, the load and energy may be reduced somewhat.

Continuity

On overhead circuits, 80–90% of the faults are of a temporary nature caused by lightning, birds, small animals, and contact with tree limbs. If the fault is interrupted for a short time, the circuit can be successfully re-energized, restoring service to all consumers. Under these circumstances, the fault is considered temporary. Overcurrent protection of primary circuits is provided by reclosing circuit breakers, automatic circuit reclosers (Figure 11.15), fuses (Figure 11.16), and sectionalizers, which divide the primary circuit into a number of sections. The time–current characteristics and operating characteristics of these devices are coordinated so that service is restored to all consumers following a temporary fault, and a minimum number of consumers are interrupted for a permanent fault.

Reclosing circuit breakers and automatic circuit reclosers have both an instantaneous overcurrent characteristic and a time-delay overcurrent characteristic. Initially, these devices trip on the instantaneous characteristics, interrupting the fault current quickly enough to prevent the blowing or melting of downstream fuses. If the fault persists when the circuit is reclosed, these devices switch to a time-delay trip characteristic. This permits downstream fuses to blow and isolates a permanent fault before the breaker or recloser trips. If the fault persists after the reclosing device has reclosed a second or third time, the device trips and "locks out," since the fault is now in the section immediately beyond the device. Sectionalizers count pulses of fault current, and open automatically while the circuit is de-energized after one, two, or three pulses of current,

SERVICE TO RESIDENTIAL, SUBURBAN, AND RURAL LOADS

Figure 11.13. Pole-top capacitor installation.

depending on the setting. In this way, a permanent fault is isolated before an upstream reclosing device goes to lock out. Sectionalizers have no interrupting capacity and no time-current characteristics.

In total underground circuits, there are no temporary faults, and the reclosing principle is not applicable. Each lateral circuit is protected by the substation breaker. The time-current characteristics of the fuses and the feeder-breaker overcurrent relays are coordinated so that the minimum number of consumers is subjected to a service interruption due to a primary cable fault. The faulted cable is then isolated by the use of load-break elbows or pad-mounted switches as necessary, and service is restored to all consumers by closing the appropriate tie switches.

There is increasing interest in the utility industry in the use of distribution automation as a means to improve service reliability by reducing outage time. Several systems now in operation involve the remote control of distribution sectionalizing and tie switches, with remote readout of fault indicators to identify the faulted section. This type of system also exists with computer control.

Equipment on distribution circuits also requires overcurrent protection. Examples of such equipment are distribution transformers, regulators and capacitor banks. The protective equipments used are fuse cutouts and, in some cases, current-limiting fuses. The time–current withstand curves of the various equipment are coordinated with the protective characteristics of the fuses.

For distribution transformers, two types are offered. The first is the self-protected type, consisting

Figure 11.14. Single-phase step-type voltage regulator.

Figure 11.15. Distribution class circuit reclosers.

of a primary fuse inside the transformer. This fuse blows only when there is a fault within the transformer. Because a blown fuse means the transformer must be changed out, special care must be taken to avoid blown fuses due to overload or secondary faults which must be cleared by secondary breakers.

The other transformer is the conventional type. Here primary protection is provided by a fuse cutout. Because a conventional transformer utilizes no secondary breaker, the fuse size can be smaller than in the self-protected transformer. This allows better transformer protection (fault removal with less damage) and also some overload and secondary fault protection.

Where the available fault currents are excessive, the cover may be blown off the transformer, with attendant fire hazard. Even though a fuse may interrupt the fault, this may not happen fast enough to prevent the cover blowing off. Essentially, the cover blows when rapidly increasing internal pressure cannot be relieved quickly enough by normal means. To protect against this condition, a current-limiting fuse is often used to limit the fault current magnitude and duration (as well as the arc energy) to a value that will not blow the cover. Current-limiting fuses are used to protect other equipments for similar reasons.

Overvoltage protection on the distribution system is obtained by proper application of distribution surge arresters (Figure 11.17).

Figure 11.16. Open-type fuse cutout with load break capability and surge arrester.

Figure 11.17. Surge arrester.

SERVICE TO RESIDENTIAL, SUBURBAN, AND RURAL LOADS

More Equipment Information

Distribution Transformers. The distribution transformer steps the voltage down to utilization levels near the electrical service entrance to the building. In overhead distribution systems, it is common to have the distribution transformers pole-mounted, with each transformer serving several loads in the same area. The transformer is connected between one line and the neutral of a three-phase, four-wire primary feeder through a fuse cutout. It is also protected by a distribution surge arrester.

In new installations, the distribution system will generally be underground and the distribution transformers will be either pad-mounted or installed in below-ground vaults. The connections for these transformers are similar to overhead transformers, except that the fuse cut-out is replaced by an internal fuse and a disconnectable cable termination. Figure 11.10 shows a single-phase, pad-mounted transformer.

Automatic Circuit Reclosers. Automatic circuit reclosers are circuit interrupters (Figure 11.15) for clearing the temporary faults on primary distribution circuits. Originally, they were single-phase devices and intended for pole mounting. Three-phase reclosers are now available, and both three-phase and single-phase reclosers are available for use in small substations, as well as on pole tops. Distribution-class reclosers have a maximum interrupting rating of 25 times rated current, and power-class reclosers have a maximum interrupting rating of 40 and 60 times rated current. They have an operating cycle that consists of one or two instantaneous openings, followed by one or two time-delay openings, or any combination of instantaneous or time-delay. If a recloser opens for the fourth time, it locks open.

The instantaneous operations are fast enough to prevent the blowing of fuses through which the same fault current is flowing. If the fault persists, the time-delay before the third or fourth openings of the conventional cycle gives the proper fuse time to blow.

Cut-outs. Cut-outs are devices for clearing persistent faults in overhead distribution systems. They can be obtained with load break capability. The cut-out also serves to isolate sections of lines and equipment for maintenance and repair.

Surge Arresters. The surge arrester protects lines and equipment against voltage surges which could damage insulation. These overvoltages are either system generated or lightning generated.

The surge arrester is connected from phase to ground and provides a conducting path to ground when the voltage exceeds a predetermined level. This limits the voltage impressed across the protected equipment.

Arresters are used extensively on primary distribution systems, since this part of the system is subject to the most exposure and wider service interruption can result from isolating a fault. Figure 11.17 illustrates a surge arrester.

Separable Connectors. These devices combine switching, connection, and fault-closing capabilities all in one small package (Figure 11.9). They have the added characteristic of occupying only a small amount of space, which is at a premium in underground installations.

The advantage provided by these connectors is that they enable energized equipment to be discon-

Figure 11.18. Single-phase watt-hour meter.

Figure 11.19. Programmable meter for time-of-use metering.

nected. This avoids the need to shut down the entire lateral when repairing equipment. As a result, down time and the number of customers affected by an equipment failure are substantially reduced.

Metering

The watt-hour meter (Figure 11.18) at the consumer's premises is sometimes referred to as the electrical cash register. It measures the useful component of energy supplied. This quantity is proportional to the amount of fuel expended to provide the energy. A related meter, called the demand meter, measures the maximum energy delivered in any interval of time (15 or 30 minutes usually) in the billing period. This quantity is related to the amount of generating equipment required to serve the load.

Customarily, residential consumers have been billed for energy only. Commercial and industrial customers, using much larger quantities of energies, are usually billed for both demand and energy.

At present, some new techniques include time-of-day or peak-load pricing and automatic meter reading and control (AMRAC* System). The former area involves meters that can be programmed to record on-peak kilowatt-hours separate from the off-peak usage. Figure 11.19 shows an example of a programmable meter. The AMRAC system utilizes the distribution line as a carrier communication path to read meters, control switching devices, and even control portions of a customer's load. The addition of this type of system is one step closer to a fully automated distribution system.

SUMMARY

Distribution is a flexible term that signifies the part of the system delivering the power to the customers. It includes the distribution transformers and their secondaries, the primaries, the distribution substations, and, sometimes, the subtransmission circuits. Different types of distribution systems are required, depending on whether the consumer falls into the industrial, commercial or residential category.

Downtown commercial areas are usually fed by means of 480Y/277-volt or 208Y/120-volt secondary networks. Primary feeders, the secondary network, and the network transformers and protectors are all underground.

The distribution to existing urban residential customers is usually by 11,000–35,000-volt overhead primary feeders, pole-top transformers, and overhead or underground secondaries. There is a steady trend towards the use of underground for serving new residential customers. Most of the primary feeders are radial. The current rating of many primary feeders can be limited by the permissible voltage drop from the first to the last transformer; the longer the distance between them, the less load the feeder can carry. The voltage drop can be partly overcome, and the feeder rating raised by feeder voltage regulators and/or by capacitors on the feeder.

REFERENCES*

1. Distribution Data Book, General Electric Publication GET-1008.
2. "Edison Electric Institute Statistical Year Book of the Electric Utility Industry/1979," November 1980.
3. Power Distribution Systems Course, General Electric Publication GEZ-6391.

* GE publications may be ordered from Publication Distribution, General Electric Co., 705 Corporations Park, Scotia, N.Y. 12302.

* Trademark of General Electric Company.

12
PROTECTIVE RELAYING

INTRODUCTION

Protective relaying is that area of power system design concerned with minimizing service interruption and limiting damage to equipment when failures occur. The function of protective relaying is to cause the prompt removal of a defective element from a power system. The defective element may have a short circuit or it may be operating in an abnormal manner. Protective relaying systems are designed to detect such failures or abnormal conditions quickly (commensurate with system requirements) and to open a minimum of circuit breakers to isolate the defective element. The effect of quick isolation is threefold: (1) it minimizes or prevents damage to the defective element, thus reducing the time and expense of repairs and permitting quicker restoration of the element to service; (2) it minimizes the seriousness and duration of the defective elements interference with the normal operation of the power system; and (3) it maximizes the power that can be transferred on power systems.

The second and third points are of particular significance because they indicate the important role protective relaying plays in assuring maximum service reliability and in system design. The effect of quick fault isolation on power transfer capability is illustrated in Figure 12.1. This figure shows the power that can be transmitted across a system without the loss of synchronism as a function of fault clearing times. It is apparent from these curves that fast fault clearing times (t^1) permit a higher power transfer than longer clearing times (t^2). On present-day bulk power systems, fault-clearing times are in the (t^1) region and, therefore, power transfers are nearly at a maximum. High-speed clearing of faults can often provide a means for achieving higher power transfers and thereby defer investment in additional transmission facilities.

The art of protective relaying is constantly changing and advancing. There are, however, certain simple and fundamental principles of relay operation and application that are timeless. It is the purpose of this chapter to first examine these fundamental principles and then show how they are applied to the protection of specific system elements. The discussion will be chiefly concerned with the protection of the bulk power system where protective relays play a most important part in system design and operation.

DEFINITIONS AND REQUIRED CHARACTERISTICS

A protective relay is a device which, when energized by suitable inputs, responds to these inputs in a prescribed manner to indicate or isolate an abnormal operating condition. Inputs are usually electrical but may be mechanical, thermal, or other quantities or combination of quantities. This chapter will be primarily concerned with relays having electrical inputs. These inputs may be a current, a voltage or a combination of current, and voltage. The relay may respond to a change in magnitude in the input quantity, it may respond to the phase angle between two quantities or it may respond to the ratio of the quantities.

Protective relays with electrical inputs fall into two basic categories—electromechanical and static.

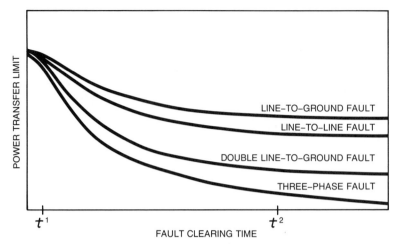

Figure 12.1. Power transfer versus fault-clearing time.

Electromechanical

The electromechanical relay consists of an operating element and a set of contacts. The operating element obtains its input information from instrument transformers in the form of currents and/or voltages, performs a measuring operation, and translates the result into motion of the contacts. When the contacts close, they either actuate an alarm or trip the necessary circuit breakers to isolate the faulted element. The relay usually includes some form of visual indicator to show that it has operated.

The operating elements for electromechanical relays can be classified, according to their construction, into four basic types: plunger, hinged armature, induction disk, and induction cup.

The plunger and hinged armature types both operate on the principle of magnetic attraction. As shown in Figures 12.2 and 12.3, an armature is attracted into a coil or to the pole face of an electromagnet, resulting in the closing of a set of contacts. These relays may be used with either alternating current or direct current, and they are generally actuated by a single input quantity, either current or voltage. They operate with no inherent time delay and, therefore, they are used for functions that require instantaneous operation.

Figure 12.4 and Figure 12.5 illustrate the construction of the induction disk and induction cup-type of relays. The principle of operation here is magnetic induction where torque is developed in a moveable disk or rotor in the same way torque is produced in an induction motor. This principle can only be used with ac quantities.

In the induction disk relay, a metal disk of copper or aluminum rotates between the pole faces of an electromagnet. This type of relay is generally actuated by a single input quantity and it operates with an inherent time delay. When actuated, it takes a finite time for the disk to rotate and close the contact.

In the induction cup relay, a metal cylinder with one end closed like a cup rotates in an annular air gap between the pole faces of electromagnets and a central core. This type of relay may be actuated by one or more input quantities and it operates with no inherent time delay. When actuated, the cup rotates in high speed to close a contact which is attached to the cup shaft.

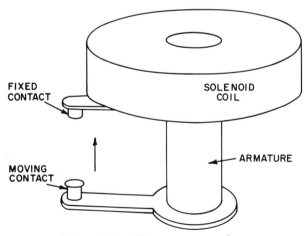

Figure 12.2. Plunger construction.

DEFINITIONS AND REQUIRED CHARACTERISTICS

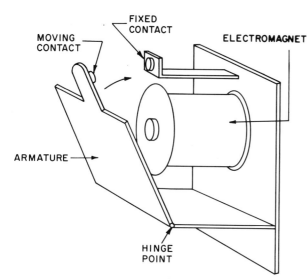

Figure 12.3. Hinged armature construction.

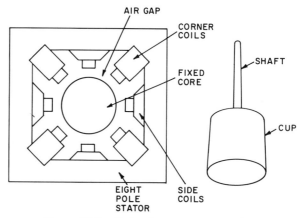

Figure 12.5. Induction cup construction.

Static

The development of highly reliable solid-state devices such as transistors, diodes, integrated circuits, operational amplifiers, and computer logic circuitry has led to the design of static or solid-state relays. As with electromechanical relays, static relays utilize electrical inputs of current, voltage, or a combination of both. The solid-state circuitry is designed to provide the various functions of level or magnitude detection, phase-angle measurement, frequency measurement, amplification, timing, pulsing, etc. The solid-state circuits react instantaneously to the input quantities to produce outputs with the required characteristics. Static relays can operate extremely fast, with response times as low as one-quarter cycle.

To illustrate how a static delay could be used to measure the phase angle between voltage and current, refer to Figure 12.6. The voltage and current sine waves at the top are supplied to separate squaring amplifiers whose function is to convert the sine wave into a square wave which is zero during the negative half cycle and provides a constant signal during the positive half cycle. These square waves,

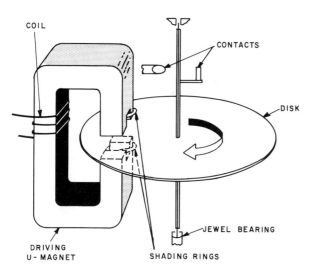

Figure 12.4. Shaded pole induction disk.

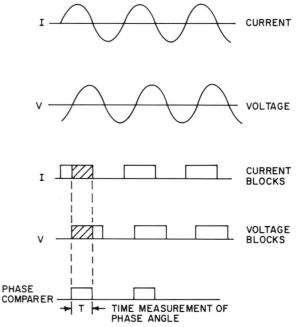

Figure 12.6. Typical waveforms used in static relay measuring functions.

commonly called blocks, are applied to a comparer circuit in such a way that an output is obtained only when both signals are present. The duration (time) of their overlap or coincidence is used to measure the phase angle between current and voltage. For example, if it is desired to produce an output when the voltage lags or leads the current by 60°, the half-cycle current and voltage blocks must overlap or be coincident for 120°. In terms of the time T in Figure 12.6 for the case where the current leads the voltage, the duration of overlap or comparer output will be 5.56 msec. An accurate timer connected to the comparer output and set to operate in 5.56 msec, will produce an output when current lags or leads the voltage by 60°.

Required Characteristics

There are three characteristics required by any protective relay to perform its function properly. These are sensitivity, selectivity, and speed. The relay must be sensitive enough to operate under conditions of minimum generation when a short-circuit fault would cause the minimum current to flow through the relay.

The selectivity of a protective relay is its ability to recognize a fault and trip a minimum number of circuit breakers to clear the fault. The relays must select between faults in their own protected equipment for which they should trip and faults in adjoining equipment for which they should not trip. Some protective relays or protective arrangements are inherently selective while others are not. This characteristic will be discussed further under the heading "Basic Relay Types."

The relay must also operate with the proper speed. Of course, speed is essential in clearing a damaged element of a power system since it has a direct bearing on the damage done by a short circuit and, consequently, the cost and the delay in making repairs. Speed of operation also has direct effect on the general stability of the power system. During a short-circuit type of fault, the various sources of generation tend to lose synchronism. The less time that a fault is allowed to persist, the less the effect on the synchronism or stability of the system.

There are two other characteristics usually associated with relays and protective systems. These are dependability and security, which collectively define protective system reliability. These characteristics may be defined as follows:

Dependability: The ability of a relay or relay system to provide correct operation when required.

Security: The ability of a relay or relaying system never to operate falsely.

The overall reliability of any relay system depends on the inherent reliability of the relays and on the proper application, installation, and maintenance of the equipment.

BASIC RELAY TYPES

While there are numerous relays and combinations of relays available for the protection of system elements, there are actually very few fundamental differences between the various relays. The operating characteristics of available protective equipment have been essentially derived from only a few basic relay types. The sections that follow will describe these basic relay types and will note their advantages or disadvantages where pertinent.

Overcurrent Relay

The overcurrent relay is the simplest type of protective relay. As the name implies, the relay is designed to operate when the current in a system element exceeds a specified level. There are two basic forms of overcurrent relays, the instantaneous type and the time-delay type.

The instantaneous overcurrent relay is designed to operate with no intentional time delay when the current exceeds the relay setting. In general, the operating time of this type of relay can vary quite widely. It may as low as 0.016 sec or as high as 0.1 sec. The operating characteristic of this relay is illustrated by the instantaneous time curve of Figure 12.7.

The time overcurrent relay has an operating characteristic such that its operating time varies inversely as the current flowing in the relay. This type of characteristic is also illustrated in Figure 12.7. The diagram shows the three most commonly used time overcurrent characteristics—that is, inverse, very inverse, and extremely inverse. These curves differ by the rate at which relay operating time decreases as the current increases.

Both types of overcurrent relays are inherently nonselective in that they can detect overcurrent conditions (faults) not only in their own protected

BASIC RELAY TYPES

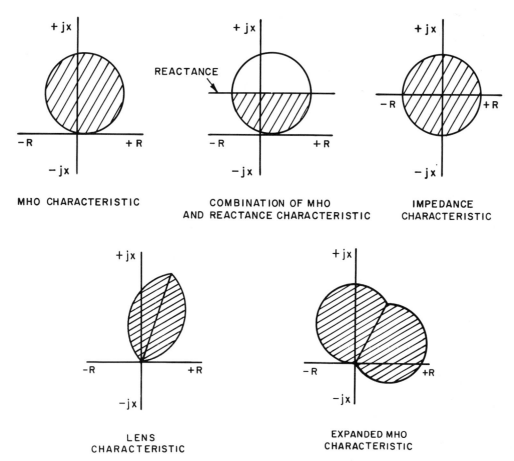

Figure 12.10. Distance relay characteristics.

tion for any problem where it is important to obtain selectively between different types of disturbances or between normal conditions and disturbances.

Pilot-Relaying Schemes

Pilot-relaying schemes are high-speed protective systems used for the protection of high-voltage transmission lines. These systems utilize protective relays at the line terminals and a communication channel between terminals to achieve simultaneous high-speed tripping of all line circuit breakers for faults anywhere on the protected line. In these schemes, the protective relays determine whether a fault is internal or external to a protected line, and the communication channel (called a pilot) is used to convey this information between line terminals. If the fault is internal to the protected line, all terminals are tripped in high speed. If the fault is external to the protected line, tripping is blocked. The location of the fault is indicated either by the presence or the absence of a pilot signal. If the presence of a signal blocks tripping (indicating an external fault), it is called a blocking pilot system. If the presence of a signal is required to cause tripping (indicating an internal fault), it is called a tripping pilot system.

Pilot Channels. There are three basic types of pilot channels used for protective relaying purposes. These are (1) the wire-line channel, (2) the power-line carrier channel, and (3) the microwave channel.

1. Wire-line channels consist of a twisted pair of copper wires in telephone-type cable. These channels are completely isolated from the power conductor and provide a direct metallic connection between the terminals of the protected line. The wire pilot may be leased from the telephone company or it may be installed and maintained by the electric utility.

2. The power-line carrier channel is the most widely used pilot for protective relaying. In this

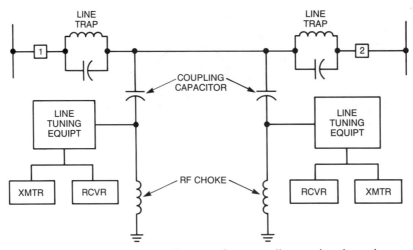

Figure 12.11. Functional diagram of a power-line carrier channel.

type of channel, the pilot signal is coupled directly to the same high-voltage power line that is being protected. The frequency of this carrier channel is in the range of 30 kHz to 300 kHz. Figure 12.11 illustrates functionally how this is accomplished.

Line traps tuned to the carrier frequency are located at the line terminals. These line traps present a high impedance to the carrier frequency and a low impedance to the power frequency. They function to contain a carrier signal within the terminals of the line being protected and they serve to isolate the carrier channel from power system faults outside of the protected line. The transmitter and receiver at each terminal are coupled to the power line through line-tuning equipment and the coupling capacitor. The radio frequency (rf) choke acts as a low impedance to the 60-cycle power frequency, but as a high impedance to the carrier frequency. This protects the equipment from high voltage at the power frequency and, at the same time, limits the attenuation of the carrier frequency.

3. The microwave channel is an ultrahigh frequency radio channel which propagates a signal through the atmosphere between line-of-sight antenna locations. The basic microwave channel is subdivided or multiplexed so that it can be used for a large number of different functions at the same time. When one of the subcarrier channels of a microwave scheme is used for protective relaying, it is usually modulated by frequency-shift audio tone equipment. In this type of audio equipment, a continuous tone of one frequency is transmitted under normal or nonfault conditions. This signal is referred to as a guard frequency. When it is desired to transmit information to another location, the audio tone transmitter is keyed and its output is shifted from the guard frequency to a trip frequency. The protective relaying scheme operates in conjunction with this audio tone equipment which, in turn, utilizes the microwave subcarrier channel to propagate the tone signals to the remote terminals of the protected line. In most instances, several different audio tone frequencies can be used simultaneously over the same subcarrier channel.

Types of Pilot-Relaying Schemes. There are four basic pilot-relay schemes in use today: (1) current differential, (2) directional comparison, (3) phase comparison, and (4) transferred tripping.

1. Current-differential protection for transmission lines is applied only over a wire pilot channel. Figure 12.12 shows one form of wire pilot current-differential relaying called the opposed voltage scheme. In this scheme, the three-phase currents at a line terminal are converted by a mixing transformer into a single-phase voltage which is applied to the relay and the pilot wire. With identical equipment at both ends of the line, current entering and leaving the line unchanged in magnitude and phase angle produces equal voltages at the end of the pilot wire. With the connection shown in Figure 12.12, these voltages will oppose each other, no current will flow in the pilot wire, and the relays will not operate. When an internal fault occurs, current will enter both line terminals, the voltages produced and applied to the pilot wire will aid each other, and

BASIC RELAY TYPES

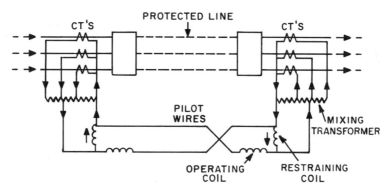

Figure 12.12. One form of wire pilot relaying.

current will flow in the pilot wire, thereby causing relay operation.

Another form of pilot wire relaying is the circulating current scheme. In this type of scheme, a current circulates over the pilot wire, thereby restraining or blocking relay operation for faults external to the protected line or with load current flowing through the line. When an internal fault occurs, very little or no current flows over the pilot wire and tripping is permitted.

2. Directional comparison protection is normally applied with a blocking-type pilot. In the blocking type of protective scheme, protective relays at the line terminals use the pilot channel to block tripping for faults external to the protected line. When relays at a line terminal respond for external short circuits, a signal is transmitted to block tripping at all terminals. When an internal fault occurs, no blocking signal is transmitted and the relays at each terminal are permitted to trip their breakers.

In the directional comparison scheme, distance relays and directional relays are used to detect internal and external phase and ground faults. The application of this type of scheme with a power line carrier channel is illustrated functionally in Figure 12.13. At each terminal, there are two sets of relays. One complement of relays indicated by T is used to detect faults internal to the protected line and operate only for faults in the direction of the associated arrow. To assure that these relays will detect all internal faults, they are set to reach beyond the remote terminal. When these relays detect a fault, they will attempt to trip unless they receive a blocking signal from another line terminal.

The second complement of relays at each terminal (indicated by CS) is set to see external faults in the direction of their associated arrows. When these relays detect a fault, they will key the transmitter and send a blocking signal.

To illustrate how this scheme operates, consider the faults F_1, F_2, and F_3 in Figure 12.13. First, consider the internal fault F_2. For this fault, the tripping relays (T) at each terminal will operate but neither

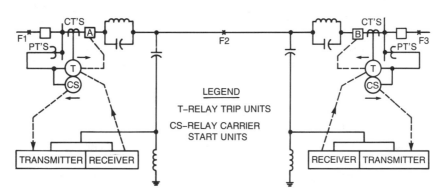

Figure 12.13. Directional comparison relaying over a power line carrier channel.

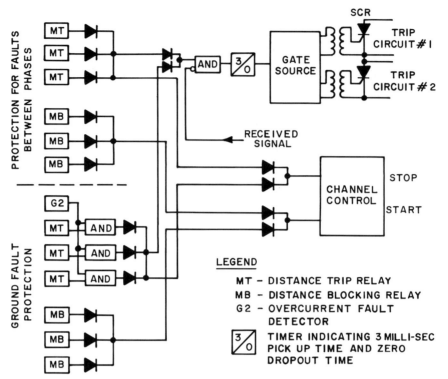

Figure 12.14. Simplified logic diagram of a static directional comparison pilot relaying scheme using phase and ground distance relays.

of the blocking relays (CS) detect this fault. Consequently, no blocking signal is transmitted and both terminals will be permitted to trip.

For the external fault at F_1, only the tripping relay at B and the blocking relays (CS) at A will operate. When the CS relay at A detects a fault, it immediately transmits a signal that blocks the T relays at B from tripping the breaker. The fault of F_3 would produce the same type of operation, except the blocking signal would come from the transmitter at B.

Figure 12.14 shows a simplified logic diagram of a static directional comparison pilot-relaying scheme using phase and ground distance relays. The AND function just before the 3/0 timer, known as a comparer, is the key logic that makes the decision whether or not to trip. To trip, there must be an input from the tripping system (MT) and there must not be a received blocking signal. The circular lower input to the AND function represents a NOT function. This type of static scheme is capable of initiating a trip signal to a circuit breaker in 12.5–21 msec (three-quarters to one and one-quarter cycles on a 60-Hz basis).

3. The phase comparison scheme is used in a special form of the blocking system. In this scheme, a phase-comparison relay and a communication channel detect line faults by comparing the phase position of currents entering and leaving the transmission line. This comparison is accomplished by the transmission of a signal from each terminal every half cycle (60-Hz basis). When a fault occurs external to a protected line section, the currents entering and leaving a line section are essentially in phase as shown in Figure 12.15. Under this condition, terminal A will transmit a signal during one-half cycle, while terminal B will transmit a signal on the alternate half cycle. In effect, a continuous signal is received at all terminals, and tripping is blocked. When a fault occurs on the protected line, the currents at the line terminals will be 180° out-of-phase, as shown in Figure 12.16. Under this condition, both terminals transmit a signal during the same half cycle. During the other half cycle, no signal is transmitted from either terminal and, therefore, tripping is permitted.

4. In the tripping type of protective schemes, protective relays at the line terminals utilize the pi-

BASIC RELAY TYPES

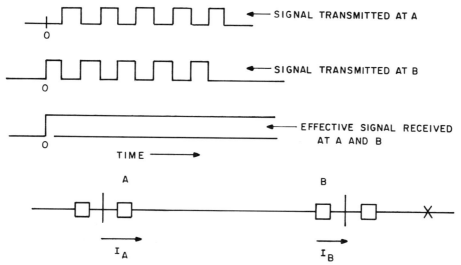

Figure 12.15. Phase comparison relaying. Diagram showing relationship between currents and signal at line terminals for external faults.

lot channel to initiate tripping for faults in the protected line. There are two basic relaying schemes used in a tripping system. These are direct-tripping and permissive-tripping schemes.

In the direct-tripping schemes, distance and directional overcurrent relays at the line terminal are set to detect faults only in the protected line section. When an internal short circuit occurs, these relays trip their respective breaker and send a trip signal to all other terminals. Receipt of the trip signal will directly trip a breaker, even though the relays at the terminal do not detect a fault. Figure 12.17 shows functionally how this scheme operates. The term underreaching, as used in Figure 12.7, refers to relays set to detect faults on 80–90% of the line section.

In the permissive-type tripping scheme, the trip circuit is always supervised by a local protective relay. To trip a breaker in this scheme, the supervising relay must see a fault and a trip signal must be received from one other terminal or from all other terminals, depending on the scheme being used.

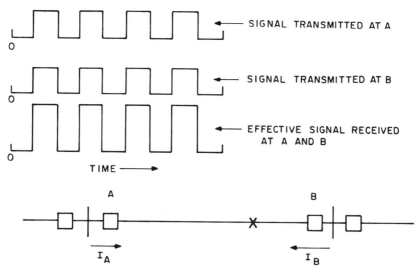

Figure 12.16. Phase comparison relaying. Diagram showing relationship between currents and signals at line terminals for internal faults.

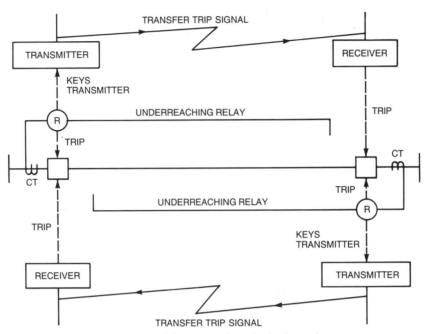

Figure 12.17. Direct transferred tripping scheme.

Static Relays

Before 1958, nearly all of the protective relay designs and schemes were based on electromechanical principles. There were a few exceptions and these utilized electronic circuits with vacuum tubes.

Since 1958, considerable progress has been made toward the development of static relays utilizing semiconductor devices. For the most part, the initial development work has been in the area of transmission line relaying, mainly because these devices and circuits are most attractive from the economic point of view when used to perform high-speed complex functions. Moreover, static transmission line relays are capable of providing the improved performance necessary to meet the stringent requirements of modern power systems—in particular, the extra high voltage (EHV) systems. Improved performance in this instance refers to faster operating times, improved reliability, greater accuracy, and reduced maintenance.

Solid-state relaying equipment is available for most of the conventional transmission line protection schemes employing distance and directional overcurrent functions. There are static mho, reactance, and impedance relays; static-phase comparison relays; and directional overcurrent relays for both phase fault and line-to-ground fault protection. In addition, special schemes have recently been developed for particular applications. For example, a static relaying scheme, which utilizes positive and negative sequence quantities in distance and directional overcurrent functions, has recently been developed for applications involving series capacitor compensated lines. Further developments can be expected to continue until static relays are available for the protection of all system elements.

FUNDAMENTAL APPLICATION PRINCIPLES

The general philosophy of relay application is to divide the power system into zones that can be adequately protected by suitable protective equipment and can be disconnected from the power system in a minimum amount of time and with the least effect on the remainder of the power system. The protective relaying provided for each zone is divided into two categories: (1) primary relaying and (2) backup relaying. Primary relaying is the first line of defense when failures occur, and is connected to trip only the faulted element from the system. If the fault is not cleared by the primary protection, backup relaying operates to clear the fault from the system. In general, backup relaying disconnects a greater portion of the system to isolate the fault.

FUNDAMENTAL APPLICATION PRINCIPLES

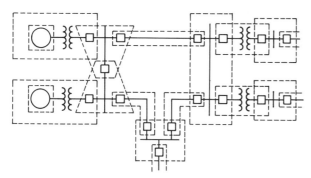

Figure 12.18. One-line diagram of system showing primary protective zones.

Primary Protection

Figure 12.18 illustrates how a power system is divided into primary protective zones. The dashed lines indicate a separate zone of protection around each system element (generator, transformer, bus, transmission line). The zone encompasses both the system element and the circuit breakers that connect the element to the system. If a failure occurs in any primary zone, the protective relays will operate to trip all of the breakers within that zone. If a breaker is omitted between two adjacent elements, both elements will be disconnected for a failure in either one. This latter arrangement is illustrated by the unit generator–transformer connection in the power plant.

As shown in Figure 12.18, the primary zones are arranged so that they overlap around circuit breakers. The purpose of this overlap is to eliminate the possibility of "blind spots" or unprotected areas. A fault in an overlap area will cause tripping of all circuit breakers in two primary zones. This extensive tripping is necessary because a fault in the overlap is, in effect, a circuit-breaker fault, and there is no assurance that the involved breaker will operate correctly.

On bulk-power generating and transmission systems, primary protection is designed to operate at high speed for all faults. Slower protection may be used in less important system areas but, in general, any system area will benefit by the fastest possible primary relaying.

Backup Protection

Backup protection is provided for possible failure in the primary relaying system and for possible circuit-breaker failures. Causes that may contribute to relay failure are:

1. Ac current or voltage supply to relays (failure in current or potential transformers or wiring)
2. Loss of dc control supply
3. Failure of auxiliary devices
4. Relay failure.

Causes that may contribute to breaker failure are:

1. Loss of dc supply
2. Open or short-circuited trip coil
3. Mechanical failure of tripping mechanism
4. Failure of main contacts to interrupt.

Any backup scheme must provide both relay backup as well as breaker backup.

Ideally, the backup protection should be arranged so that anything that may cause the primary protection to fail will not also cause failure of the backup protection. Moreover, the backup protection must not operate until the primary protection has been given an opportunity to function.

As a result, there is time delay associated with any backup operation. When a short circuit occurs, both the primary and the backup protection start to operate. If the primary protection clears the fault, the backup protection will reset without completing its function. If the fault is not cleared by the primary protection, the backup relaying will time out and trip the necessary breakers to clear the fault from the system.

There are two forms of backup protection in common use on power systems. These are remote backup and local backup.

1. *Remote Backup.* In remote backup relaying, faults are cleared from the system one station away from where the failure has occurred. For example, consider the simplified system shown in Figure 12.19. In a remote backup scheme, time delay

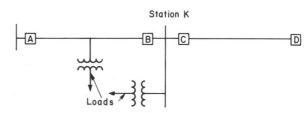

Figure 12.19. Illustration of remote backup relaying.

relays at A provide backup protection for line C-D. If, during a fault on line C-D, there is a failure of protection or circuit breaker at C, these remote relays will trip their respective breaker to isolate the fault. The relays and breaker at A provide relay and breaker backup for breaker C. Similarly, the relays and breaker at D provide backup for breaker B and both A and D provide backup protection for the bus at station K.

When distance relays are used for line protection, backup clearing times for faults near breaker C will generally be in the range of 0.25–0.5 seconds, while clearing times for faults near breaker D will be between one and three seconds. When backup is provided with time overcurrent relays, the backup times will generally be higher.

Remote type of backup is commonly used in less important areas of a power system. It provides an economical means for achieving back-up protection because the relays at A can be used in dual functions: they can provide primary protection of line A-B and back-up protection for line C-D. However, there are several shortcomings to remote type of backup.

Remote backup is inherently slow and, when called on to operate, often trips more breakers than actually necessary to clear a fault. For instance, in the above discussion, it is pointed out that for some fault locations, fault clearing times could be 3 seconds or higher. Moreover, it should be apparent from Figure 12.19 that when the backup relays at A operate, the loads on line A-B and the bus are disconnected.

Present-day systems have become increasingly complex due to a greater number of interconnected generating stations, a multiplicity of short-circuit current paths, and higher line loadings. As a result of this expansion, there are system locations where remote relays cannot be set to operate for faults on other system elements. Consider the same system as in Figure 12.19, with lines and generation added to the intermediate bus shown in Figure 12.20. These added lines and generation will contribute short-circuit current to faults on lines A-B and C-D and, consequently, will make it difficult for the remote backup relays at A and D to see faults on the adjacent lines, especially the faults near the remote buses. In effect, in-feed of fault current at the intermediate station makes a fault appear farther away to the remote relays. The greater the in-feed, the farther away the fault will appear. While it is some-

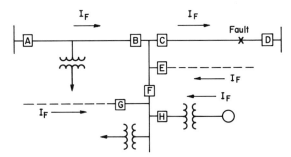

Figure 12.20. Illustration of remote backup relaying with intermediate feed.

times possible to increase the sensitivity or reach settings of remote relays so that they will see faults with in-feed, there will be situations where such settings will not be feasible because the required sensitivity or reach would cause relay operation on load currents or on minor synchronizing power swings.

Where remote backup is not acceptable or applicable because of the above shortcomings, a local type of backup is used.

2. *Local Backup.* In local backup relaying, faults are cleared locally in the same station where the failure has occurred. This type of backup provides both relay and breaker failure backup and is illustrated for a single bus arrangement in Figure 12.21. This functional diagram shows the basic plan

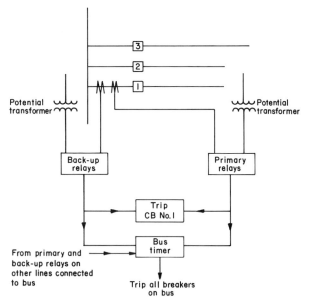

Figure 12.21. Local backup scheme.

APPLICATION PRACTICES

of this scheme for one of the lines connected to the bus. As shown, redundant relaying is used at each line terminal.

The primary protection shown in this diagram can be any form of distance-carrier relaying, pilot-wire relaying, phase comparison-carrier relaying, some form of microwave relaying, distance relays, or simple overcurrent relays. Relay backup may be identical to the primary protection or it may be one of the other types. Ideally, the primary protection and relay backup are supplied from separate current and potential sources. If so, a failure in any one set of relays, current, or potential supplies will not cause complete loss of protection.

For faults on the protected line, both the primary and the backup relays will operate to trip the line breaker. Relay backup may be just as fast as the front line relays. When either of these relays operates to initiate tripping of the line breaker, it also energizes a timer to start the breaker backup function. If the breaker fails to clear the fault, the line relays will remain picked up, permitting the timer to time out and trip the necessary other breakers on the associated bus section.

In addition to the above coverage, this scheme is designed so that no single failure in the dc control circuits can nullify all protection. Auxiliary relays are used where necessary to separate the dc circuits and all dc control and trip circuits are fused separately.

On important EHV systems, it is not uncommon for utilities to use two station batteries and to design the circuitry so that the primary protection is electrically isolated from the backup protection.

To illustrate the application of this type of backup protection, assume that a local backup scheme is used at breaker C in Figure 12.20. If a fault occurs on line C-D close to breaker C, the scheme would operate as follows to clear the fault. First, if it is assumed the primary and backup relays at C are distance relays, they will operate in high speed to clear the fault. If breaker C should fail to clear the fault for some reason, the bus timer can be set to operate to trip breakers B, E, and F in about 0.15–0.20 seconds.

With local backup protection in Figure 12.20, the fault will be cleared from the system in less time than that provided by remote backup, and the loads, generation, or lines are not unnecessarily disconnected from the system. Moreover, since the backup relays are located on the faulted line, they can easily detect faults on that line and there are no problems due to the effect of fault current infeed.

APPLICATION PRACTICES

Special requirements of the various system elements and the application practices involved are considered in this section. While the application practices are standardized to some degree, variations are to be expected for such reasons as personal preference, past experience, and special economic considerations.

It should be noted that this discussion will be limited to the practices involved in providing protection for the bulk power system and to the most commonly used protective schemes, equipment connections, and system arrangements.

Generator Protection

Protective relaying practices for generators are fairly well standardized throughout the utility industry. Protection is provided for the detection of short circuits and for the detection of abnormal operating conditions. Electrical protection for most all of the following conditions is normally provided on important system generators:

1. Stator short circuits
2. Grounded field
3. Loss of excitation
4. Unbalanced phase currents
5. Motoring or reverse power
6. Loss of synchronism
7. Abnormal frequency
8. Overexcitation
9. External faults.

In discussing this protection, only the unit generator-transformer arrangement will be considered. In this arrangement, a generator and a step-up transformer are connected to the system as a unit (see Figure 12.22). In this approach, there is no generator breaker, the step-up transformer is usually connected grounded wye-delta, and the neutral of the wye connected generator is high-resistance grounded through a distribution transformer. This arrangement is widely used with all types of generators from 25-MW combustion turbine generators up to 1500-MW steam turbine-generators.

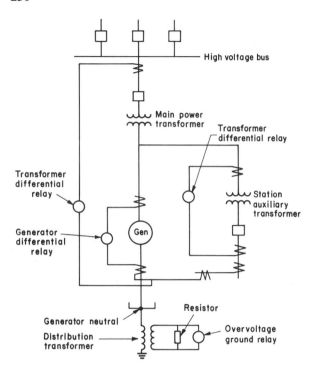

Figure 12.22. Unit generator-transformer protection.

Stator Short-Circuit Protection. Protective relaying is provided for the detection of phase and ground faults in the generator stator winding. In this and all following discussions, phase faults refer to faults between two- or three-phase conductors or windings while ground faults generally refer to a fault between a phase conductor or winding and ground.

1. Phase Faults. Percentage current differential relaying is used to provide phase-fault protection for the generator and the step-up transformer. As shown in Figure 12.22, differential relaying is provided for the generator and a separate differential relay for the transformer. The transformer differential protection is usually connected to include the generator within its zone, thereby providing primary protection for the transformer and backup protection for the generator. It also supplies backup protection for the station auxiliary transformer. Three differential relays, one for each phase, are provided for each elements.

2. Ground Fault Protection. With high-resistance grounding of the generator neutral, the primary fault current for a single line to ground fault at the terminals of the generator will be limited to 10 amperes or less. With this low-fault current level, current differential relaying cannot detect ground faults in either the generator or in the step-up transformer.

Where high-resistance grounding is used, ground faults are detected with a sensitive overvoltage relay connected across the resistor in generator neutral. A ground fault anywhere in the generator zone will cause a voltage to appear across the neutral resistor, thereby operating the overvoltage relay.

When any of the phase or ground fault protection operates, it is connected to energize a hand-reset multicontact auxiliary relay. The auxiliary relay is connected to trip the main generator breaker—the field breaker—to shut down the prime mover and to sound an alarm.

Field Ground. The exciter and field circuit of a generator is an ungrounded system. A single ground fault in the field will not affect the performance or operation of the generator. However, a second ground fault in the field could, in effect, short circuit part of the field winding and thereby cause unbalanced air-gap fluxes in the generator. This unbalance can cause severe vibration which can be damaging to the generator. Therefore, it is common practice to provide relay equipment to detect a ground fault in a generator field or in the exciter armature and leads.

One type of protection used for this purpose consists of a dc source and a voltage relay connected in series between the negative terminal of the field winding and ground. The generator rotor is grounded through a brush riding on the rotor shaft. A "ground" fault anywhere in the field (from the winding to the rotor) or in the exciter armature will impress a voltage on the relay, thereby causing relay operation.

This protection should be connected to trip the generator and field breaker and to shut down the prime mover.

Loss of Excitation. Loss of excitation can be caused by a complete short circuit or an open circuit in the field winding. When a generator loses excitation under full load, it will lose synchronism and operate as an induction generator. It will continue to supply some power to the system and it will receive its excitation from the system in the form of vars. The vars taken by the generator can equal generator rating. During this mode of operation, generator damage can occur due to high peak currents in the stator, excessive rotor heating, and pulsating-shaft torques. In addition, other system generators may lose stability and thereby cause a complete system collapse.

As noted above, the generator will lose synchronism when there is a loss of excitation. When this occurs, there will be a variation in generator current and voltage and therefore, a variation in impedance as seen at the generator terminal. The relay equipment used for loss of excitation protection is a distance relay connected at the generator terminals and set to detect the variation in impedance seen during the loss of excitation condition. This protection is usually connected to trip only the generator and field breakers. The prime mover is not shut down since it may be possible to remedy the problem quickly and restore the generator to service.

Unbalanced Phase Currents. Unbalanced currents in a generator can be caused by impedance dissymmetries, unbalanced loads, and unbalanced faults, such as phase-to-phase faults, all ground faults, and open phases. When any of these conditions cause unbalanced currents in a generator, large 120-Hz currents may be induced in the rotor. These induced currents can cause rapid overheating and damage to the generator rotor.

The currents induced in the rotor are proportional to negative-sequence currents in the stator. Therefore, a negative-phase-sequence time overcurrent relay is used to detect unbalanced current conditions in a generator. This protection is usually connected to trip the generator from the system without a complete shut-down, since it may be possible to quickly remedy the unbalanced condition.

Motoring Protection. A steam turbine requires protection against overheating when its steam supply is cut off and the generator runs as a motor. The turbine manufacturer usually supplies a number of temperature devices to detect this condition and to remove the generator from the system. It is commonly the practice to supplement these devices with a power-directional relay connected at the generator terminals to detect reverse generator power.

Power-directional relaying is also used to protect against motoring of combustion turbine or diesel generators. No immediate harm is done to a diesel if it is driven by its generator, although a substantial load is put on the power system. Combustion turbines are normally taken out of service by cutting off the fuel and allowing the motoring relay to shut down the machines.

Reverse power relays used for this purpose should have sufficient time delay so that the generator breaker will not be tripped because of momentary power reversals during severe oscillations between machine and system that may be caused by system disturbances.

Loss of Synchronism Protection. Protection for a loss of synchronism is not applied to all generators. This protection is only used when the system electrical center during a loss of synchronism appears in the generator or in the generator step-up transformer. The electrical center is the point on the system where the voltage is zero when the generator is 180° out-of-phase with the system.

Protection for this condition is provided by a distance-relaying scheme that detects the variation in impedance as measured at the machine terminals during a loss of synchronism. This protection is set to trip the generator when the machine voltage is 90°, or less out-of-phase with the system voltage. This protection is usually connected to trip only the generator breaker, thereby permitting quick resynchronization of the generator when conditions stabilize.

Abnormal Frequency Protection. Operation at abnormal frequencies can be detrimental to steam turbine-generators. While both the generator and turbine are limited to the degree of off-frequency operation that can be tolerated, the turbine is more restrictive since there are mechanical resonances that could cause turbine damage in a relatively short time for small departures in speed. Turbine manufacturers provide off-frequency capability limits which give the permissible operating time at different frequencies. The effects of off-frequency operation are cumulative. That is, the operating time accumulated during all off-frequency events should not exceed the permissible operating time specified by the capability limit.

When abnormally low frequency conditions occur on a power system, the primary protection for the steam-turbine generator is the system load-shedding program. The load-shedding program should be designed and implemented to bring system frequency close to normal quickly. Backup protection for the steam turbine generator is provided by a use of several frequency relays. The pick-up frequency and time settings of these relays are adjusted to protect the turbine in accordance with the manufacturer's capability limits.

Overexcitation Protection. Generators and transformers are designed to operate continuously at a specified maximum excitation (flux) level. This level is 105% excitation for a generator and 110% excitation for a transformer. Excitation or flux level is usually defined in terms of volts per Hertz (V/Hz). When these excitation levels are exceeded, the equipment may be driven into saturation and, flux may enter parts of the equipment not designed to carry high flux-density levels. This may result in the breakdown of lamination insulation and/or the creation of hot spots caused by increased eddy current levels, both of which can cause equipment failures.

Protection for this condition is provided by the use of voltage relays having a pickup that is a function of volts/Hertz. Two or more of these relays may be used to protect both the generator and the step-up transformer, and the relays are connected to trip only the generator and field breakers.

External Fault Backup Protection. External fault or system backup protection is usually provided on generators. The purpose of this protection is to disconnect the generator for uncleared phase and ground faults on the high-voltage system.

Two types of relays are used for phase fault backup: a voltage-restrained time overcurrent relay or a distance relay with a timer. The overcurrent relay is generally used when the connected transmission line is protected by overcurrent relays. The distance relay is used when the transmission line is protected by distance relays.

Ground fault backup is provided by a time overcurrent relay located in the neutral of the transformer high-voltage winding. These backup relays are connected to operate a hand reset lockout relay which trips and shuts down the generator.

Power Transformer Protection

There are three types of protection available for power transformers. These are (1) gas detection and analysis, (2) fault pressure, and (3) differential relaying.

Gas Detection and Analysis. This protection is designed to detect incipient faults in transformers. Incipient faults can be defined as the beginning stages of a failure where no actual arcing is involved. The sign of an incipient fault in an oil-filled transformer is the generation of combustible gases. These gases are produced by the breakdown in insulating material (oil or paper) due to a localized heating in the transformer.

The generation of gas in a transformer can be detected by two means: (1) gas-detector relay and (2) measurement of dissolved gases in the oil.

The gas-detector relay is basically a float-operated gauge mounted in a liquid-filled float chamber which is connected by tubing to the high point of a cover on an oil-filled transformer. Gas generated by an incipient fault will rise to the transformer cover and pass through the tubing to the float chamber. Gas accumulates in the chamber and lowers the liquid level. When 200 cm^2 of gas have accumulated, the float will activate an alarm. The gas can be removed from the chamber and tested to determine the levels of combustible gas.

The rate of gas accumulation indicates the magnitude of the fault. If gas accumulates quickly with resultant operation of the relay, the transformer will be taken out of service.

When measuring dissolved gases in oil, monitoring equipment is programmed to analyze periodically a sample of oil from the transformer. The concentration of nine different gases are determined through chromatographic analysis and the data is displayed on a paper chart recorder. The transformer will be removed from service if there is an increase or change in gas content.

Fault Pressure Relay. This type of relay is mounted on the transformer tank and operates when an internal fault creates a sudden rate-of-pressure rise within the oil. Theoretically, this protection is capable of detecting a turn-to-turn fault in a transformer. The fault pressure relay may be connected to sound an alarm or to trip the transformer from the system.

Differential Relaying. High-speed percentage differential relays are used for the detection of phase and ground faults in transformers. However, a special form of differential relay must be used to prevent incorrect relay operation when the transformer is energized.

When energizing a transformer, there will be a magnetizing-inrush current whose magnitude can be eight to twelve times the related load current. The inrush current appears in only one winding of the transformer and will cause operation of a conventional differential relay. The standard approach

APPLICATION PRACTICES

for preventing relay operation during the energization period is to utilize the high level of harmonics present in the inrush current. The harmonic components in the inrush current are filtered out and introduced into a circuit that restrains the relay from operating. This type of relay is called a harmonic restrained-percentage differential relay and it is the type most widely used for high-speed transformer protection.

It is common practice to use gas detection, fault pressure, and differential relaying on all important system transformers. When differential type relaying is not applicable or difficult to implement, only gas detection and/or fault pressure relays may be used to provide primary protection for small transformer banks and for phase shifting or wide-range voltage-regulating transformers.

Other Protection Considerations. Where small transformer banks are energized from a high-voltage line, it is common practice, for reasons of economy, to omit the high-voltage circuit breaker and to protect such transformers with fuses and overcurrent relays. This type of protection is not considered adequate since fuses and overcurrent relays will only detect severe internal faults (fault currents above twice full load current).

It has also become common practice (again for reasons of economy) to omit high-voltage circuit breakers when connecting large power transformers on high-voltage transmission lines or when terminating transmission lines in a transformer. In such a case, percentage differential relaying is used on the power transformer, and remote tripping is used to trip the breaker at the far end of the line whenever the transformer relaying operates. The reason for using remote tripping is that many transformer faults would not cause enough current to flow to operate the relays at the far end of the line. If carrier current or microwave pilot relaying is used for the line, the same channel is used to send a special transferred-tripping signal to the distant end. A wire pilot and an audio tone channel could be similarly used. If a pilot channel is not available, the practice is to use a high-speed grounding switch to throw a short circuit on the line whenever the transformer differential relaying operates. This short circuit causes the remote line protective relays to trip their breakers.

Overheating Protection. It is not the usual practice to disconnect a power transformer automatically when it overheats, whether the transformer is in an attended station or not. Replica-type relaying accessories are available that can be used to sound an alarm.

On low-voltage systems, it is the usual practice to have overcurrent relaying on one side or another of a transformer for backup protection against short circuits in the equipment supplied by the transformer. If sensitive enough, this will also serve to prevent serious overloading.

Bus Protection

It is an almost universal practice to use differential relaying in one form or another to protect buses against short circuits. Two types of relays are used (1) conventional current percentage differential relaying and (2) high-impedance differential relaying.

1. Conventional current differential relaying is generally used on low-voltage buses where high speed clearing of faults is not essential. In general, time delay must be used with this form of relaying to prevent incorrect relay operation during external faults. Such relay misoperation can occur due to a reduction in current transformer accuracy for a short circuit just outside the zone of protection. For this condition, the current transformers in the circuit energizing the fault will carry all of the current from the other circuits connected to the bus. This total current might cause the current transformer core to saturate. The saturated current transformers will produce a distorted and reduced output, while the current transformers in the other circuits (with a lower current level), will usually maintain their accuracy and produce nearly full outputs. The currents in the differential circuit will be unbalanced and a large error current will flow in the differential relay-operating coil. Relay misoperation can be prevented for this condition by desensitizing the relay or by adding time delay.

2. High-impedance differential relaying is a high speed scheme designed to operate correctly even when one or more current transformers saturate. This scheme uses standard bushing current transformers and a high-impedance voltage relay connected in a differential circuit, as shown in Figure 12.23. In this approach, discrimination between internal and external faults is made on the basis of the voltage magnitude that appears across the relay. During internal faults, the voltage that appears across the relay will be much greater than the volt-

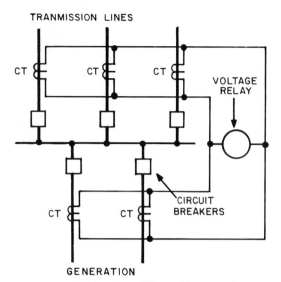

Figure 12.23. Bus differential protection.

age that appears for external faults, even with complete current transformer saturation. Therefore, to prevent operation for external faults, it is only necessary to set relay pick-up well above the maximum voltage obtained for an external fault, assuming complete saturation of the current transformers in the faulted circuit.

Air-core current transformers and a sensitive voltage relay can also be used to eliminate the current transformer saturation problem. In this approach, the current transformer outputs are connected in series and total output impressed on the relay. This scheme also discriminates between external and internal faults by the voltage that appears across the relay. During internal faults, the voltage will be high; while for external faults, the voltage should theoretically be zero.

In general, the high-impedance relay scheme is the most widely used because it provides the required high-speed protection using conventional current transformers.

Line Protection

Because most of the short circuits in power systems occur on overhead transmission or distribution lines, their protection has received the greatest attention over the years. In terms of damage and cost of repair, a fault on an overhead line is usually much less objectionable than a fault in any other part of the system. It is the greater frequency of line short circuits and their possible effect on service that makes the protection of lines so important.

While there are a variety of line-protective schemes and practices used by utilities to meet particular system requirements, this section will be primarily concerned with the application of the basic type of relaying equipment used for line protection. These are overcurrent relaying, distance relaying, and pilot relaying.

Overcurrent Relaying. Overcurrent relaying is applied principally on subtransmission systems, on radial distribution circuits, or on simple loop distribution circuits. Faults in such circuits do not generally affect system stability and, consequently, do not require high-speed relaying.

Many systems still use overcurrent relaying for high-voltage transmission lines to obtain effective and satisfactory protection. However, as the demand grows for faster fault-clearing times, overcurrent relaying is being replaced by distance or pilot relaying.

Overcurrent relaying is the simplest and least expensive form of line protection. It is also the most difficult to apply and requires readjustment as the system configuration changes. Overcurrent relays cannot discriminate between load and fault current and, therefore, when these devices are used for phase-fault protection, they are only applicable when the minimum short-circuit current is greater than the maximum full-load current.

There are two forms of overcurrent relays: the time overcurrent relay and the instantaneous overcurrent relay. Both forms are used for line protection.

When the time overcurrent relay is used for line protection, it is usually set and adjusted to provide primary protection for a line section and remote backup for an adjacent section. For example, in Figure 12.24, a time overcurrent relay at A provides primary protection for line A-B and backup for line C-D. To provide both of these functions, the relay at A must be set to pick up for a fault near breaker B and near breaker D for the fault and system conditions that produce minimum current in the relay.

Since the relay at A can operate for faults on line C-D, it must be adjusted to be selective with the primary relays at C. The relays at C must be permitted to operate and trip first for a fault on line C-D before the relays at A operate. Selectivity is achieved by utilizing the inherent time delay char-

APPLICATION PRACTICES

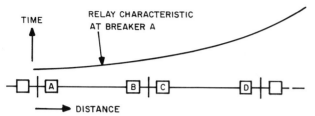

Figure 12.24. Time–distance characteristic of an inverse-time overcurrent relay.

acteristic of this type of relay. The time overcurrent relay has an operating characteristic such that operating time varies inversely as the current in the relay. This characteristic is illustrated in Figure 12.24 as a function of distance for the relay at A. As seen in Figure 12.24, the virtue of the inverse time characteristic is that there is less time delay for faults in the primary zone than is necessary for backup. The farther a fault is from the relay location, the greater the circuit impedance to the fault and the lower the fault current; the lower the fault current, the greater the operating time.

Figure 12.25 shows the application of time overcurrent relays to a series of radial lines. This figure illustrates how time coordination is achieved between inverse time overcurrent relays at each breaker location. For the fault shown, the relay tripping breaker 1 operates quickly; the relay at 2 has enough time delay to permit tripping of breaker 1 if it can; the relay at 3 is selective with the relay at 2 by having a still longer time delay; and finally, the relay at 4 has the longest time delay of all and will not trip its breaker unless all of the other breakers fail to trip. This figure shows that for faults near the power source, the time required to clear a fault becomes longer. This is a principal disadvantage of time overcurrent relaying.

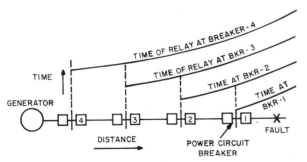

Figure 12.25. Operating time of overcurrent relays with inverse-time characteristics.

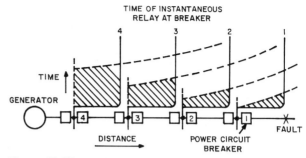

Figure 12.26. Reduction in tripping time by use of instantaneous relays in conjunction with inverse time relays.

Time overcurrent relays have different time current characteristics: inverse, very inverse, and extremely inverse. It is advisable to use relays whose time current curves have the same degree of inverseness, or the problem of obtaining selectivity over wide ranges of short-circuit current levels may be difficult.

Instantaneous overcurrent relaying is used only for primary relaying to supplement inverse-time relaying. This type of relaying is applicable only if there is a substantial increase in the magnitude of fault current as the short circuit is moved from the far end of a line toward the relay location. This increase should be at least two or three times. Since the instantaneous overcurrent relay operates without time delay, it must be set so that is can never detect an external fault. The zone of protection of an instantaneous relay is established entirely by adjustment of sensitivity and is terminated short of the far end of the line. The instantaneous relay is usually set so that its pick-up is at least 25% higher than the maximum current the relay will see for a three-phase fault at the end of the line. With this setting, the instantaneous relay will protect about 80% of the line section.

Instantaneous overcurrent relays are frequently added to inverse time relaying to effect a reduction in tripping time. This is illustrated in Figure 12.26 where the two characteristics are superimposed. The shaded area indicates the reduction in time achieved by using instantaneous relays. A reduction in short-circuit current magnitude shortens the distance over which the instantaneous relays will operate and, in some instances, this distance may be reduced to zero. This is usually of no great importance, since the primary objective is fast tripping under the maximum short-circuit conditions.

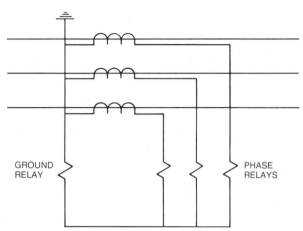

Figure 12.27. Complete overcurrent relay protection for a line.

The preceding discussion has described in a general way how overcurrent relays are applied and coordinated. Complete protection for a line is provided by the use of three overcurrent relays for phase-fault protection and one overcurrent relay for ground-fault protection as shown in Figure 12.27. Each phase-fault relay receives a current from a current transformer in a phase conductor while the ground relay only senses a current that would flow in the neutral connection of the wye-connected current transformers (zero-sequence current). Under balanced or nonfault conditions, no current flows in this neutral connection. During a ground fault, a zero-phase sequence current will flow in the neutral and is used to detect all faults involving ground. The ground overcurrent relay would be applied, set and coordinated as described above. If the system loads are balanced, this relay can be set quite sensitively and is not restricted by load currents as are the phase relays. If the system loads are unbalanced, there will be a current flowing continuously in the neutral connection and the pick-up of the ground relay will have to be set above this current level.

Toward the ends of primary distribution circuits, fuses are sometimes used instead of relays and breakers. In the region where the transition occurs, it is frequently necessary to use overcurrent relays having extremely inverse characteristics to coordinate with the fuse characteristics.

The extremely inverse relay characteristic has also been found helpful in permitting a feeder to return to service after a prolonged outage. After a feeder has been out of service so long that the normal "off" period of all intermittent loads (such as furnaces, refrigerators, pumps, water heaters, etc.) has been exceeded, reclosing the feeder throws all of these loads on at once without the usual diversity. In this case, the total inrush current may be four times the normal peak-load current. This inrush current decays very slowly and will be approximately 1.5 times normal peak current after as much as 3 or 4 seconds. Only an extremely inverse characteristic relay provides selectivity between this inrush and short-circuit current.

When overcurrent relays are used for phase-fault protection the minimum fault current must be greater than the maximum load current. If this criteria cannot be met, overcurrent relays will not be applicable and some form of distance relaying will have to be used.

Distance Relaying. Distance relaying is the most versatile and widely used form of transmission line protection. It has replaced the overcurrent type of protection on most high-voltage transmission circuits because it:

1. Provides faster protection
2. Makes coordination simpler
3. Is less affected by changes in generation and system configuration
4. Permits higher line loadings
5. Is less affected by power swings.

Distance relays are generally used to provide primary and back-up protection for three-phase and phase-to-phase faults. Overcurrent relays are still generally used for ground fault protection because this type of protection is less susceptible to the type of problems encountered by phase overcurrent protection, and it can generally provide the necessary tripping speed. However there is a trend toward the use of distance relays for ground fault protection.

The application of distance relays is fairly simple. They are generally applied to provide primary protection for a line section and remote backup for an adjacent line section. Three separate sets of relays, arranged into three protective zones, are used to achieve this protection. A first zone and a second zone are used to provide primary protection for a line while the third zone is used for remote backup of an adjacent line.

The first zone relays are set to provide high-speed protection for about 90% of the line. These

APPLICATION PRACTICES

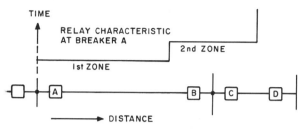

Figure 12.28. Time–distance characteristic of distance relay for primary relaying.

relays operate without time delay and they must be set so that they will never operate for faults beyond the end of the line.

The second-zone relays provide protection for the remaining 10% of the line and they are set to reach beyond the end of the line so that relay operation will be assured for any phase fault in the 10% region. A typical impedance setting for this zone is 120% of the protected line. Since this zone will see faults in the next line section, it must be time coordinated with the primary relays in the next line section. To achieve this coordination, the second-zone tripping is delayed from 0.2 to 0.5 seconds by means of a separate timer.

The time distance characteristics of the first and second-zone relays is illustrated in Figure 12.28. It should be noted that this step characteristic is achieved with two sets of relays.

The third-zone relays are set to provide backup protection for an adjacent line section. In Figure 12.28, third-zone relays at A would provide backup for line C-D. The third-zone relay is set to reach beyond breaker D so that relay operation is assured for phase faults in line C-D. Tripping by the third zone is delayed between 1 to 3 seconds to obtain coordination with the relays on line C-D. The setting and coordination of three zones for several line sections is illustrated in Figure 12.29.

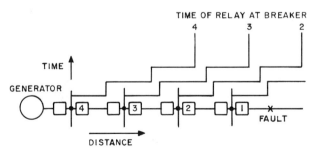

Figure 12.29. Time–distance characteristic of distance relays.

In the section on basic relay types, four types of distance relays were introduced: the impedance, the reactance, the mho, and the lens-type relays. The mho type relay is generally used for the protection of long lines, while the reactance type is most applicable to short lines. In the past, the impedance-type relay was used extensively for the protection of both long and short lines. At the present time, however, it is seldom used for line protection, since both the mho and reactance types provide adequate and more desirable coverage for most line configurations. The lens type is generally used to restrict the tripping area of a mho unit when these relays are applied on long transmission lines.

Ground-fault protection is usually provided with overcurrent relays, as discussed in the preceding section. In some instances, distance relays are also used for ground-fault protection.

While the operation of a distance relay is relatively independent of the magnitude of short-circuit current flowing in the relay, it is affected by the magnitude of any current that may be fed into the system between the relay and the fault. As was pointed out in the section on backup protection, this in-feed of fault current makes a fault appear to be farther away than it really is. This effect is important and must be considered when applying distance relays to a tapped line or when setting the reach of backup relays.

On many systems, distance-type relaying will provide adequate protection and will be sufficient to preserve stability. However, when stability becomes a critical factor and high-speed reclosing of the line breakers is necessary to maintain stability, straight-distance relaying is not usually applicable. This is because high-speed reclosing requires the breakers at both ends of the lines to be tripped simultaneously at high speed for all faults. This is not possible with only distance relaying. For such applications, pilot relaying is the only solution.

Pilot Relaying. Pilot relaying is recommended for the protection of all bulk power transmission lines. It is inherently selective and it is capable of providing simultaneous high-speed tripping of all circuit breakers for all faults.

In general, pilot-relay schemes are designed to provide primary protection for a line. Backup protection may be provided by a separate set of relays, or the relays used in the pilot may also be used to provide a backup function. When separate backup relays are used, they may be step-distance relaying, as discussed in the preceding section or they may be another pilot-relaying scheme. On extra-high-voltage systems, it is becoming common practice to use two pilot-relaying schemes for the protection of important lines and where stability is a problem.

The selection of a pilot-relaying scheme is a function of such factors as line length, line configuration (two or multiterminal line), fault current level, reliability requirements, and channel availability. In the sections that follow, these factors will be considered in discussing the application of the various types of pilot relaying.

Wire-Pilot Relaying. Wire-pilot relaying is generally used for the protection of short lines; that is, lines ranging in length from 0.2 miles (.38 km) to about 30 miles (48 km). The relaying schemes that can be used with this pilot are (1) ac wire pilot relaying and (2) any of the blocking or transferred tripping schemes using the frequency-shift audio-tone equipment.

Ac wire-pilot relaying is capable of providing low cost, high-speed protection for short lines, and power cables. These schemes are generally applicable where the one-way length of the pilot wire is 10 miles (16 km) or less. The distance limitation is usually a function of the attenuation of the signal due to the distributed capacitance and the series resistance of the pilot-wire pair. If a leased wire is used, its characteristics should be checked because it may be longer than the actual distance between the terminals of the power line.

The opposed voltage type of ac pilot-wire relaying is generally applicable only on two-terminal lines. The circulating current type has a wider range of application and is applicable on two and three terminal lines. Both of these schemes provide primary protection only for a line. Separate phase and ground relaying are required to provide the necessary backup protection.

Any of the blocking schemes (directional or phase comparison) or any of the tripping schemes can be used with frequency-shift audio-tone equipment over pilot wire. The most widely used form of protection has been the permissive type of transferred tripping. This scheme can provide economical protection for short lines and is applicable where the length of the pilot wire circuit is between 20 and 30 miles (32 and 48 km). This approach has been used over longer distances on a leased pilot wire where the audio-tone signal is amplified, repeated, or transmitted via another carrier.

The blocking and tripping schemes can be designed to provide only primary relaying or they can be designed to provide both primary and backup protection. If they provide only primary protection, it is necessary to provide some form of backup relaying.

A major factor of concern with all types of wire pilot relaying is the reliability of the pilot wire. The wire pilot may be exposed to lightning, mutual induction with the power line, contact with the power line, differences in ground potential, and interference by maintenance personnel. All of these factors can cause the loss of the pilot wire when it is needed for line protection. The wire pilot must be designed and protected for the natural and electrical hazards and supervising equipment can be applied to detect the loss of the pilot wire.

Power Line Carrier Relaying. Usually, the blocking type of protective system is used with the power line carrier channel. The schemes that fit in this category are the directional comparison scheme and the phase-comparison scheme.

The directional comparison scheme is the most versatile and most widely used scheme for line protection. It is applicable on short or long lines, on multiterminal lines, and on series capacitor-compensated lines. It can be used for single-phase relaying and it has the sensitivity and speed to meet system requirements. In most instances, it can be readily modified as the line configuration and/or the system changes.

Directional comparison protection is available with electromechanical and solid-state relays and the scheme can be designed to provide only primary protection or to provide both the primary and backup functions. On important high-voltage lines, it is common practice to use separate primary and backup relaying connected in a local backup scheme.

Phase-comparison relaying in its simple overcurrent form is not as versatile or as sensitive as directional comparison relaying. In general, phase-comparison relaying is best applied where fault current levels are two to three times larger than load current. When high-fault currents are available, this scheme can provide reliable protection on two and three terminal lines.

The advantages of the overcurrent form of phase-comparison relaying is simplicity. It does not require a potential source and it is immune to incorrect operation on power swings and zero-sequence

APPLICATION PRACTICES

mutual induction. Because of these advantages, this form of phase comparison is often used where there are a number of lines on the same right-of-way or as an alternate pilot scheme on extra-high-voltage lines.

The sensitivity and versatility of phase-comparison relaying can be improved by combining it with directional comparison relaying. In these combined schemes, directional comparison relaying may be used to provide three-phase-fault protection while phase comparison is used for the detection of all other faults; or directional comparison may be used to detect all phase faults and phase comparison used to detect only ground faults.

There are instances where the transferred-trip schemes have been used with the carrier current pilot for line protection. This type of protection is not considered as reliable as the blocking system since it necessitates transmitting a trip signal through a fault. There is some question as to whether or not the trip signal can get through the fault 100% of the time, especially if the fault is located near a line terminal.

Aside from providing the protective function, the power line carrier channel can be used for other services. For example, simplex telephony is a standard part of carrier current pilot-relaying equipments. Also, where a line terminates in a power transformer without a high-tension breaker, remote tripping for transformer faults can be provided with little extra equipment. In certain cases, the coupling capacitors and line traps can be used in common by still other services, such as telemetering, supervisory control, etc. When other services share the relaying channel, the relay equipment automatically takes preferential control when faults occur.

Microwave Pilot Relaying. In general, the microwave pilot cannot be justified economically for protective relaying functions alone. Because microwave installations are expensive, it is feasible only when used to provide a number of services, such as telephony, telemetering, supervisory control, load-frequency control, etc. In such a case, the addition of a relay pilot to a microwave channel is not expensive.

While both the blocking and the tripping type of protective systems are applicable over the microwave pilot, in most instances, the tripping schemes are used with this pilot. Tripping schemes are used because they are slightly less complex than the blocking schemes and since the microwave channel is independent of the power circuit, these schemes can be used without interference from the fault.

The microwave channel generally utilizes audio tone equipment to achieve a tripping pilot. The protective relays key the audio tone equipment and the tone signal is propagated over the microwave channel to the other line terminals.

In general, the microwave pilot is not considered to be as reliable for relaying purposes as carrier current, especially when numerous repeat stations must be used in the microwave system. Furthermore, when the microwave system is widely used as a relaying pilot, a microwave failure may affect the quality of protection for a large part of the power system.

Other Considerations

Series Compensated Lines. Series capacitor compensation of transmission lines is used to improve stability limits, to improve voltage regulation, to provide a desired load division, and to maximize the load carrying capability of the system. Series capacitors introduce special problems for line-protective relaying. Under fault conditions, they introduce an impedance discontinuity (negative inductance), subharmonic currents, and when the capacitor protective gaps operate, they impress high-frequency currents and voltages on the system. All of these factors can cause incorrect operation of the conventional relaying schemes.

Special relaying schemes have been developed for the protection of series compensated lines. These schemes are more complex and costly than standard forms of relaying and the same type of scheme must be used for primary and backup protection.

Single-Phase Relaying. The conventional line-protective schemes discussed in the preceding sections are designed to trip all three phases of a transmission line for all types of faults. In the single-phase tripping and reclosing schemes, the scheme is designed to trip and reclose only the faulted phase on single phase to ground faults and to trip all three phases on multiphase faults. Since most transmission line faults are single-phase faults, this approach tends to minimize the effect of faults on system voltages and operation and it provides additional stability margin since power can still be transmitted over the two sound phases.

Single-phase tripping and reclosing schemes are more complex and costly than conventional schemes because additional protective relaying is required to detect faults on individual phases. Similar schemes are generally used for both primary and backup protection.

Transmission Line Reclosing. Since most transmission line faults are of a transitory nature, it is the practice to reclose automatically transmission lines remote from a generating station after they have been tripped for a fault. In most instances, a single high-speed reclosure will be attempted only after the simultaneous, high-speed tripping of all line breakers by the primary pilot-relaying system. If the fault persists, the line relays will trip the line out again and, usually, no further attempts will be made to reclose the line automatically. Some utilities, however, will attempt to reclose the line automatically two or three times with time delay between each attempt. If all attempts at automatic reclosure fail, a system operator may attempt to reclose the line manually after some time interval. Of course, if this fails, it would indicate a permanent fault and the line is taken out of service for repairs.

This practice of high-speed automatic reclosure of a line provides several benefits since it:

1. Returns a line quickly to service, thus maintaining the integrity of the power system.

2. Minimizes the effect of a line outage on critical loads.
3. Permits higher loading of the transmission system.

The utility industry has used this practice for years with excellent results. Experience has shown that successful reclosure can be achieved 80–90% of the time.

High-speed reclosing is not used in all instances. In some cases, a reclosure into a persistent fault may cause system instability and, therefore, would not be desirable.

In addition, high-speed reclosing after tripping for phase faults is not recommended on lines leaving a generating station. High-speed reclosures into persistent phase faults can cause mechanical stresses and possibly shaft fatigue in steam turbine-generators. Reclosing duties on steam turbine-generators can be minimized by:

1. Delaying reclosure for a minimum of 10 seconds.
2. Using selective reclosing, that is, reclosing only on single-phase to ground faults.
3. Using sequential reclosing. Reclose initially at the remote end of the line and block reclosing at the generating station if the fault persists. This approach is only applicable if the line is long and/or if there is no generating station at the remote end.

SUMMARY

The type of protective relays and practices used for power systems may vary widely between utilities. Factors such as service reliability requirements, system stability, equipment and/or line damage considerations and economics may determine the type of relaying used. In general, as these requirements and considerations become more stringent, so do the requirements for protective relaying. To meet present and future system requirements, the development of solid-state relays is continuing and expanding into the area of microprocessors and computers to meet the need for complex relaying functions and specialized protective characteristics.

REFERENCES

For further information on protective relaying, see the references listed below. Unfortunately, very few books have been published that offer a good, basic understanding of this subject. However, many publications and papers covering all aspects of protective relaying are available from the relay manufacturers.

1. *Guide for A.C. Motor Protection,* ANSI/IEEE Standard Publication C37.96-1976.
2. *Guide for Protection of Shunt Capacitor Banks,* ANSI/IEEE Standard Publication C37.99-1980.
3. *Guide for Protective Relaying Application to Power System Buses,* ANSI/IEEE Standard Publication C37.97-1978.
4. *Guide to Protective Relay Applications to Power Transformers,* ANSI/IEEE Standard Publication C37.90-1978.
5. Mason, C. R., *The Art and Science of Protective Relaying,* Wiley, New York, 1956.

13
STABILITY

INTRODUCTION

Stability is a condition of equilibrium between opposing forces. In a power system, when the forces tending to hold the machines in synchronism with one another are in equilibrium with the forces tending to pull them out of synchronism, the system is stable. Stability is desirable under all operating conditions, both normal and abnormal. These forces may properly be expressed as torques since only rotary motion is involved. The average rotational velocity is constant, although there may be momentary excursions above and below synchronous speed. Thus, torque and power will be used interchangeable as long as both are treated as per unit quantities.

When any synchronous machine is operating stably, there is equilibrium between the power input and the power output; that is, between the power tending to accelerate and the power tending to decelerate its rotor. The power may be either mechanical or electrical. In a generator, the accelerating power is the mechanical input from its prime mover, which tries to speed up the rotor. The decelerating power is the sum of the electrical output and the losses. In a motor, the accelerating power is its electrical input, while losses and the mechanically-driven load are the decelerating power that act as a brake. These power components tend to produce either acceleration or deceleration of the rotor, and may do so momentarily in any given machine when there is a temporary excess of either. However, there can be no sustained acceleration or deceleration if stable operation is to continue. For approximate stability studies, it is assumed that the losses are negligible.

The simplest form of power relationship for every synchronous machine operating stably is

Mechanical power = electrical power.

Power-system stability is determined by the generator's ability to develop synchronizing or restoring torques when disturbed from an equilibrium (above equation) by some disturbance. Following a disturbance the power-balance equation requires another term; that is

Mechanical power = electrical power
                          + acceleration power.

During a nearby fault, the electrical power output is reduced and the amount of this reduction goes into accelerating the rotor. Something must be done to reverse this, i.e., decelerate the rotor, before the rotor position has advanced too far and pulls "out of step" or is unstable. Stability can be lost if the electrical load on a machine is suddenly increased too much. For instance, the sudden tripping of one large generator may impose an excessive momentary load on another nearby generator. To preserve the power balance, the acceleration power must decrease by an amount equal to the increased electrical power imposed on the generator. The resulting "deceleration" can cause the generator to fall behind too far and pull out of step with other generators in the system.

The maximum amount of power that can be transferred between machines, or between groups of machines, without loss of synchronism between them following a disturbance is called the power limit, or stability limit. It is the critical value of power transfer below which the system is stable and above which it is unstable.

For convenience of analysis, stability problems are generally divided into two major categories (IEEE Task Force on Terms and Definitions, 1982). (Note that stability is actually one characteristic and that any subdivision into classifications is only for convenience of analysis.) The two categories are as follows:

1. Steady-state instability occurs when the power system is forced into a condition for which there is no equilibrium condition. For example, the power output of a generating plant may be slowly increased until maximum electrical power is transferred. At this point, either an increase or decrease in generator angle will result in a reduction in power transferred. Any further increase in generating plant output will cause classical steady-state instability and loss of synchronism between that plant and the rest of the system. Dynamic instability, another form of steady-state instability, is characterized by hunting or steadily growing oscillations, ultimately leading to a loss of synchronism.

2. Transient instability applies to a system's inability to survive a major disturbance, thus causing an abrupt and large transient change in the electrical power supplied by the synchronous machines. The occurrence of a fault or the sudden outage of a transmission line carrying heavy load from a generating plant will cause a severe momentary unbalance between the input power and the electrical load on one or more generators. If the input/output power unbalance is large enough or lasts long enough, the result will be transient instability.

A complete understanding of power system stability and the factors that either preserve or threaten stability, cannot be gained without first reviewing certain fundamental concepts (Clarke, 1950). The next few sections cover these necessary fundamentals.

Transfer of Power

Consider the system of Figure 13.1, where $\bar{E}_{A1}$, $\bar{E}_{B1}$, and $\bar{E}_{C1}$, are the balanced three-phase voltages at point 1, separated by an impedance $\bar{Z}$ in each phase, from the balanced three-phase voltages $\bar{E}_{A2}$, $\bar{E}_{B2}$,

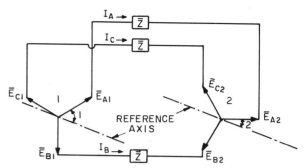

Figure 13.1. Simplified representation of current, voltage, and impedance in a three-phase system.

and $\bar{E}_{C2}$ at point 2. $\bar{E}$, $\bar{I}$, and $\bar{Z}$ are treated as phasor quantities.

Current flows in the circuit between points 1 and 2 in response to a difference between the voltages at the two points. A difference in magnitude of the voltages causes only reactive current to flow between them when $\bar{Z}$ is a pure reactance $+jX$. A difference in phase position of the voltages, that is, an angle between them, causes inphase (active) current to flow. As a result, angular displacement between voltages is necessary to the transfer of power. This can be demonstrated by the following simple experiments.

This experiment can be performed on a synchronous machine which has a horizontal shaft and can be operated as a motor or a generator. To begin, carefully paint a well-centered arrow on the end of the shaft. With the machine running as a motor at no load, stroboscopically illuminate the end of the shaft that has the arrow. The arrow, not visible under ordinary light, can be made to appear stationary with the proper adjustment of the stroboscope. Load the "motor" by adding a mechanical load, noting that the arrow will appear to move, i.e., rotate through a small angle and then remain stationary at a new position. This angle is called the load angle. If this is a four-pole motor on a 60-Hz voltage (1800 r/min), the observed angle (mechanical degrees) must be doubled to obtain electrical degrees.

Next operate the synchronous machine as a generator. To do this, the machine must be driven mechanically. The prime mover (in this case it could be a dc motor coupled to the synchronous machine), increases the applied torque (or input power) to the synchronous generator. The actual speed will not change but a transient will occur so the arrow will appear to move to a new position. This will be in the direction opposite to that observed as a motor. The

STEADY-STATE STABILITY

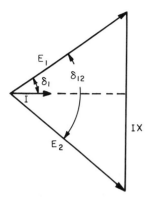

Figure 13.2. Angular relationships between current and voltages of Figure 13.1.

arrow will move with shaft rotation as a generator, and opposite to mechanical rotation when operated as a motor.

This experiment indicates that increasing the load, on either a motor or a generator, will increase the load angle. The machine appears to be connected to the load through a flexible connection. When load is suddenly changed, the arrow's position will oscillate. The oscillation will be damped if the system is stable and the arrow will reach a final steady-state position as referred to above. The stability of a given generator is then determined by whether or not this flexible connection is "broken."

The power transfer between two points, as between 1 and 2 of Figure 13.1, is given in terms of the phasor diagram of Figure 13.2 by the expression

$$P = \frac{E_1 E_2}{Z} \sin \gamma \sin \delta_{12} + \text{loss terms} \quad (1)$$

where, E_1 = voltage magnitude at point 1; E_2 = voltage magnitude at point 2; δ_{12} = total angle between voltages E_1 and E_2; Z = total impedance magnitude between the points 1 and 2 (this may include machine internal impedances depending upon the location of 1 and 2); and $\bar{Z} = Z\underline{/\gamma}$ total phasor impedance.

In most systems $Z(= R + jX)$ is predominantly inductive reactance, and R can be neglected with little error. The expression for power then becomes

$$P = \frac{E_1 E_2}{X} \sin \delta_{12}. \quad (2)$$

This relation is shown in Figure 13.3. The maximum value occurs at an angular displacement of 90° and is

$$P_{\max} = \frac{E_1 E_2}{X}.$$

Equation 2 is the simplest form of the fundamental equation of power flow in an electrical system. An appreciation of this equation and its significance is basic to an understanding of all stability problems.

Voltages and Reactances

What voltages and what reactance are represented by the symbols E_1, E_2, and X in equation 2 above? In the derivation of this general expression, the voltages E_1 and E_2 were maintained at the terminals of the impedance Z (of which only the X part will be used). The expression for power is valid so long as E_1 and E_2 are on either side of X, and δ is the angle between them. The X used in stability calculations almost certainly includes internal machine reactances. Since there are several values of machine reactance, it is necessary to choose that value of machine reactance that is appropriate for the purpose.

STEADY-STATE STABILITY

The general classification of steady-state stability refers to a system's ability to withstand *small* changes or disturbances from the equilibrium state without the loss of synchronism between two or more synchronous machines in the system. Two subclassifications that were mentioned earlier, "classical steady-state stability" and "dynamic stability," will be discussed in the sections to follow.

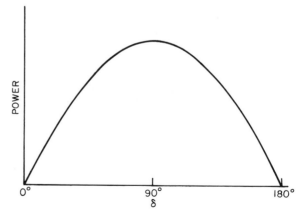

Figure 13.3. Power versus angular difference between voltages.

The classical definition of steady-state stability applies to the ability of interconnected synchronous machines to remain in synchronism while *small and gradual* changes are occurring on the system. It is assumed that automatic controls, such as continuous-acting voltage regulators on synchronous machines are not active.

The above definition is of historical significance since the earliest voltage regulator systems were of the noncontinuous-acting types whereby deadbands prevented control action for very small changes in the sensed voltage. The classical definition does have meaning today for those instances when a generator's excitation system is operating in the manual mode; that is the terminal voltage regulator is out of service.

The classical treatment of steady-state stability theory (Crary, 1947; Kimbark, 1948; Kimbark, 1956), is also of practical value because it provides insight into the interconnected operation of synchronous machines without the added complication of control system behavior. Ignoring control system actions and considering gradual changes in system load permits the analysis of the dynamic power systems in terms of easy-to-understand steady-state equations of machines in a static transmission network.

In the sections that follow, reference made to steady-state stability will refer to this classical definition. A later section entitled dynamic stability will discuss the importance of continuous-acting controls, particularly the excitation systems in modern power systems, and how damped or growing oscillation can determine the stability or instability of a system.

Effect of Saturation

Experience has indicated that ignoring saturation leads to unduly pessimistic estimates of the ability of the machines in a system to remain in synchronism when gradual and small disturbances occur. For this reason, the effect of saturation is usually included in analyzing a system for steady-state stability. The usual method of estimating saturation effects is to utilize an equivalent reactance X_{eq} instead of X_d in the machine equations. In effect, X_{eq} is treated as a saturated value of X_d. The voltage behind X_{eq} can no longer be regarded as being proportional to field current. This representation by a voltage behind X_{eq} constitutes an equivalent unsaturated machine that will have the same behavior at its terminals for small system changes as the actual, physical saturated machine. For the purpose of steady-state stability analysis, the saturated machine is being replaced by an equivalent generator consisting of a constant voltage behind a linear reactance, X_{eq}.

There are various techniques for estimating the proper value of X_{eq} which are discussed in texts on stability (Crary, 1947; Kimbark, 1948; Kimbark, 1956). These methods involve not only the saturation characteristics of the generator but also the characteristics of the system to which the generator is connected and the terminal conditions of the generator. For this reason, "exact" values of X_{eq} for a given machine are very difficult to calculate, and generally the value of X_{eq} is estimated. The practice is to use the following relationship to estimate the value of X_{eq}.

$$X_{eq} = \frac{1}{\text{SCR}}$$

where SCR = short-circuit ratio.

Figure 13.4 represents a motor and a generator connected directly together. In this diagram, E is the terminal voltage common to both machines. E_G

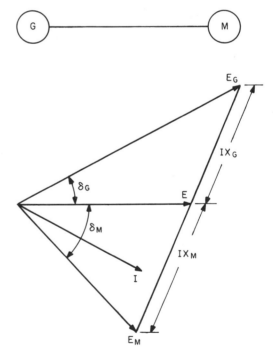

Figure 13.4. Representation of angles between internal and terminal voltage and current of motor connected directly to generator.

STEADY-STATE STABILITY

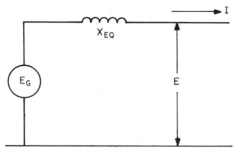

Figure 13.5. Illustration of internal voltage back of generator-equivalent reactance.

is the voltage back of the equivalent synchronous reactance, X_G, of the generator; and E_M is the voltage back of the motor equivalent synchronous reactance, X_M. While the voltages E_G and E_M may be considered fictitious, in that they are not directly measurable, they do nevertheless represent excitation that must exist inside the machines to fulfill measurable terminal conditions. Figure 13.5 illustrates this concept of an "internal" or "generated" voltage E_G, back of the machine equivalent reactance X_{eq}, as viewed from the machine terminals.

Neglecting losses, the power being transferred from the generator to the motor of Figure 13.4 is as follows:

$$P = \frac{E_G E_M}{X_G + X_M} \sin(\delta_G + \delta_M). \quad (3)$$

This must equal both the prime-mover input to the generator and the power output from the motor to its mechanical load. To illustrate a method of calculation and to gain some concept of typical quantitative values, assume that the following constants apply to the system of Figure 13.4.

G	= 10,000 kVA.
M	= 5000 kW at 0.80 PF and is loaded to rating.
Generator X_d	= 1.20 per unit on its own rating.
Motor X_d	= 0.90 per unit on its own rating.
E	= rated terminal voltage,
	= 1.0 per unit.
X_{eq}	= 0.75 X_d.

Selecting 10,000 kVA as base,

$X_G = 0.75 \times 1.20,$
$\quad = 0.90$ per unit on 10,000 kVA.

$X_M = 0.75 \times 0.90 \times \dfrac{10{,}000 \times 0.80}{5000}$ on 10,000 kVA,
$\quad = 1.08$ per unit.

$I = \dfrac{5000}{0.80 \times 10{,}000},$
$\quad = 0.625$ per unit.

$E_G = |1.00 + 0.60(0.625 \times 0.90)$
$\qquad + j\,0.80(0.625 \times 0.90)|,$
$\quad = |1.338 + j\,0.450|.$

$E_G = 1.412$ per unit,

$E_M = |1.00 - 0.60(0.625 \times 1.08)$
$\qquad - j\,0.80(0.625 \times 1.08)|,$
$\quad = |0.595 - j\,0.540|.$

$E_M = 0.804$ per unit.

$\delta_G = \arctan \dfrac{0.450}{1.338},$
$\quad = \arctan 0.337,$
$\quad = 18.59°.$

$\delta_M = \arctan \dfrac{0.540}{0.595},$
$\quad = \arctan 0.908,$
$\quad = 42.23°.$

$P = \dfrac{1.412 \times 0.804}{0.90 + 1.08} \sin(18.59° + 42.23°),$
$\quad = \dfrac{1.135}{1.98} \sin 60.82°,$
$\quad = 0.573 \times 0.873,$
$\quad = 0.500$ per unit,
$\quad = 5000$ kW.

This figure checks the initially assumed 5000-kW load on the motor since all losses have been neglected. The power limit for this system with constant excitation is simply

$$P_{max} = \dfrac{1.412 \times 0.804}{0.90 + 1.08}$$
$$= 0.573 \text{ per unit, or } 5730 \text{ kW.}$$

The mechanical load on the motor could be increased to 5730 kW, at which value the torque on its shaft would just equal the maximum electrical torque which the system is capable of transferring. Any further increase in motor load would pull it out of synchronism.

In most practical power systems, the motor and generator are not connected directly terminal to ter-

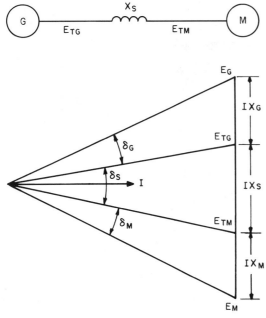

Figure 13.6. Same as Figure 13.4, with external reactance between motor and generator.

minal, but through some intervening system reactance, as represented by X_S in Figure 13.6. In the phasor diagram, E_G and E_M represent, as before, the voltages behind the equivalent synchronous reactances X_G and X_M, respectively. E_{TG} is the generator terminal voltage, while E_{TM} is that of the motor. For the purpose of drawing the phasor diagram, the power factor is indicated as unity near the midpoint of the line. Actually it might be anything consistent with practical system operation. Here the power transfer between the generator and the motor is

$$P = \frac{E_G E_M}{X_G + X_S + X_M} \sin(\delta_G + \delta_S + \delta_M). \quad (4)$$

The presence of X_S in this equation tends to reduce the inaccuracy introduced through the simplifying assumption: $X_G = 0.75\, X_d$; since in typical cases X_{dG} is comparable in magnitude to X_S of the lines and transformers, whose characteristics are generally known to be fairly accurately. For the same reason, this masking effect by the system reactance makes it permissible in practical problems to disregard the difference between X_d and X_q existing in salient pole machines.

Often steady-state stability problems can be resolved into cases where a single machine is connected into a system so extensive that it may be regarded as infinite by comparison with the machine being investigated. An example might be the case of a small municipal power plant having a single tie with a large integrated power system serving the entire surrounding area. In such cases, analysis of the problem is simplified by the assumption that the large system has zero equivalent reactance, an infinite inertia, and its operating voltage remains constant regardless of the system disturbance. Figure 13.7 illustrates an example of this type. Here the machine, G, is connected, through an external reactance, X_S, representing transformers and tie line, to the infinite system represented by sustained rated voltage $E = 1.0$. The phasor diagram as drawn indicates unity power factor at the machine terminals. In this case, since the voltage of the infinite system equals 1.0,

$$P = \frac{E_G}{X_G + X_S} \sin(\delta_G + \delta_S). \quad (5)$$

This is a very useful means for comparing the stability characteristics of several different machines, or for evaluating the relative merits of alternative transmission line voltages or transformer arrangements that might be used for connecting an isolated plant into a large system. Even where the large system is not so great that its reactance is entirely negligible, this concept of "infinite system" permits a rough appraisal of the situation without actually determining its reactance and terminal voltage characteristics. Unless this appraisal indicates

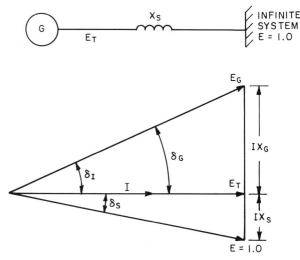

Figure 13.7. Generator connected to infinite system through external reactance.

STEADY-STATE STABILITY

that stability is questionable, it will not be necessary to make a more detailed study.

Excitation

Up to this point, it was assumed that the voltages that have governed the transfer of power between machines have been those internal generated voltages corresponding to the excitation in the machines. These voltages were assumed to remain constant in magnitude, as is essentially the case when the machine fields are manually controlled. Under this condition it will be apparent from inspection of Figures 13.4, 13.6, and 13.7 that as the power transfer increases, thereby widening the angle between these sustained voltages, the machine terminal voltages necessarily decrease. To an operator attending a machine, this drop in terminal voltage is the signal for increasing the excitation so as to restore the terminal voltage to normal. Increasing the excitation increases the internal voltage; and to conform to the principle of maintaining equilibrium between accelerating and decelerating forces, the angle between these internal voltages must decrease. The stability limit appears to be somewhat higher than we had previously found, since the load can now be increased again until we reach the previous angular displacement. Note however, that this increase in excitation, under hand control, does not take place until after some change in load has occurred. There will be some time delay even if the operator sits watching the voltmeter with his hand on the rheostat; and while increased excitation under hand control may raise the stability limits in a few cases, this factor can not be depended on and is disregarded in determining practical stability limits.

From the foregoing discussion, it can be reasoned that the maximum power that a machine under hand control of excitation can handle with stability is a function of the load and power factor at which it is already operating. The lightly loaded or underexcited machine operates at a relatively low internal voltage. Therefore, if its field remains constant, the maximum load it can carry without losing synchronism is less than that of a machine already heavily loaded or operating in the over-excited region. Quantitative values can readily be determined in any specific instance from the now familiar power equation. Figure 13.8 shows the relation between initial load and total permissible load for a typical machine operating at 0.80 PF and connected

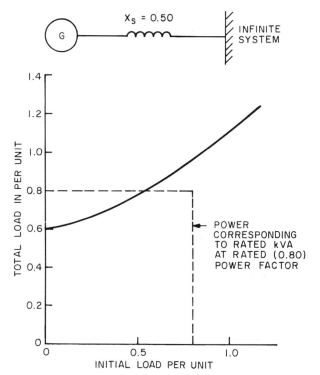

Figure 13.8. Effect of excitation on load-carrying ability.

through 0.50 per unit reactance to an infinite system.

Automatic voltage regulators at the machine terminals may be regarded as robot operators who constantly keep an eye on the voltmeter and a hand on the field rheostat. The principal difference is that the regulator is always on the job, whereas the human operator may be elsewhere at the crucial moment. To represent the continuously acting voltage regulator, it is necessary to express mathematically, by differential equations, the overall performance (including the regulator, excitation system, generator, and system transient characteristics) and solve them for small changes. Modern automatic voltage regulators, in conjunction with adequate excitation systems, do offer a means for increasing system power limits above those obtainable under hand control. Figure 13.9 will serve to illustrate the upper limit of what might be accomplished in this regard. Here two equal machines, G and M, are connected together through external system reactance, X_S. The constants are as indicated on the diagram. Without voltage regulators, and with the excitations adjusted to give unit voltage midway in the system when maximum power is reached, the maximum power is

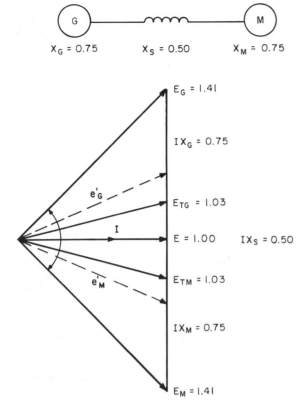

Figure 13.9. Effect of automatic generator voltage regulators on load-carrying ability.

$$P_{max} = \frac{1.41 \times 1.41}{2} = 1.00 \text{ per unit.}$$

Under this condition, the voltage at the terminals of each machine is 1.03 per unit. The angle between E_G and E_M is 90°. This corresponds to the maximum power capability shown in Figure 13.3.

With active voltage regulators, the maximum power will be somewhat greater than that achievable with hand control. As the power transfer increases the angle between E_G and E_M increases. As that angle increases without a change in the magnitude of E_G and E_M, voltage E_{TG}, E_1 and E_{TM} decrease. The voltage regulators sense the reduction in E_{TG} and E_{TM} and increase E_G and E_M to restore the terminal voltages to the desired values, 1.03 pu in the example.

If the voltage regulators were of perfect accuracy, infinitely fast, and the excitation system could sustain 3.0 pu field current ($E_G = E_M = 3.0$ pu) indefinitely, the power limit would approach 2.12 per unit. This would occur when the angle between E_{TG} and E_{TM} was 90°. The angle between E_G and E_M would be about 150° at that point. These "ideal" voltage regulator and exciter characteristics are unachievable. Modern exciter/voltage regulators do, however, permit higher stable power transfers than with hand control of excitation (Byerly and Kimbark, 1974).

An alternate view is that the combined voltage regulator and dynamic excitation system response are instrumental in holding constant (in magnitude) a voltage phasor called e'_G, which is located between E_G and E_{TG}, as shown dashed in Figure 13.9. The machine reactance between E_{TG} and e'_G is the transient reactance X'_d which is generally about one-third the value of X_d. The theoretical power limit under these conditions occurs when the angle between e'_G and E'_M approaches 90°.

If we assumed $E = 1.0$ pu at the mid-point and $X'_d = 0.25$ pu for both machines the theoretical power limit would be approximately 2 per unit. The terminal voltages would be about 1.11 pu and $e'_G = E'_M = 1.41$ pu. The excitation levels required to sustain this power would be $E_G = E_M = 2.23$ per unit and the angle between E_G and E_M would be about 126°. Some utilities have reported stable operation for angles up to about 120°.

Application to Complex Systems

To illustrate fundamental considerations, this discussion has been concerned with the simple case of two machines and the even simpler case of one machine connected to an infinite bus. How are the more complex stability problems handled? For example, how is the case of an interconnected system having dozens of machines located in many different plants and spread over several states solved? There are several possible answers.

Assume that the solution is to be undertaken longhand. Multimachine equations can be written similar to the two-machine equations developed at the beginning of this chapter. The use of equations involving more than two machines, however, becomes very laborious. Fortunately, various simplifying assumptions can usually be made that will permit breaking down the problem to a point where it can be handled either as an equivalent two-machine problem or as one machine against an infinite bus. The simplifying process may represent as much art as it does science, since an understanding of system behavior and good engineering judgment will govern the choice among those factors that can be neglected, those that can be combined and repre-

STEADY-STATE STABILITY

sented as equivalents, and those that must be examined in their entirety. Two examples will illustrate these distinctions.

1. Assume that the problem concerns the possible loss of synchronism of one particular machine in a large generating station in the event that its field excitation is lost or is seriously reduced. Here, because of the low intervening reactance, the remainder of the machines in the station can all be lumped together as an equivalent single machine. The problem is then solved as a simple two-machine problem. If however, this station is connected to the system through "stiff" (i.e., low-reactance) ties, then the rest of the system will require consideration. In this case, the remote machines are combined into an equivalent machine, taking into account the reactances of the circuits that link them together. Then, they are combined with the previously determined equivalent for the remainder of the machines in the station being considered. This final combined equivalent is so large that it may be regarded as an infinite system. If so, the generator with weakened field can be studied as if it were connected directly to an infinite bus, which is maintained at normal voltage. Taking into account the rest of the system might not appreciably change the constants determined for the equivalent machine representing the machines in this particular station. In this case, it would have been valid to neglect entirely the effects of the outlying machines.

2. Consider a system in which a tie line joins two areas, each of which is closely knit electrically; the areas are reasonably well balanced as to load and generation, so that normally there is little interchange of power over the tie line. It would be essential that the system operator know the maximum power that can be transferred in the event of a generator failure or an unexpected power demand in either area. If there is a major generating station at or near each terminal of the tie line, a first approximation might be to neglect the remainder of the system and represent each of the terminal stations as an equivalent machine. The power limit could then be determined for these two equivalent machines, connected through a reactance representing that of the tie line. If this calculation indicates a potential power-transferring ability well in excess of that required in an emergency (for example, when the largest generator is out of service at the time of peak load in its area), then further study might be unnecessary. The reason is that such an extreme simplification of the problem neglects the help that other machines would give in maintaining voltage, and therefore it gives a pessimistic result. If, however, the power-transfer ability is found to be marginal, then a more accurate representation would be necessary to answer the question. This would require that all the machines in each area be factored into the equivalent, taking account of the system reactances interlinking them. These larger equivalents would then be considered as acting at the terminals of the tie line, and the power limit calculated accordingly. Obviously, a problem like this one cannot be simplified beyond that involving two equivalent machines; the infinite system approach is not applicable unless one of the systems is many times larger, electrically, than the other.

From these examples, it is evident that the best method of attacking steady-state stability problems in complex systems by long-hand calculation is to simplify them until they are amenable to calculation, either as equivalent two-machine problems or as single machines against an infinite bus.

The analysis of large, complex systems for stability is usually accomplished by using a digital computer problem with sufficient capability for representing generator (Young, 1972), excitation system (IEEE Committee Report, 1968), and turbine control system behavior (IEEE Committee Report, Dec. 1973), as well as the power system behavior.

Dynamic Stability

The analysis of classical steady-state stability assumed that the automatic voltage regulator had so much deadband and was so slow, that for all practical purposes the machine had constant field voltage when a small disturbance occurred. The automatic voltage regulator could change the excitation level from one operating condition to another, but, for purposes of stability, this type of control was no better than "close-hand control." When a system with this type of control was steady-state unstable, it experienced a monotonic increase in machine angle until the machine pulled out of step.

With the introduction of continuously acting voltage regulators, the nature of the system's response to a small disturbance changed to an oscillatory behavior. Now, the problem was whether the system had positive damping. This concern with system damping is often labeled as "dynamic stability." The assumption was made in the past that the net damping of the system would be positive, and that the primary source of negative damping, if any, was the hydraulic turbine governor response to speed errors. Excitation systems that were prevalent had sufficient deadband and were slow enough so that they had little effect upon damping. The sources of positive damping (generator amortisseur and field damping, load damping, and prime-mover damping) more than balanced the sources of negative damping. With the introduction of continuously acting excitation systems, there came another important source of negative damping.

In many systems, damping has not been a problem. But in some areas of the world, due to the wide geographic dispersion of load and generation, the size of the ties, and the presence of both hydro and steam generation, the problem of damping is significant both now and in the future. To add positive damping, devices such as excitation or hydro-turbine supplementary stabilizing (Byerly and Kimbark, 1974; DeMello and Concordia, 1969) have been added.

TRANSIENT STABILITY

Certain fundamental concepts of power system stability were developed at the beginning of this chapter. Transient stability was defined as that particular subdivision of overall stability that is associated with relatively severe and sudden changes in system conditions, such as faults or switching operations. Stability was said to be a single characteristic of the power system; and subdivisions of stability, such as steady-state and transient, are purely arbitrary for convenience of analysis. As a result, the fundamental concepts already developed remain valid throughout the discussion that follows.

The expression for power

$$P = \frac{E_1 E_2}{X} \sin \delta_{12}$$

was developed in general terms. Later, by specifying the particular values of voltages and reactances appropriate to steady-state conditions, certain conclusions regarding steady-state stability were drawn. This same expression can be used in the study of transient stability. The procedure will be first to investigate the particular values of voltage and reactance appropriate to transient conditions, and then to apply the fundamental equation, employing these values. Finally, the effects on transient stability by other characteristics of the system will be discussed.

Transfer Impedance

When a fault occurs on a circuit that interconnects two parts of a system, the ability to transfer power between those two parts of the system is reduced. In the extreme case of a three-phase fault having negligible fault impedance, the power-transferring ability is reduced to zero just as effectively as if the circuit breaker had been opened. Faults that do not involve all three phases will impair the power-transferring ability to a lesser degree. Also, the more impedance in the fault, the less it reduces the power-transfer ability.

Figure 13.10a represents a transmission line between two systems having a fault at F. Figure 13.10b is an equivalent circuit, in which X_A and X_B are, respectively, the line impedances on either side of the fault and X_F is the fault impedance. The "transfer impedance" between points A and B is the ratio of voltage at A to current at B with the voltage at B set at zero (or vice versa, since this is a

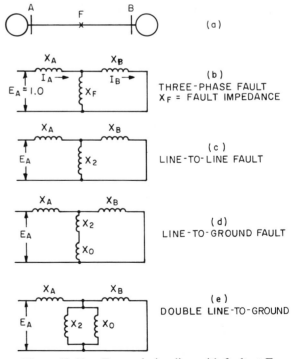

Figure 13.10. Transmission line with fault at F.

TRANSIENT STABILITY

linear circuit). This transfer impedance, X_T, may be determined as follows:

By definition,

$$X_T = \frac{E_A}{I_B},$$

letting

$$E_A = 1.0 \text{ per unit}$$

$$X_T = \frac{1}{I_B}.$$

But

$$I_A = \frac{1}{X_A + \dfrac{X_B X_F}{X_B + X_F}}$$

and

$$I_B = \frac{X_F}{X_B + X_F} I_A$$

$$= \frac{X_F}{X_A(X_B + X_F) + X_B X_F}.$$

Hence,

$$I_B = \frac{X_F}{X_F(X_A + X_B) + X_A X_B}.$$

Therefore,

$$X_T = \frac{1}{I_B}$$

$$= X_A + X_B + \frac{X_A X_B}{X_F}. \tag{6}$$

Obviously, if there was no fault, X_F would be infinite and X_T would simply equal the total line impedance, $X_A + X_B$. Similarly, it will now be apparent how the power-transferring ability becomes zero (i.e., X_T becomes infinite), if X_F is zero. While the impedance is not zero in most faults, it can be negligibly low. In a case where the bare conductors of a transmission line are brought into metallic contact, the fault impedance would be zero. In most faults, arcing occurs so there is a fault resistance. It is the usual practice to neglect fault resistance in making studies of system stability, since the total impedance is predominantly reactive. This does not mean that all faults result in infinitely high transfer reactance, since not all faults involve the three phases. The majority of faults are line-to-ground or line-to-line and are less severe than a three-phase fault.

In deriving equation 6 through the use of the equivalent circuit of Figure 13.10b, no restriction was made as to the exact nature of X_F. In the case of a three-phase fault, X_F does represent the fault impedance. The methods of symmetrical components, however, permit the utilization of X_F for representing the effects of the negative- and zero-phase-sequence components of system reactance for unbalanced faults. In each case (Figure 13.10c, d, and e), equation 6 can be used by substituting X_2, $X_2 + X_0$, or $(X_2 X_0)/(X_2 + X_0)$, respectively, for X_F.

The significant point concerning the transfer reactance between any two points in the system is that the transfer reactance is always greater during a fault than the normal through reactance without fault; and, any fault, regardless of type or severity, effectively reduces the synchronizing power that can be transmitted.

Switching

Faults are not usually permitted to remain on power systems any longer than is required for relays to detect their presence and location, and then for circuit breakers to isolate the afflicted apparatus or line section. This removal of a portion of the system from service necessarily increases the load angles between machines or generating/load areas of the system (unless the fault happened to be on a stub feeder), and reduces the capacity for transferring power. If the two areas of the system are interconnected through a multiplicity of circuits, the outage of any one of them may not seriously impair the system stability. If, however, only a single tie line links the two areas, then a fault that requires opening this tie line will completely destroy any possibility of continued transfer of power. Perhaps the most typical case is that in which the areas of the system are interconnected through a double-circuit tie line (either two single-circuit lines or one double-circuit line). Here it is essential, for intelligent system operation, to know the stability limit in the event of a fault and the subsequent opening of either circuit.

From this description it is evident that the value of X to be used in the expression

$$P = \frac{E_1 E_2}{X} \sin \delta_{12}$$

is not one value, but three. Prior to the fault, X is the through reactance, with both lines in service; during the fault it is the transfer reactance; and after the fault has been isolated, it is the through reactance of the remainder of the system. Both the latter figures are always greater than the value of through reactance, which was used in the steady-state calculations. (Again, the special case of a stub feeder fault, the isolation of which does not increase the through reactance, is an exception.)

Voltages

In using the expression

$$P = \frac{E_1 E_2}{X} \sin \delta_{12}$$

to find the power transfer between any two points, the only restriction is to be consistent and select values that correspond to one another. If voltages E_1 and E_2, at two given points in the system, are to be used, the X must be the reactance between these same two points, and δ_{12} must be the angle between E_1 and E_2 at those points. In the usual case of steady-state stability, the voltages used are those behind the equivalent synchronous reactance in the machines. The internal generated voltages correspond to normal field current, but are modified to account for the saturation that occurs near the pull-out point.

For the first half second or so following a transient disturbance, the synchronous machine is assumed to behave as though it possessed a constant voltage behind its transient reactance. The machines field L/R time constant is large so that the net air gap flux is assumed to be constant as a first approximation. In practice, this assumption is not made in modern computer studies of machine dynamics.

During a fault-initiated transient disturbance, the machine may supply unusual amounts of reactive power to a system, which is accompanied by high armature reaction that causes the machine (net air gap) flux to decrease. This same situation also reduces the machine terminal voltage, and causes the automatic voltage regulator to react so as to increase the machine's excitation. This, in turn, increases the flux. If the excitation system is able to overcome the demagnetizing effect of the excess armature reaction during the time period of interest (about one-half second after the fault), the net effect of these two actions is to maintain, on the average, constant flux in the machine. Using the concept that the net air gap flux (rather than the field currents, as in the steady-state case) remain constant, the voltages used are those behind transient reactance. Correspondingly, the angle is that between these voltages, and the machine reactances are the transient values X'_d.

The expression for electrical power that applies during a transient period is as follows:

$$P = \frac{e'_1 e'_2}{X} \sin \delta$$

where

e'_1, e'_2 = constant voltages behind transient reactances X'_{d1} and X'_{d2} for machines at two ends of the system.

X = $X'_{d1} + X_T + X'_{d2}$

X_T = transmission impedance as defined in equation 6 and shown in Figure 10.

δ = angle between phasors e'_1 and e'_2.

Inertia

The analysis of the transient reponse in a power system involves the behavior of the generator rotors. It is actually the rotor motion that determines the stability of the system. The rotor motion is determined by its inertia and the torques exerted on it. These torques are the prime mover input, the electrical output torque, and damping torques. The prime mover and electrical torques are the most significant and will be discussed first. The damping torques are often neglected for studies of first-swing transient stability unless a refinement is considered necessary.

Before a transient disturbance, the machine rotors run at a constant synchronous speed under steady-state conditions. The mechanical input torque is balanced by the electrical output torque and losses. When a sudden disturbance (fault or switching) occurs, the electrical conditions suddenly change. In the case of a fault, the electrical torque suddenly drops to a lower value, leaving an unbalanced torque on the rotor. The excess of mechanical input torque over electrical output torque acts on rotor inertia to accelerate it. The calculation of these torques and the rotor response gives the solution to the transient stability performance.

The calculation of the torques require an expression for torque for each component. They are:

1. P_i or T_i for mechanical input. Note that power and torque are being used interchangeably since machine speed rarely varies more than 5% from synchronous during a transient (postfault) period. Actually $P_i = \omega T_i$ so when $\omega = 1$ (in per unit), $P_i = T_i$.

2. $P_{max} \sin \delta$ for electrical torque. This term will be more complex if the saliency torque is included. It is nearly true in the case of round-rotor machines. It is this term that is nonlinear and requires step-by-step solution of the equation. Each step is for a short interval of time, equal to or less than one cycle of the system's frequency.

The accelerating torque (T_a) on the rotor is the difference between 1 and 2 above. When the equation is reduced to per unit terms, the result is

$$T_a = \frac{H}{180f} \frac{d^2\delta}{dt^2} \text{ per unit torque}$$

$$H = \frac{(.231)(WR^2)(\text{r/min})^2 \times 10^{-6}}{\text{base kVA}}$$

f = frequency (usually 60 Hz in the United States)
δ = rotor angular position in electrical degrees
H = per unit inertia constant in kW seconds per kVA or seconds. If the accelerating torque acting on the rotor is equal to rated torque and is held constant, the speed will change 100% in $2H$ seconds.

Equal-Area Criterion

A qualitative concept of how the effects of inertia modify the now familiar power equation can be gained with the aid of Figure 13.11. Since transient reactances, X_d', and voltages e' behind X_d' are used to determine the power relationship, this will be called the "transient power-angle" characteristic or transient P-δ curve for short. Here two machines, G and M, are interconnected through parallel tie lines. One line is faulted at F. Each of the three curves is a representation of the relation

$$P = \frac{e_G' e_M'}{X} \sin (\delta_G + \delta_M).$$

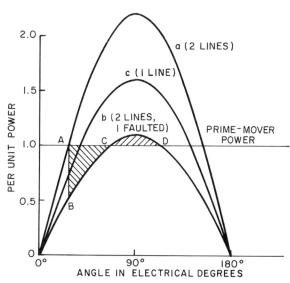

Figure 13.11. Transient torque- (or power-) angle curves for two-circuit line with a fault on one circuit.

In all three, e_G' and e_M' are, respectively, the voltages behind transient reactance in G and M, while ($\delta_G + \delta_M$) is the total angular separation between them. Curve a represents the predisturbance transient P-δ curve of the system with no fault. Its crest value is not the steady-state power-transfer capability. Instead, the crest value is the maximum synchronizing power (or torque) capability of the two machine/transmission system.

Curve b was determined using the transfer reactance (X_T) before the faulted line was disconnected. This curve represents the transient (synchronizing) P-δ characteristic during the fault. Without specifying the type of fault, it is apparent that the transfer reactance is great enough to reduce the power-transferring ability. Curve c represents the condition after the faulted line is switched out. The system does not regain its prefault ability to transfer power because, with only one line remaining in service, the through reactance is higher than it was originally.

The analysis will require consideration of all these curves. Operation prior to the fault is shown on curve a. Conditions immediately after the occurrence of the fault are shown on curve b. When the fault is tripped off, the synchronizing power capability is up to the level represented by curve c.

Where and how these transfers are accomplished can best be seen by reference to Figure 13.11.

The horizontal line represents the constant power input supplied by the prime mover. Its constancy is assured in spite of any momentary fluctuations of the angle δ, because the controlling mechanism is responsive only to changes in speed. The small change in speed, together with the time delays inherent in the turbine control system (IEEE Committee Report, Dec. 1973), result in minor changes of prime-mover power input during the time when our basic assumptions relative to transient stability remain valid. Prefault (steady-state) operation is represented by the intersection of the prime-mover power line with curve a at point A. At the instant the fault occurs, the ability to transfer electrical power is suddenly reduced to point B on curve b. The prime-mover power remains constant and now exceeds the power that the machine can transfer electrically at the initial value of δ. The excess in driving torque causes acceleration of the rotor, thereby increasing its speed, storing added kinetic energy, and increasing the angle δ. This will carry the point of operation to C, where the accelerating and decelerating forces are again in balance, but the rotor speed is greater than nominal and the rotor angle δ will continue to increase.

The rotor was accelerated from B to C by an amount depending on its inertia constant, H, and on the integrated excess of input over output power represented by the area A B C. It follows that at point C it is running too fast and contains excess kinetic energy. This kinetic energy will move the operating point past C toward point D. In this region, the electrical output exceeds the mechanical input and the rotor tends to slow down while the angle δ continues to increase, but at a decreasing rate. From Figure 13.11, it is apparent that at point D the excess kinetic energy has not all been dissipated, and the angle δ will continue to increase. This is true because the area enclosed above the horizontal line between C and D is less than the area A B C below the horizontal line. Stability is now lost, because to the right of point D a region of net accelerating torque is entered. As a result, δ will increase, and at some point far beyond the region of validity of our basic assumptions, the machine speed will have increased to the point where an operator or an overspeed device will shut down the machine.

To illustrate the phenomenon clearly, the situation has been oversimplified by assuming that the

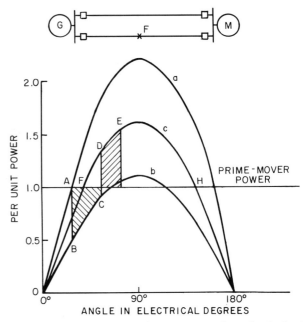

Figure 13.12. Same as Figure 13.11, when the faulted circuit is tripped off.

fault remained on the system. By so doing, curve c of Figure 13.11 has not been used. Actually, there are some systems that are stable with sustained faults. This fact in no way decreases the general desirability of fast fault clearing, based on such other considerations as conductor burning at the point of fault, reduced system voltage, etc.

A more realistic case is that illustrated in Figure 13.12. The initial conditions are identical with those of Figure 13.11, except that this time the effects of switching will be observed. Starting from the prefault conditions at point A, the occurrence of the fault immediately shifts the operation to B on curve b. An excess of mechanical input to the generator over its electrical output accelerates the rotor, thereby storing excess kinetic energy and the angle δ increases. Assume that the faulted line is switched out at point C. This shifts the operation to curve c, starting from point D since the angle δ cannot change instantaneously. The net torque is now decelerating, and the previously stored kinetic energy will be reduced to zero at point E when the shaded area below DE equals that above BC. The excess kinetic energy is zero at E, because the rotor speed is rated speed.

Again, at least momentarily, the angle is too large for steady-state operation since the electrical output at point E exceeds the input from the prime mover. The rotor will decelerate along curve c,

TRANSIENT STABILITY

passing through points D and F, oscillating in smaller and smaller swings (positive damping is assumed), with steady-state operation occurring at F. If the system inherently has sufficient damping for the rotor to experience a decayed oscillation, and will reach the new steady-state operating point at F in about 3–10 seconds.

Damping

The rate at which the oscillation dies out is determined by the damping torques. These damping torques may be due to factors such as friction, windage, hysteresis, and eddy-currents. All these factors are present to some degree in every electrical machine. Friction and windage are purposely held to the lowest practicable levels by careful design, since they always represent power loss. On the other hand, there is little hysteresis or eddy-current loss in the rotor during normal operation because it rotates at exactly the same speed as the armature flux. But as the rotor oscillates above and below synchronous speed, the resultant hysteresis and eddy-currents produce important damping torques.

Returning to Figure 13.12, there are two additional points to be made. A machine can operate with transient stability, even though momentarily swinging past the peak of the power-angle curve—i.e., even though δ is greater than 90°. In terms of Figure 13.12, this means that point E could be moved to the right along the middle curve as far as required for the area below DE to equal that above BC. This point might well be past the peak of the curve, but obviously could not be beyond the next intersection of the curve with the constant prime-mover power line at H; at this point there would again be a reversal of the net torque, which would carry the machine further out of step with the system. The final point of stable operation, F, will necessarily be to the left of the crest of the curve—i.e., at an angle less than 90°.

Figures 13.11 and 13.12 have been explained in terms of a generator; the prime-mover power was regarded as positive (in such a direction as to produce acceleration) and electrical power as a negative (in such a direction as to produce deceleration). The explanation could have been given in terms of a motor, in which case the electrical power would have caused acceleration and the mechanical shaft load would have caused deceleration. Consequently, both Figures 13.11 and 13.12 are general, and with proper choice of algebraic sign along the vertical axis, they can apply to either a generator or a motor.

In either case, prompt isolation of a fault is desirable—the sooner the fault is removed, the smaller will be the area below the horizontal line. For this reason, there is a better chance that the available area above the line will be adequate to balance it, resulting in a stable condition.

Methods of Calculation

Transient stability can be estimated by hand, using an analog computer or a digital computer. Hand calculation is usually limited to a two-machine case where the equal area criterion or standard curves (Crary, 1945, 1947) can be used. Analog computers, used in conjunction with manual sliderule or desk calculator calculations, formerly provided a tool for the calculation of power system stability. This method was very cumbersome and generally impractical for representing the transient saliency, excitation system, and prime-mover characteristics. The digital computer can easily include all of these features to give an accurate calculation for several seconds (Stott, 1979; Concordia and Brown, 1971). It then gives results in not only the "transient stability" range but also the "steady-state" stability region. Transient and steady state were separations that were previously made for ease of calculation. Using the digital computer, this distinction is unnecessary, and a system is either stable or unstable as is the actual power system.

For making a stability study, the system to be studied must be defined in terms of impedances, network configurations, and rotor inertias. A system configuration must be selected, i.e., all generators and lines in service for a heavy load or some out of service for a light load. If the disturbance to be studied is a short circuit or fault, the type of fault must be selected. The switching time and method must be determined from the protective relay settings and circuitry.

Since the torque or power equations for the synchronous machine are nonlinear, some appropriate step-by-step method (Stott, 1979) of solving nonlinear equations is used. First, an increment of time is elected; for example, 0.05 seconds (three cycles at 60 Hz) might be used. The maximum rotor movement per step is an indication of the accuracy. The shorter the time interval selected, the more accurate the results and the greater the number of calculations. The net accelerating torque on a rotor is assumed to be constant over this small time interval, and, in the next step, the angular displacement of the rotor is calculated in this given time interval. Smaller time intervals tend to increase the accuracy of the results, but also increase the calculating time and cost of the calculations.

The first step in the stability calculations is to solve for the initial load flow in the selected system to determine the initial conditions of bus and generator voltages, generator flux linkages, rotor angles, etc. The next step is to apply the disturbance, fault or switching, and again solve for the power flows under this new circuit condition. Since this takes place in "zero" time, the rotor angular position has been unchanged from the first step. Comparing the results of these two steps will give the difference between the driving power or torque on any given generator rotor and the electrical load or torque. (Note that a motor could be used as easily as a generator, with only the power flow being reversed.) That difference is the net accelerating, or retarding torque on the machine rotor. The rotor inertia is known; so, the angular change for a given time interval can be calculated. The rotor positions of all machines are adjusted to the angular position at the end of this first time interval and the procedure is repeated. The circuit changes, such as faulted line switched off, or the line reclosed, are to be incorporated at the proper time. Calculations are continued until the stability characteristics are determined. The result is a swing curve, as shown in Figure 13.13, where the angular position of each rotor is shown as a function of time. This curve reveals the stability (or instability) of the system under the assumed conditions. Several other sets of assumed conditions will usually be tried to determine system performance under all operating conditions.

Figure 13.13 is a plot of angular position versus time (commonly known as a "swing curve") for a representative group of four machines. Note that although the curves end at 0.65 second (when the machines have not regained their new steady-state angular positions), the curves have been carried far enough to conclude that this is a first-swing transiently stable case. The reason is that three of the machines have remained together, while the fourth has turned back toward the others. The tendency for all the machines as a group to move somewhat ahead in angular position is typical. It is caused by the constancy of prime-mover input, in conjunction

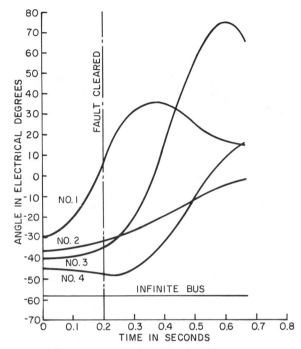

Figure 13.13. Representative swing curves.

with a reduction in system load occasioned by the fault.

A system of machines is not necessarily "stable" merely because they all survive the first swing without losing synchronism. For some systems, the post-first swing oscillations can be severe and continue to grow until synchronism is lost between several machines. This is a "dynamic instability" (De Mello and Concordia, 1969) or oscillatory (controller aggravated) steady-state instability. Studies to determine stability for such systems must be carried out for 3–7 seconds or more of simulated real time.

Factors in Transient Stability

Practical use has been made of the methods of determining transient stability, not only to guide the operation of specific existing systems, but also to yield results in general terms that may be employed in the design of new systems and in the development and the improvement, with respect to stability, of existing systems. The transient power limit in any given case is determined by many factors. Some of the principal factors are indicated in the paragraphs that follow and in their accompanying illustrations; also included is an evaluation of their

TRANSIENT STABILITY

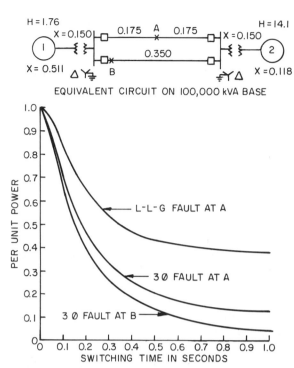

Figure 13.14. Effect of switching time on maximum power for various types of faults.

effectiveness in improving transient stability. (See Byerly and Kimbark's *Stability of Large Electric Power Systems* for several good papers on this subject.)

Switching Time. The faster a faulted section can be isolated from the system, the greater will be the power that the system can transmit with stability. In fact, rapidity in clearing faults has been the most effective single factor in improving transient stability. This is easily understandable when we recall the development of the equal-area criterion and Figures 13.11 and 13.12.

The relation between power limit and fault-clearing time, for both three-phase and double line-to-ground faults, is shown for a typical system in Figure 13.14. This figure emphasizes the real importance of modern high-speed relay and circuit breakers whenever a problem of transient power limits arises.

Types of Faults. Figure 13.15 shows the relative severity of various types of faults in reducing the transient power limit of a typical simplified system. While three-phase faults are the most severe, the fact is that they are the least frequent in occurrence. Recognizing this, some operators are willing to accept the possible loss of stability in the rare event of a three-phase fault and base their system designs on the more realistic possibility of being able to maintain stability for double line-to-ground faults. As Figure 13.15 indicates, a switching time for three-phase faults of approximately 0.05 second, or three cycles, would be required to provide a 0.60 per-unit power limit for this particular system. This same power limit, however, can be obtained with five- to six-cycle switching, provided operators are willing to use double line-to-ground faults as the basis of design and to accept the risk of a system outage in the event of a three-phase fault.

Single line-to-ground faults are usually the least severe, and their severity is subject to control if we are willing to ground the system neutral through impedance. Generally, such faults do not constitute the basis for system transient-stability design. If these faults are not promptly detected and isolated, they may develop into the more severe double line-to-ground or three-phase faults; equally careful attention should be given to ensuring their prompt isolation.

Inertia. The inertia of the rotating parts has a bearing on the extent to which any net excess of accelerating or decelerating power will change the angular position of a machine rotor with respect to

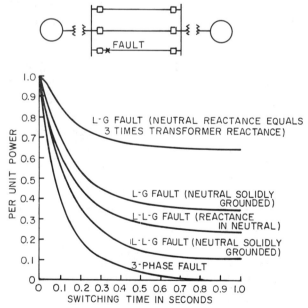

Figure 13.15. Effect of impedance in the neutral ground.

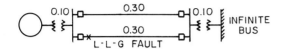

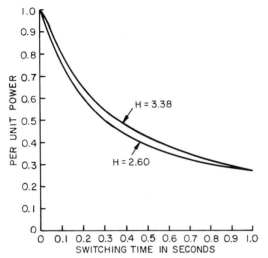

Figure 13.16. Effect of generator inertia.

the rest of the system and, influence the transient-stability limit (Concordia and Brown, 1971). Figure 13.16 shows a quantitative concept of the gain in transient-power limit that may typically be realized through an arbitrary increase of 30% in the inertia of a machine which is connected through transformers and two parallel lines to an infinite bus. Again, the data are presented as power limit versus switching time, for a double line-to-ground fault at the generator end of one of the lines. As Figure 13.16 shows, this 30% increase in inertia raises the power limit about 5% for usual switching times.

Other Factors. Curves similar to those in Figures 13.14, 13.15, and 13.16 might be drawn to illustrate the effect of the following factors on the transient-stability limit:

1. The transient reactance of machines and the reactance of transformers (Concordia and Brown, 1971).
2. The number of parallel transmission lines, also the number of sectionalizing points.
3. Fault location along transmission lines or stub feeders.
4. Simultaneous versus sequential switching at the two ends of the faulted line section.
5. Excitation system performance (Byerly and Kimbark, 1974; DeMello and Concordia, 1969).
6. Dynamic behavior of the system loads (IEEE Committee Report, Mar./Apr. 1973).
7. Performance of fast valving or braking resistor schemes if used (Byerly and Kimbark, 1974; Kimbark, 1969).
8. Performance with controlled reactive compensation on the line (Hauth, et al., 1982; Miller, et al., 1982).

In general, the gain that may be realized from these factors will only be a few percent. Even so, these few percent can amount to a large amount of power when considering bulk transmission.

Without detracting from the importance of these factors, fast relays and fast circuit breakers are the most significant considerations in system design as related to transient stability.

Several years ago, total switching time (relay time plus breaker time) was 10–15 cycles or more, as contrasted to the shorter times available today. The "fault on" period (see Figure 13.12 and the equal-area criterion) was also more important than it is today. Since this period could not be shortened, emphasis was placed on limiting the fault current. Today, the primary concern is to remove the fault from the system as quickly as possible. A further point of contrast is that the outage of a faulted line section used to be accepted as permanent since it affected transient stability. This is no longer the case due to the acceptance of high-speed reclosing, a technique which returns lines to service in time to restore the power-transferring ability to its prefault value.

High-Speed Reclosing

The usefulness of high-speed reclosing in raising the transient stability limit is illustrated in Figure 13.17. Start as before with steady-state operation at point A, which is the intersection of the system power-angle curve with the constant prime-mover power line. The occurrence of the fault immediately reduces the system power-transferring ability to B, and the angle δ increases along curve b to C. At this point, the simultaneous tripping of the breakers at the ends of the faulted line section permits partial recovery to point D on curve c. It appears from

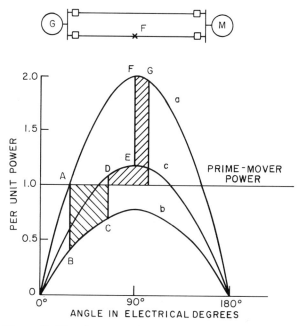

Figure 13.17. Same as Figure 13.12, with high-speed reclosure.

Figure 13.17 that if nothing further happened, the area enclosed below curve c and above the horizontal line would be less than the previously accumulated area below the line, indicating a case of transient instability. Assume, however, that by the time point E is reached, the faulted line section will have been deenergized long enough for the transitory fault path to become deionized, and for a successful reclosure to be accomplished at E. Operation now changes to point F on curve a; and when δ has increased only a few degrees further, to G, the equal-area criterion will have been satisfied, leaving a wide margin of transient stability. Finally, following oscillation about point A on curve a, the system will settle into steady-state operation at A, the initial starting point. Not only is the transient stability limit increased by automatic high-speed reclosing, but, by returning the line to service promptly, without waiting for the attention of an operator, the original steady-state limit is restored.

When high-speed reclosing is under consideration as a means for increasing the transient stability limit of a system (Byerly and Kimbark, 1974; Kimbark, 1969), it should be recognized that there is a risk that stability may be endangered rather than benefited if a line is reclosed on a persisting fault. If reclosure is necessary to maintain stability, that is, if the system would be unstable without reclosure, then the risk of possibly closing in on a persistent fault may be worth taking. Faults, i.e., short circuits, may be permanent or temporary. While reclosure is not possible for a permanent fault, a temporary fault may be reclosed to maintain stability. As an example, lightning may cause a temporary fault. The voltages resulting from the lightning current may flash over an insulator string. When the arc is over and the arc products dispersed, the line may be successfully reenergized. However, if the line voltage is reapplied too quickly, the arc path will again break down and be a conductor.

A temporary fault due to lightning occurs when an arc path across a transmission line insulator string is ionized. As soon as the transmission line is deenergized, the arc no longer exists but the arc path is still ionized. If line voltage is restored by reclosing the breaker in one cycle, the arc will be reestablished. However, if the line voltage is restored after the arc path has recovered, i.e., the ions removed from this area by buoyancy, wind, or recombination, the reclosing will be successful. The insulation between line conductor hardware and the tower must be sufficient to withstand the line voltage successfully. This establishes the minimum time between breaker opening and breaker reclosing. A dead time of 20 cycles has been successfully used on 345-kV lines, 25 cycles on 500-kV lines, and 30 cycles (one-half second) on 700-kV lines.

High-speed reclosing can be useful in preserving the postdisturbance steady-state stability. This can be true particularly if the faulted line is in a critical high-capacity transmission circuit whereby its loss would cause other lines to be overloaded to the point that the system could split up. The loss of synchronism of one plant or several units may be less severe than the system split that might occur if reclosing were not practiced.

There is another risk associated with the practice of high-speed reclosing. The risk is that of accumulated mechanical fatigue of turbine-generator shafts (Bowler, et al., 1980; IEEE Working Group on the Effects of Switching on Turbine-Generators, 1980) to the eventual point of shaft failure. Mechanical fatigue is generally a factor whenever severe faults on the power system are near a power plant. The fatigue damage associated with high-speed reclosure on a permanent fault can be very severe and may exceed the duty produced by generator terminal faults or synchronizing out of phase at 120°.

The rotating mechanical system of a turbine-generator is basically a torsional spring mass system capable of oscillating at several natural frequencies above and below 60 Hz. These oscillations involve differential motion between the generator section and the turbine sections, and between the several turbine sections.

The impulse torque and energy associated with a fault may have a frequency component equal to one of the torsional natural frequencies of the turbine-generator shaft with its several concentrated masses. When this impulse occurs, the torsional mechanical system absorbs some of the fault energy as mechanical spring energy causing oscillations in shaft torque. Reclosing onto the fault may cause additional energy absorption to the detriment of the torsional mechanical system.

For this reason, it is recommended that for transmission lines near generating plants reclosing after any fault be delayed sufficiently to allow the torsional response to decay. Delay of 10 seconds or longer may be required.

Single-Pole Switching and Reclosing

The majority of faults do not involve all three phases and most faults occurring on transmission lines are temporary. These facts are the basis for further refining high-speed operations on each phase so that the one or two line conductors not involved in the fault can remain in service.

The gains in transient stability afforded by single-pole switching are appreciable only in the case of single-circuit lines. Even here the gain may be small if the generating capacity in each system is large relative to the capacity of the tie line. For this type of situation, single-pole switching is a special tool of limited application. In addition to requiring that the circuit breakers be equipped with an independent operating mechanism for each pole, a relaying system is necessary; this system must be capable of distinguishing between the different kinds of faults. Only those breaker poles are opened that are necessary to isolate the faulted conductors. Because most transmission systems have directly grounded neutral, there is still some gain in stability over the case of three-pole switching, even if only a single conductor remains in operation.

In a quantitative evaluation of single-pole switching, the methods of calculation previously used are used again. Up to the point of isolating the fault, the procedure is identical with that for three-pole switching—using a value of transfer reactance appropriate to the type of fault being considered. After the one or two poles of the terminal breakers have opened, symmetrical components must be used to determine the correct representation for the one or two open conductors. A single, open conductor in a three-phase system may be represented by inserting the parallel value of the negative- and zero-phase-sequence reactances in series with the positive-phase-sequence network. Two open conductors may be represented by inserting, in series with the positive-phase-sequence, the negative- and zero-phase-sequence reactances connected in series. Single-pole switching must result in some gain in transient stability to be used.

The time required for deionization of the arc path is increased in the case of single-pole switching by the capacitance coupling with the one or two adjacent energized conductors. In typical cases, the amount of the increase may be on the order of 50% over the deionization time required for three-pole switching. Studies have shown that extremely fast reclosing is of less importance when line-to-ground faults are cleared by single-pole switching.

HIGHER-TRANSMISSION VOLTAGE

The power-angle relationship, $P = (E_1 E_2/X) \sin \delta_{12}$ is fundamental to all stability problems. Assume a single generator (or station) is to transmit power over a distance to a system bus. In this case let E_1 be the generator internal voltage and E_2 the voltage of the system bus (more accurately the voltage behind the system equivalent reactance as viewed from the bus, the Thevinin equivalent voltage). Thus, the voltages E_1 and E_2 cannot be raised to increase the power flow P. The angle δ_{12} must vary between 0° at no load, to less than 90° at peak load. In this case, the $\sin \delta_{12}$ is between 0 and 0.9. (The value of 0.9 was used to emphasize that 1.0 cannot be used). The power equation has been reduced to $P = K/X$ where the constant K is $E_1 E_2 \sin \delta_{12}$. The only way to increase P is to reduce X. Let $X = X_g + X_t + X_b$ when X_g is the generator reactance and X_b the system equivalent reactance behind the bus. Since X_g and X_b are fixed in value, X_t, the transmission reactance, is the only variable. Using two parallel lines will reduce X_t to $0.5X_t$. If a line for one voltage level is compared with a line of another voltage level (same number of conductors per phase) the series reactance is substantially the same number of ohms per unit of line length. The reactance of the line, when viewed from the low-voltage winding of a step-up transformer is reduced by the square of the transformer turns ratio. Thus, to increase capacity or power transfer ability, one can use higher voltage transmission. The transformers will insert a series reactance but this increase is much less than the reduction in "effective" line reactance.

It is clear that increasing transmission voltage will increase the ability of the system to transfer power, thus raising the stability limit. The transmission line reactance X_t increases with its length so that the stability limit is reduced as the line length is increased. Raising the transmission voltage is the usual method of increasing stability limit for long lines. Usually at least two lines are desired from the standpoint of reliability.

Anything that reduces the series X of the power-angle equation will increase the stability limit. Multiple-phase conductors (bundle conductors) reduce the series reactance and are effective. Another approach is to cancel the series reactance with a negative series reactance or with a series capacitor.

SERIES CAPACITORS

Since distance is one of the principal limits to the amount of power that can be transmitted over a line with stability, considerable attention has been given to methods for making lines appear to be "shorter," electrically, then they actually are. The use of multiple (bundles) conductors for reducing the inherent inductance is one approach. This approach has the added advantage of reducing corona—the primary reason bundled conductors were initially considered for transmission lines.

The use of series capacitors is often considered (Miller, et al., 1982) for EHV lines, the final choice being dictated by the economics of the particular application. By reducing the reactance of a transmission circuit, a series capacitor improves stability and voltage regulation. In some cases, a series capacitor provides the desired load division between transmission circuits.

Because of the large ratio between system fault current and normal load current, it is not economical to apply series capacitors on the basis of their being able to withstand fault current. Therefore, protective equipment is provided that short circuits the series capacitor whenever the current flowing through it exceeds approximately twice normal value. In analyzing the stability characteristics of a system employing a series capacitor, the presence of the series capacitor is neglected during the "fault-on" period, since it is then short circuited. Consequently, the gains in system stability are realized during normal, steady-state operation, prior to the occurrence of a fault and, again, following the isolation of the fault, as long as its by-pass equipment is promptly reopened.

The introduction of series capacitors in a power system may produce many problems for the turbine generator. These problems are collectively called subsynchronous resonance (SSR) (see IEEE Subsynchronous Resonance Working Group, 1980). The subsynchronous resonance problem arises because capacitors and inductors are energy storage devices, and may produce an oscillating dynamic system. When both are present, large energy oscillations may be induced by transmission line faults; these oscillations may also occur spontaneously. The frequency of these oscillations are less than 60 Hz because the capacitive reactance is always less than the inductive reactance of the compensated lines. Subsynchronous postfault line currents create subsynchronous components in the generator air gap electromagnetic torque. If the frequency of these components is at or near a turbine generator torsional frequency, then a reinforced resonance phenomenon may occur. When resonance is a factor, large torsional response and fatigue damage may occur because of fault application; it may also occur spontaneously as a form of dynamic instability.

Because of the potential effects of subsynchronous resonance, a study to determine compatibility of the series capacitors with the turbine generators is always advised. After any potential SSR problems have been recognized and solved, the improvement attainable in transient stability may be determined. The series capacitor control and protective equipment must be recognized in the stability analysis. (See Chapters 3 and 7 in Miller, et al., 1982.)

SUMMARY

Stability is that property of a power system that makes it possible to maintain synchronism between all of its synchronous machines.

System disturbances caused, for example, by short circuits or switching, may temporarily upset the balance between input and output of one, several, or most of the machines. When the input to a particular machine exceeds the output, the difference causes the rotor to accelerate. The excess of energy input over output is stored in the rotor as kinetic energy. If the rotor contains enough WR^2 to absorb this energy without having its angle (with respect to the system) advanced too far, it will probably remain in synchronism after the disturbance subsides. If the difference between input and output is too great, or if it lasts too long, the rotor will be driven out of synchronism with the system.

The methods that have been suggested to improve stability include increased generator WR^2, reduced generator reactance, high-speed excitation, and many others. By far the most effective method is reduction of fault duration through the use of a high-speed relays and circuit breakers. A combination of these methods is usually applied in modern power systems.

REFERENCES

1. Bowler, C. E. J., Brown, P. G., and Walker, D. N., "The Evaluation of the Effect of Power Circuit Breaker Reclosing Practices on Turbine-Generator Shafts." IEEE Power Engineering Society, 1980 Winter Meeting paper #F80-199-0.
2. Brown, P. G., and Concordia, C., Effects of Trends in Large Steam Turbine Driven Generator Parameters on Power System Stability," *IEEE Transactions*, 71TP74-PWR, January 1971.
3. Byerly, R. T., and Kimbark, E. W., Editors, *Stability of Large Electric Power Systems* (book with 59 excellent papers on the subject of power system stability), IEEE, New York, 1974.
4. Clarke, Edith, *Circuit Analysis of AC Power Systems*, 2 Vol., Wiley, New York, 1943 and 1950.
5. Crary, S. B., *Power System Stability*, 2 Vol., Wiley, New York, 1945 and 1947.
6. DeMello, F. P., and Concordia, C., "Concepts of Synchronous Machine Stability as Affected by Excitation Control," *IEEE Transactions*, **PAS-88**, 316–329 (April 1969).
7. Hauth, R. L., Miske, Jr., S. A., and Nozari, F., "The Role and Benefits of Static Var Systems in High Voltage Power System Applications," paper 82WM 076-8, IEEE Winter Power Meeting, February, 1982.
8. IEEE Committee Report, "Computer Representation of Excitation Systems," *IEEE Transactions*, **PAS-87**, 1460–1468 (June 1968).
9. IEEE Committee Report, "Dynamic Models for Steam and Hydro Turbines in Power System Studies," *IEEE Transactions*, **PAS-92,** 1904–1915 (Dec. 1973).
10. IEEE Committee Report, "System Load Dynamics-Simulation Effects and Determination of Constants," *IEEE Transactions*, **PAS-92,** 600–610 (Mar./Apr. 1973).
11. IEEE Subsynchronous Resonance Working Group. "Proposed Terms and Definitions for Subsynchronous Oscillations," *IEEE Transactions*, **PAS-99** (2), 506–511 (March/April 1980).
12. IEEE Task Force on Terms and Definitions. "Proposed Terms and Definition for Power System Stability," *IEEE Transactions*, **PAS-101** (7), 1894–1898 (July 1982).
13. IEEE Working Group on the Effects of Switching on Turbine-Generators, "IEEE Screening Guide to Planned Steady-State Switching Operations to Minimize Harmful Effects on Steam Turbine-Generators," IEEE Transactions, **PAS-99**, 1519–1521 (July/Aug. 1980).
14. Kimbark, E. W., "Improvement of Power System Stability By Changes in the Network," *IEEE Transactions*, **PAS-88**, 773–781 (May 1969).
15. Kimbark, E. W., *Power System Stability, Vol. I: Elements of Power System Calculations*, Wiley, New York, and Chapman and Hall, London, 1948.
16. Kimbark, E. W., *Power System Stability, Vol. III: Synchronous Machines*, Wiley, New York, and Chapman and Hall, London, 1956.
17. Miller, T. J. E., et al., *Reactive Power Control in Electric Systems*, Wiley, New York, 1982.
18. Stott, B., "Power System Dynamic Response Calculations," *Proceedings of IEEE*, **67,** (2) 219–241 (February 1979).
19. Young, C. C., "Equipment and System Modeling for Large-scale Stability Studies," *IEEE Transactions*, **PAS-91,** 99–109 (Jan/Feb, 1972).

14
SYSTEM OPERATION

INTRODUCTION

The object of system operation is to generate and deliver the power required by the utility customers and interconnections as economically and reliably as possible while maintaining the voltage, frequency, and time error (integrated frequency) within permissible limits.

As systems become larger and more complex, the electric utility industry devotes greater efforts to apply automation technology to the solution of system operating problems. The relevant developments in automation technology are associated with analog and digital computers, data collection and supervisory control equipment, communication systems, and displays.

SYSTEM OPERATION TASKS

The tasks facing the system operator can be divided into three basic categories: planning, control, and accounting.

Operations Planning

Operations planning is concerned with looking ahead to the next hour, day, week, or month. Tasks include:

1. Load forecasting
2. Maintenance scheduling
3. Generation spinning reserve determination
4. Generation unit commitment scheduling
5. Evaluation of possible interconnection transactions
6. Fuel selection
7. Hydro-thermal generation coordination
8. Selection of load-shedding procedures.

The public concern over air and water pollution resulting from power generation has also made pollution an important factor in operations planning and control.

Operations Control

In operations control, the second-by-second or real-time calculations and decisions are made. Examples are automatic generation control (AGC), generation dispatch, var dispatch, and transmission control. Generation dispatch involves allocating generation on a continuous basis so as to achieve minimum fuel cost. AGC involves maintaining generation at such a value as to keep frequency and the scheduled net interchange with neighbors at desired values.

Operations Accounting

Operations accounting is concerned with data collection and analysis, after-the-fact evaluations, preparation of reports, and billing. Examples are:

1. Collection of system and unit production statistics
2. Interconnection billing
3. Evaluation of system and unit performance
4. Analysis of abnormal conditions.

264 SYSTEM OPERATION

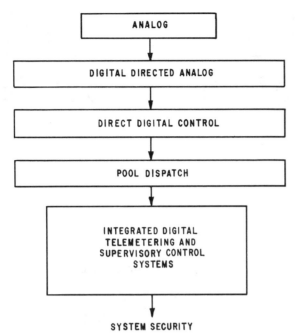

Figure 14.1. Progress in application of system operations computer.

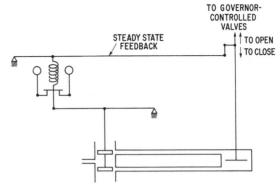

Figure 14.3. Schematic diagram of speed-governing mechanism with speed regulation.

ROLE OF AUTOMATION

Control functions received the first consideration in the application of automation equipment. The advent of the process computer led to a rapid advancement in application of automation to other system functions. Figure 14.1 shows the progress made in applying computers to system operation.

Originally, analog control systems were applied for both load-frequency control and economic dispatch. Then process computers were used in conjunction with analog systems in digitally directed analog control systems. Subsequently, the process computer was used to calculate all economic dispatch and load-frequency control functions on an instant-by-instant basis, resulting in direct digital control systems.

These developments were then applied on a pool basis. Finally, with the advent of high-speed supervisory control and data acquisition (SCADA) systems, it is feasible to integrate digital telemetering and supervisory control systems with process computers to permit the effective solution of problems involving system security.

CONTROL FUNCTIONS

A simplified description of turbine governors is useful to an understanding of the automatic generation control.

Most speed governors do not attempt to hold constant speed, but are designed to permit the speed to drop as the load is increased (Figure 14.2). This speed regulation is necessary for stable load division between turbine-generator units. If two or more machines are operated in parallel, an isochronous (constant speed) governor may be used on only one of the turbines, since units equipped with isochronous governors tend to "fight" against each other due to slight differences in their characteristics. Governors typically have a speed regulation of 5–6% from zero to full load.

Figure 14.3 shows schematically the essential elements of a speed governor with speed regulation. A turbine equipped with such a governor would operate well in parallel with similar type units since each unit would have a definite steady-state valve position for each value of speed error. It is likely, however, that the resulting split of power between

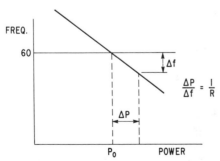

Figure 14.2. Steady-state characteristic of typical governor.

CONTROL FUNCTIONS

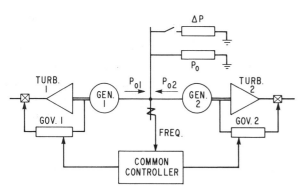

Figure 14.4. Common controller.

units would not be satisfactory from the standpoint of overall system economy, or perhaps from a system security standpoint. Furthermore, the frequency has no way of getting back to 60 Hz.

To accomplish the dual function of obtaining the desired power split between units and of restoring frequency to normal, some type of common control is necessary. In its simplest form, this control looks at frequency and signals the individual governors to reset their operating points to the desired levels. This is illustrated in Figure 14.4. The common controller generally acts more slowly than the governors. The point of control in the governor is the speed changer (Figure 14.5), which will act to change the speed set-point and the power delivered at rated speed. The steady-state characteristic is shifted so that the new steady-state operating point is the intersection of rated speed and the new desired power (Figure 14.6).

Many system loads are frequency sensitive. For example, motor loads drop 2% or more for a 1% drop in frequency. On the other hand, heating and

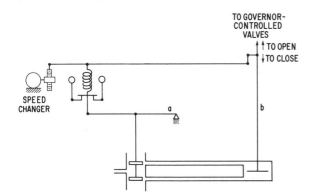

Figure 14.5. Schematic diagram of speed-governing mechanism with motor-adjusting mechanism to change the speed set-point.

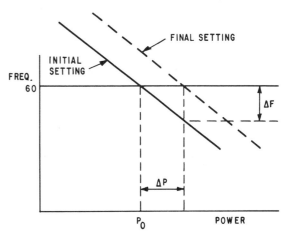

Figure 14.6. Steady-state characteristic of typical governor with speed changer.

lighting loads are insensitive to frequency. The composite load consists of a mixture of these loads, and the overall characteristic is often approximated by a 1% drop in load for a 1% drop in frequency. In Figure 14.7, the solid line illustrates the characteristic for a system load P. The dashed line is the new characteristic if the load is increased by ΔP.

Previously, in the absence of a common controller, the addition of ΔP resulted in a frequency drop of $R \times \Delta P$, where R is the slope of the governor characteristic.* Due to the frequency sensitive load, the frequency will not drop quite that far, as some of the load will be shed.

The amount of frequency drop in the presence of governor control and frequency sensitive load is of interest in understanding how load-frequency control works. The final steady-state operating point, in the absence of a central controller, after load has been added, is the intersection of the governor characteristic and the new system load-frequency characteristic (Figure 14.8). The frequency drop is equal to ΔP divided by $(1/R + D)$, where D is the slope of the system load frequency characteristic.

Tie Line and Load-Frequency Control

With this background of governor and system characteristics, the performance of an interconnected system may be discussed. The operation objectives are to maintain reasonably uniform instantaneous

---

* The standards for this particular area do not follow the conventional definition of slope. See Figure 14.8 for the definition of slope used in this section.

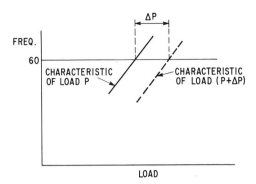

Figure 14.7. Frequency characteristic of system load.

frequency; to maintain correct time (integrated frequency); to divide the loads between systems, stations, and generators so as to achieve maximum economy and correctly control the net interchange. All these objectives can be met simultaneously. Tie line interchange schedules are set by each interconnected utility system based on mutually agreed upon schedules dictated by area regulation requirements and economy interchange agreements.

Consider a typical set of interconnected systems as in Figure 14.9. Each area acts as an independent agent in buying or selling power, and each calculates a quantity known as ACE (area control error) as its fundamental control variable. The ACE for each area is a function of deviation from desired net interchange for that area and the frequency deviation multiplied by $(1/R + D)$. The units of ACE are megawatts.

As an example, Figure 14.10 shows what happens when a load of 100 MW is suddenly applied to system A which, initially, is serving 900 MW and has 1000 MW of capacity running. Taken together, systems B, C, and D are serving 90,000 MW of load with a capacity of 100,000 MW. The example shows that the ACE of systems B + C + D = 0 and the ACE for system A is 100 MW. Area A should then pick up the 100-MW load to restore the frequency to normal and the net interchange to that which was scheduled. This 100 MW is allocated to the different generators in system A in the most economical manner.

Load-Frequency Control and Economic Dispatch Hardware

Figure 14.11 shows the initial application using an analog load-frequency control system with manual economic dispatch. The area control error and the resulting signal to each individual unit are formed by analog equipment. The settings that allocate the generation change required from each unit are set manually.

Next in the application of automation came the digitally-directed analog system (Figure 14.12). Analog circuits are involved in the instant-by-instant control of generation, while the digital computer calculates the settings necessary for the most economic allocation of generation.

In direct digital control systems, the digital computer is included in the control loop. It scans the unit generation and tie flows and generates the raise/lower pulses to be sent to the individual units, using software rather than hardware. The formation of ACE may or may not be done within the computer. This system may have analog backup as in Figure 14.13, or digital backup as in Figure 14.14. In the latter case, the backup computer normally performs other functions, such as computer-directed supervisory control. Today, most utilities use direct digital control systems based on a dual digital computer configuration.

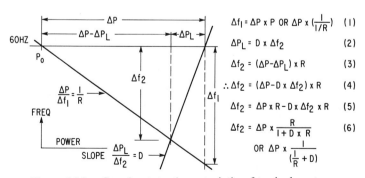

Figure 14.8. Steady-state characteristic of typical system.

CONTROL FUNCTIONS

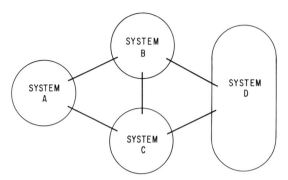

Figure 14.9. Typical interconnected systems.

Economic Allocation of Generation

The individual efficiencies of units, as well as their location in the transmission network, affect the amount of power each should contribute to the system load. Since there is a definite cost associated with starting or stopping a unit, this has an effect on which units would be running (unit commitment).

To illustrate the principles involved, a simplified example with two units, both running, and no transmission losses will be considered. The input/output curves of both units are shown in Figure 14.15. Figure 14.16 shows the incremental cost curves of both units. These curves are the slope of the input/output curves against output.

It can be shown that, for a given load, the most economical operation of the two units occurs when the incremental cost of unit 1 equals the incremental cost of unit 2. Consider a horizontal line $X - X$ in Figure 14.16. Total power output corresponding to that line is $M + N$. If $M + N$ is equal to the load requirement, then these are the correct operating points. A manual procedure to find the most economical operating points is to move the horizontal line up and down until the sum of the power outputs of the units equals the system load. This is the basis for the first attempts at economic dispatch where the operator used an incremental loading slide-rule to obtain the most economical allocation of generation. The results could be used to set the levels in Figure 14.11. The economic allocation is calculated by the computer in Figures 14.12, 14.13, and 14.14.

Reactive Power and Voltage Control

The control of voltage and reactive power are inseparable. It is a complex task and the approach to it is still based as much on operator experience as on any defined method.

The source of reactive power (vars) is machine vars and capacitor and reactor vars. The machine vars are controlled by field excitation, and the capacitors and reactors can often be switched for con-

	SYSTEM A	SYSTEMS B+C+D	SYSTEM A+B+C+D
ASSUMED STARTING CONDITIONS			
CONNECTED GENERATION (MW)	1000	100,000	101,000
INITIAL CONNECTED LOAD (MW)	900	90,000	90,900
ΔP (MW)	100	0	100
FINAL CONNECTED LOAD (MW)	1000	90,000	91,000
ASSUMED SYSTEM CHARACTERISTICS			
1/R (MW/%Δf)	200	20,000	20,200
D (MW/%Δf)	10	900	910
1/R + D (MW/%Δf)	210	20,900	21,110
COMPUTED RESULTS			
$\Delta f(\%) = \dfrac{\Delta P}{1/R + D} = \dfrac{100}{21,110} = .004737\% = .00284\,hz$ (STEADY STATE)			
$\Delta P_{GEN} = \Delta f \times 1/R$	.947 MW	94.7 MW	95.65 MW
$\Delta P_{LOAD} = \Delta f \times D$	.047 MW	4.26 MW	4.31 MW
NEW GENERATION	900 + .947 MW	90000 + 94.7 MW	90900 + 95.65 MW
NEW LOAD	1000 - .047 MW	90000 - 4.26 MW	91000 - 4.31 MW
NEW (GENERATION−LOAD)	−100 + .947 + .047 = −99 MW	0 + 94.7 + 4.26 = +99 MW	−100 + 100 = 0 MW
ACE	−99 − (−210)(−.004737) = −100 MW	99 − (−20900)(−.004737) = 0	

Figure 14.10. Calculation of area control error following increase in load in area A.

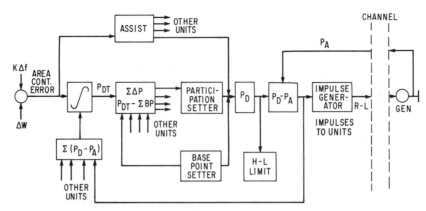

Figure 14.11. Analog load-frequency control with manual economic dispatch.

nection or disconnection. Tap-changing transformers also perform a voltage control function.

The present operating procedure is to monitor and control a number of buses in the system. The dispatcher controls these voltages based on experience, previous operating practices, and off-line load-flow studies. He employs the generator bus voltages for control, in addition to the controlled capacitors and reactors. A number of conditions of the controlling variables would yield the proper controlled bus voltages—provided that these voltages are feasible. A methodical, often long, search procedure using off-line load flows, is required before an acceptable set of variables can be found to satisfy the desired conditions.

The implementation of var dispatch for a process computer is feasible and practical methods of application are being pursued. One approach would be to allow the computer to calculate one or more sets of feasible conditions in terms of realizable generator bus voltages, switched capacitors and reactors, transformer tap settings, and other system constraints that cannot be violated using information available concerning the current state of the sys-

tem. The dispatcher could then select the condition that is most easily implemented.

This approach could be modified to allow the computer to pick the best of all the feasible operating conditions, and then automatically schedule the source of vars. Equipment and techniques to carry out this complex task adequately have not yet been fully developed and tested under actual operating conditions.

SYSTEM SECURITY

The initial justification of process-control digital computers was primarily based upon improvement to be achieved in fuel economy. Of even greater importance today is the application of computers to improve the reliability of system operation.

Automation technology cannot be used as a substitute for good system design. Sound design prac-

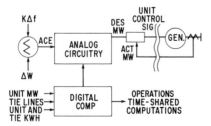

Figure 14.12. Digital-directed analog system.

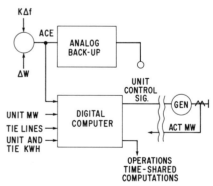

Figure 14.13. Direct digital control system with analog back-up.

SYSTEM SECURITY

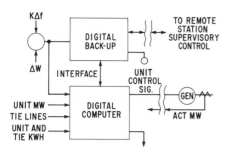

Figure 14.14. Direct digital control system with digital back-up.

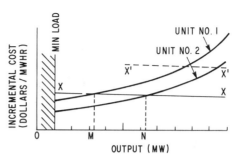

Figure 14.16. Incremental cost curves.

tices must be applied to selecting generator size and location, transmission size and location, and the determination of the control system and the protective system. Only then can rational policies regarding maintenance scheduling, line loading, generation loading and shedding, and separation rules be formed.

In applying automation technology to system security, it is appropriate to examine the time scale for control action. For fault removal, control action must be accomplished within cycles or milliseconds to be effective. For fast load changes, action must be taken over a period of seconds. New unit commitment can be accomplished in minutes or hours, depending on the nature of the prime mover.

Overall operation for reliability means trying to keep the system in a normal state, i.e., all customer and interconnection demands are met, no apparatus or line is overloaded, and the consequences of an unexpected contingency are minimal. If the system departs from this state for any reason, the object is to restore it in minimum time.

System Operating States

To analyze power systems and to design appropriate control systems, it has been found helpful to classify the operating states of the system into several categories. These states, ranging from normal operation through several abnormal states, are depicted in Figure 14.17, which also indicates the state transitions that can take place. The discussion that follows reviews the definition of these operating states.

Normal State

The power system is in the normal or secure state if no emergency conditions exist and no likely contingencies would cause an emergency condition to exist.

Alert State

The power system is in the alert or insecure state if one or more likely contingencies would create an emergency condition. The dividing line between

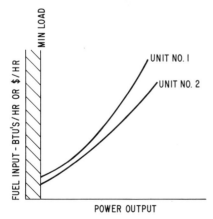

Figure 14.15. Input-output curves.

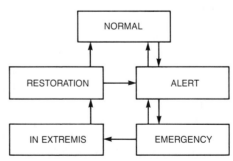

Figure 14.17. Power system operating states.

normal and alert states is dependent upon what contingencies are considered "likely." Normally, all single contingencies are used as the basis for assessing the security of the system and, thereby, flagging alert conditions. An alert state may also be the result of severe weather conditions that increase the likelihood of transmission line outages.

Emergency State

The power system is in an emergency condition when critical operating contraints are being violated. When a violation occurs, the integrity of the system is jeopardized. Examples of critical operating constraints include:

1. Thermal loading limits of transmission lines, cables, transformers, and other current-carrying equipment
2. Line or corridor loading limits based on stability or voltage collapse limitations
3. Voltage limits at transmission, distribution, or power plant auxiliary substations
4. Transient or oscillatory responses that may lead to loss of synchronism of parts of the system.

The emergency state, as distinguished from the in extremis state, exists when constraints are being violated but the system is still intact and satisfying all load demands.

In Extremis State

The in extremis state (islanding) is an emergency condition in which system integrity has been lost either by shedding of load or by separation of the system into islands. In the islanding situation, generation-load imbalances within the islands generally exist with resulting frequency excursions and further shedding of load and tripping of generation. Thus, the in extremis state is frequently associated with the blackout of some portion of the system.

Restorative State

Once an in extremis condition is brought under control, the system is in the restorative state. In this state, the transmission and distribution system are reconnected, generating units are restarted, and loads are restored.

Steps to Improve Security

The digital computer offers an opportunity to improve the security of the power system through a four-step evolutionary approach:

1. Status monitor and display
2. Contingency evaluation
3. Corrective strategy formulation
4. Automatic control.

A significant improvement in system security can be achieved by using the computer to give the operator increased information on the status of the system. In the second step, the computer predicts the effect of contingencies and planned outages, and alerts the operator to potential troubles. In the third step, the computer formulates corrective strategies. At this stage, the operator would call for the execution of the strategy he chooses. In the final stage, the computer, through the communication network, would execute automatically the computer-formulated strategies.

HIERARCHY OF ACTION CENTERS

Power Companies

In the continental United States, the electric power industry includes nearly 3500 systems that vary greatly in size, type of ownership, and range of functions. Generally, there are four distinct ownership segments: investor-owned companies, non-federal public agencies, cooperatives, and federal agencies. Most systems are vertically integrated, i.e., they perform the functions of generation, transmission, and distribution. On a higher level, most companies are organized according to a hierarchy of action centers, as illustrated in the lower portion of Figure 14.18. Note the data, control, and voice communication requirements between each location.

Overall responsibility for the bulk generation-transmission system resides with a company dispatching center. This center directs the second-by-second control of generation to maintain tie-line schedules and frequency, and is responsible for minimizing the cost of power generation within constraints imposed by security. Typically, such a center has access to telemetered values of all major tie-

HIERARCHY OF ACTION CENTERS

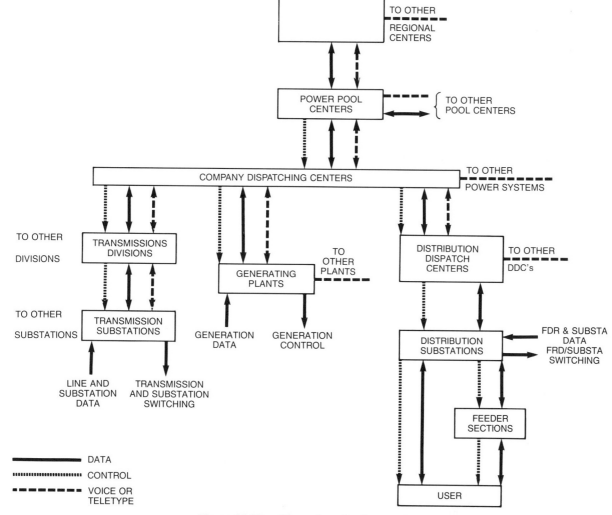

Figure 14.18. Hierarchy of action centers.

line power flows, the power output of all large units or plants, and system frequency.

Operation of the substation and transmission facilities is accomplished within transmission divisions. Division operators plan and carry out preventive and emergency maintenance and are responsible for switching operations within the division. These operators coordinate with the company dispatch center and other division centers as required.

Transmission substations within the divisions may be manned or unmanned. Unmanned substations are monitored and controlled at a Division Dispatching Office by means of SCADA equipment discussed in Chapter 10. Manned stations may or may not have telemetry equipment installed to enable automatic monitoring by the division.

Generating plants, with the exception of remote hydro- or combustion-turbine installations, are staffed with operators and maintenance personnel. The operation of large thermal plants involves the local control of many plant variables, and the high degree of automation achieved has contributed significantly to their reliability and economy. Viewed from the company dispatch center, these plants provide a point from which generation is controlled and is the source of data on the plant status.

At the company dispatch center, improvements in the degree of monitoring of conditions on the bulk power system are being applied by many utilities. Through supervisory control and data acquisition (SCADA) equipment, dispatchers in the company dispatch center can quickly see any adverse fluctuations experienced by the system. Many times, these conditions call for several voice conversations.

The number of automated distribution systems in electric utilities in the United States is growing annually. Automation at the distribution substation, feeder, and customer service levels is being studied and trial systems are either planned or implemented.

Reporting to the company dispatching center may be 3–15 distribution dispatch centers, each of which may operate in conjunction with 5–100 distribution substations. The distribution dispatch center has overall responsibility for the operation of the distribution system. Each distribution substation may supply 2–20 feeders, each of which may serve 200–1500 users. With feeder automation, each feeder may be divided into three or four sections by means of remotely controlled sectionalizing switches. Tie switches are associated with each feeder section so that various sections may be energized through more than one feeder and from more than one distribution substation. User automation includes customer load control (water heaters, central air conditioning, etc.) and automatic meter reading, which collectively are referred to as load management.

Company and Pool Levels

The company and pool levels are responsible for operation of a complete bulk generation-transmission system and for coordinating energy and capacity interchanges with neighboring companies and pools.

Various arrangements of computers may be considered with respect to AGC between the company and pool levels. A configuration such as shown in Figure 14.19 is good for flexibility and reliability of operation. At the pool level, company area control errors are formed that recognize both economic and regulation requirements. If the pool computer is out of service, the company computer can form an area control error in the usual manner.

Backup should be provided for all major computer functions. One possible arrangement of dual computers is illustrated in Figure 14.20. One processor is dedicated to scanning the process inputs and to performing the basic control and alarm functions. The second is dedicated to the study functions and those real-time functions that require relatively large amounts of active storage or whose running times are long. The bulk memory associated with both processors is kept up-to-date on the status of the system. Each central processor should be selected to conduct at least the control and security functions and, at reduced speed, a number of other functions. It is not considered necessary to back up all the study functions when one of the processors is shut down.

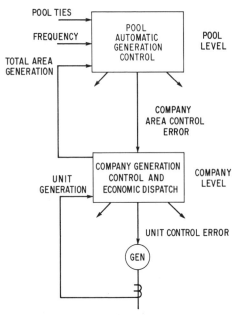

Figure 14.19. Relationship between company and pool computer center.

Regional Coordinating Centers

With the recognition that economic and reliability considerations cannot be treated completely within

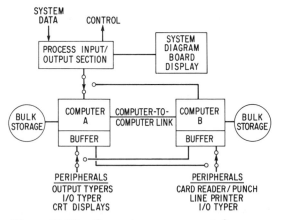

Figure 14.20. Computer arrangement for company and pool levels.

a single pool, regional coordinating centers with responsibility for large geographic areas have been formed.

At present these centers have the following functions:

1. Role as communication center
2. Analysis of future operating conditions
3. Coordinating emergency procedures and analyzing disturbances
4. Coordination between planning and operation.

Computer requirements are mainly for analysis of large networks so that a large scientific computer is appropriate. Backup is not required.

FUNCTIONAL SERVICE REQUIREMENTS

It is helpful to review the functional service requirements for utility system operation and communication. These requirements are as follows:

1. Telemetering and Control
   a. Analog telemetering
   b. Supervisory control and data acquisition (SCADA)
   c. Automatic generation control (AGC)
   d. System security
   e. Distribution automation
   f. HVDC converter control
2. Telephony (voice communications)
   a. Dispatch
   b. Maintenance
   c. Operation
   d. Executive
3. Teletype and bulk data transmission.

Telemetering and Control

Power system automation relies on many different types of telemetering and control signals. Analog telemetering and SCADA systems provide information to indicate breaker or line status, generator status, frequency, megawatt flows and voltage at selected buses, and Mvar flow at var sources. Megawatt and Mvar flows are also required from tie lines that interconnect individual utilities and power pools. All of this information is critical in maintaining the security of a power system. Frequency is also monitored at several locations, since it is one of the first indications of trouble on an interconnected system.

Automatic generation control requires raise-lower control signals to each generator unit under control, or to each member utility within a power pool. Conversely, generator megawatt output, high-low limits, and the status of various equipment pertinent to the operation of the turbine-generator are required at the company dispatch center. Telemetering and control functions call for extensive use of voice channels to accomplish the data transmission requirements.

Distribution automation provides remote control and indication of feeder sectionalizing and tie switches, remote monitoring of fault indicators, remote control of feeder capacitors and voltage regulators, and telemetering of certain voltages and currents. The functions include distribution data base, fault identification, fault isolation, feeder load management, remote control and data reporting, service restoration, volt/var control, and dispersed storage and generation.

Communication requirements for the operation of unattended HVDC converter terminals consist of communication channels for supervisory control equipment, a communication channel for voice communication between converter terminals and other remote terminals, and communication channels for control, interlocking and protection purposes between converter terminals.

Telephony

Telephony is a valuable tool for the modern electric utility system. In times of an emergency, it is indispensable; for efficient normal operation it is a necessity; and for maintenance and day-to-day company business it is a time-saving convenience. Voice communication is provided by privately owned communication systems such as microwave and power line carrier and leased telephone company facilities.

Teletype and Bulk Data Transmission

Bulk data information interchange consists of random data that must be transmitted within a specified time period. Such data may consist of forecast data, boiler/unit status or current condition, generation schedules, fuel information, etc. The data transmission requirements vary from teletype to voice grade and wide-band, high-speed data transmission.

COMMUNICATION ALTERNATIVES

Communication is required for three general areas: finance and accounting, engineering, and operations. Of the three, probably the largest requirement for communication is operations.

A review of the communication alternatives for power system operation may be examined by considering two general categories—communication for bulk power system operation and communication for automated distribution systems.

Communication for Bulk Power System Operation

A variety of communication systems are available for communication between dispatch centers, generating stations, transmission substations, and distribution substations. This upward communication can be accomplished with available art, including microwave, power line carrier, and telephone. Each may satisfy a particular need or application, and the choice depends on technical and economic considerations and the total functional requirements.

One type commonly used is microwave. This communication system has several general characteristics. Licensing is required and it is regulated by the FCC. Generally, it provides many voice channels over a medium distance. It makes use of free space propagation, and requires a path survey to consider the terrain and objects in its path. Microwave is generally not affected by line noise. It utilizes a large frequency spectrum in the 2-GHz, 6-GHz, and 12-GHz bands. Generally, it is not economically competitive for low-density channels, e.g., 8–12 voice channels. Land sites must also be considered, in addition to more complex maintenance requirements.

Telephone channels leased from the telephone company are generally economical for short distances, e.g., less than 50–100 miles (80–160 km), and it depends on the number of voice channels between locations. Leased channels are not under user control and require careful coordination between the telephone company and user. Also, where a service ties into an electric power station, interface coordination and protection is required. The telephone company does make available a wide offering of channels and a wide range of characteristics and conditioning.

Power line carrier (PLC) has been applied directly to transmission lines for many years and is suited for few channels (four to eight voice channels between locations) over long distances of 100–200 miles (160–320 km). Power line carrier is reliable, because it is as reliable as the transmission line over which it is applied. Minimum repeating is required and it is not unreasonable for PLC to be applied over a single hop of 100 miles (160 km) and greater, depending on the transmission line voltage.

Power line carrier is suited for voice communication and all of the services described earlier. However, its spectrum is limited to 30–500 kHz and on insulated shield wires, it can go as low as 8 kHz. Good application rules must be followed to make effective use of the available carrier spectrum. Also, power line carrier is susceptible to line noise and requires careful signal-to-noise ratio analysis to insure a good application.

Communication for Automated Distribution Systems

The approach to automated distribution systems requires communications upward from the distribution substation to the higher-level control centers and downward to the distribution feeder and user location. Upward communications can be accomplished with available technology, as discussed under communication for bulk power systems. However, the downward technology is new and the subject of study by many groups. The role of communication and the possible communication alternatives are significant.

A variety of communication techniques from the distribution substation downward are being considered:

1. Telephone
2. Dedicated private lines
3. Radio
4. Distribution line carrier
5. Ripple (audio signals) for load control
6. Hybrid, involving a combination of radio and distribution line carrier.

Generally, irrespective of the type of communication used between the distribution substation, the feeder, and user locations, the need for a voice-grade circuit or channel with 300-baud capability is required between the distribution dispatch center and the distribution substation.

DATA DISPLAY

A comprehensive control center involves extensive display equipment. Operator communication is very important in system operation. Pertinent data must be made available in a prompt and efficient fashion so that the operator can make correct decisions. Display systems play an increasingly important role as a means of efficient data communication.

Common display requirements in a company or pool center are:

1. Generator status
2. Tie-line status
3. Spinning reserve status
4. Frequency and frequency trend
5. Transmission system status
6. Post disturbance display
7. Alarm and logging.

Devices such as indicating lights, switches, meters, recorders, and digital readouts have been widely employed throughout the industry. Other devices are cathode ray tubes (CRTs), projection displays, and electro-luminescense displays as might be used in the case of a one-line diagram.

CRTs provide an effective method of displaying a large amount of data on request. They are especially useful for displaying substation one-line diagrams, including telemetered and computed data.

A system diagram board which portrays the major transmission circuits in general is utilized even though its size places a significant constraint on the physical arrangement of the operating center. It provides the dispatcher with a continuously updated view of the physical network so that he may readily visualize alternate paths for power.

SUMMARY

There is extensive application of automation equipment both on an area dispatch level as well as the pool dispatch level. At the area level, system operation computers are being applied to provide improved automatic dispatch of system generation, improved system security through the interface with remote supervisory control equipment, improved operator display systems, and communication with plant computers to provide up-to-date information in the central system operations computer.

The company system operations computer also communicates with a pool operations computer. The pool operations computer provides improved economic operation of the entire pool, as well as permitting coordinated system security of the pool. The pool office may utilize business and engineering computers to carry out all the tasks required at the pool level. Improved data systems are being applied to display the status of each of the critical elements of member companies that make up the pool.

Interest is growing in further automation of the distribution system to obtain the benefits of better utilization of the power system facilities and equipment, improved operation, reduced outage duration, and better information for planning.

REFERENCES

1. Cohn, N., "Power System Interconnections: Control of Generation and Power Flow," Section 15 *Standard Handbook for Electrical Engineers,* 10th Edition, McGraw-Hill, New York, 1968.
2. "Energy Control Center Design," IEEE Power Engineering Society Course Text, 77TU0010-9-PWR.
3. Kirchmayer, L. K., *Economic Control of Interconnected Systems,* Wiley, New York, 1959.
4. Kirchmayer, L. K., *Economic Operation of Power Systems,* Wiley, New York, 1958.
5. Miller, R. H., *Power System Operation,* McGraw-Hill, New York, 1970.

15
SYSTEM DESIGN

THE EARLY SYSTEMS

Thomas Edison designed generators, distribution systems, and loads. He invented the first load, the electric light, and organized the first company to generate, distribute, and sell electric energy. Not only was Edison the most comprehensive system engineer of the electric power industry, he actually created the industry. In Edison's time, the concept of system planning, as known today, was not necessary. Generally, the pioneers of the industry began by building a generating station and some feeders to potential loads. As these isolated enterprises grew, the areas they served began to touch and overlap. Gradually, some of the separate companies were combined into the ancestors of our present companies.

The separate elements of the early systems were carefully designed. When a hydroelectric site was to be developed, there also had to be transmission lines to bring the power to the load. It was a major design problem to select exactly the right voltage and conductor size to transmit the output of that plant to market at the lowest cost over a period of years. The designer could not foresee that in a few years a large system would grow up around these transmission lines and absorb them, nor that the conductor size selected so carefully to do one job would probably be inadequate for the later function.

SYSTEM DESIGN AND CHANGING CONDITIONS

The concept of designing an entire power system developed slowly. The idea that a system can and should be designed or planned has gained wide acceptance.

Progress in the electric utility industry has come about by trying new ideas that appear to have a reasonable chance of success. Those that worked came into general use; the others did not. Prior chapters have cited examples of both successes and failures. At one time, the expulsion protector tube offered the promise of making transmission lines immune to lightning. This promise was never fulfilled, and the modern line design for satisfactory lightning performance uses overhead shield wires and proper grounding of towers.

Other examples can be cited as in the development of a successful automatic reclosing relay which made the unattended substation practical. As a result of this innovation, the unattended substation could be smaller and placed nearer the load. A new method of primary distribution system design was achieved and standardized factory-fabricated substations became available.

Beginning about 1925, intensive studies of the nature of lightning were carried on in both field and laboratory. The knowledge gained from these studies of lightning, and of the ability of insulation to withstand lightning voltages, enabled much more reliable apparatus and transmission lines to be built.

As the utilities have been able to give more and more dependable service, their customers have placed more and more dependence on a continuous supply of electric power. For example, in certain oil refineries, the loss of power for much more than a minute may require a week to get back to full production. In mills and factories producing cloth, thread, paper, and other commodities, losses of production or damaging flaws in finished products

SYSTEM DESIGN AND CHANGING CONDITIONS

can result from brief service interruptions, including momentary voltage dips.

Each improvement in continuity is an invitation to attempt some new application, which in turn requires still greater freedom from interruptions. Three more examples will be cited to describe the effects of changing conditions on system design. First, the change in philosophy of selecting generating units, second, the expansion of interconnections and pooling, and, third, the race between short-circuit currents and breaker capacity.

Changing Generation Patterns

After the second World War, the electric utility industry had little difficulty in selecting the next generating unit that would be put on the system; in each case, it was the largest unit that was available to burn fossil fuel. During the 1950s this philosophy continued, with a gradual recognition that as the unit size became larger and the load was carried by the larger units, the remaining units were used more and more for peaking purposes, i.e., to carry load only a few hours a day and be shut off during the night. As this transition took place, it became more evident that large fossil fuel units were not well suited for this daily cycling or peaking duty.

In the late 1950s, utilities realized that beside purchasing large base-load capacity, they should be planning to purchase units especially designed for peaking capacity. This capacity might be in the form of pumped storage hydro, combustion turbines, diesels, etc. These peaking capacity units could also be used as area protection since they were small and could be located close to the load or in outlying districts at the end of the long transmission line. Although the initial impetus to large-scale purchasing of peaking generation, and in particular combustion turbine-generators, was given by the Northeast Blackout on November 9, 1965, operating experience coupled with numerous computer studies have shown that peaking generation is a sensible and economic addition to a utility's system. Coupled with economic installation costs, operational flexibility, and rapid start-up capabilities, the combustion turbine has operated far more often than originally planned. The combustion turbine is now an established and important member of the power generator family.

More recently, factors such as varying system load, the availability of large steam-electric plants and the responsibility of meeting load requirements

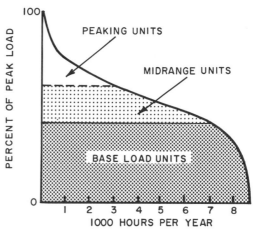

Figure 15.1. Load duration curve.

during maintenance outages, have resulted in the middle portion of the load duration curve (Figure 15.1) receiving greater attention. The combined-cycle plant which takes advantage of the low cost of heavy-duty combustion turbines and the high thermal efficiency of steam turbines has evolved to meet midrange generation requirements. Consequently, utilities now approach generation expansion on a mixed-mode basis wherein base-load units, midrange units, and peaking units are combined to provide the optimum system.

The system planner must explore the many kinds of generation available, determine which ones are right for his system, how much of each he should have, and when the installations should be made. This interplay of choice and economics will be discussed more thoroughly later in this chapter.

Large Interconnected Systems

One of the major changes occurring in past years has been the interconnection of many electric utility companies into large pools. These interconnections have greatly improved the economy of operation of the individual utilities, but, at the same time, complicated the process of system planning and design. For example, in the past, utilities planned their own supply of power and energy without any need to consider the plans of their neighbors. With interconnected systems, planning a unit addition becomes a matter of considering both the individual utility needs and the needs of the pool. Several questions need to be answered in this case. Should the utility supply energy for itself and/or other members of the pool? Should unit additions be rotated among the members of the pool? How large a unit can the pool sustain?

The Race Between Short Circuits and Breaker Capacity

Throughout most of the life of the industry, there has been a race between circuit-breaker development and the need for higher and higher interrupting capacity. The designs of many systems have hinged upon the changing fortunes of this race. As systems grow, the magnitude of possible short circuits also tends to grow. When short-circuit magnitudes approached the maximum interrupting ratings of available circuit breakers, the system designer used reactors, split his system, and adopted any compromise that worked. As a result, there has been, and will continue to be, pressure to develop higher-capacity circuit breakers. The ability of the manufacturer to supply circuit breakers with adequate interrupting capacity permits the attainment of a good system design.

SYSTEM PLANNING

The large expenditures involved on a world-wide basis to expand electric utility generation and transmission systems require that careful attention be given to improved methods for system planning. Digital computer programs are playing an important role in aiding the system planner in determining an expansion plan that provides reliable service at lowest cost, subject to environmental limitations.

The achievement of the system planner's objective is a most difficult task, in view of the many opportunities and limitations. For instance:

1. Equipment decisions have long-term effects requiring a forecast and study period of 15–25 years.
2. There are many alternate means of generating electric power: nuclear, base-load fossil, mid-range combustion turbines or hydro, and in large-, medium-, or small-size plants. Longer-range horizons might also include solar and wind-based technologies, magnetohydrodynamics (MHD), and different forms of energy storage.
3. There are several alternate means of transmitting electric power, for example by alternating or direct current, overhead or underground, and all over a wide range of voltages.
4. The planning decisions are affected by load management techniques and the load patterns that change with conservation.
5. Uncertainty exists concerning the study parameters, such as future fuel costs, interest rates on money and capital availability, equipment forced-outage rates, new technologies, environmental restrictions, and social/political influences.

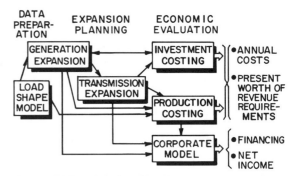

Figure 15.2. Relationships between system planning and economic evaluation programs.

To help the planner formulate and evaluate the many and lengthy expansion alternatives, computer-implemented planning and simulation methods have evolved. These methods are directed by engineers trained especially for generation and transmission system planning.

The various types of computer programs used in expansion planning are related as shown in Figure 15.2. Each program may be used separately or in combination with any of the others as required by a given problem.

A load shape model accepts several years of hour-by-hour load history (Figure 15.3) and produces a tabulation of the daily peak loads throughout a typical year for use in a generation expansion program. Typical daily load shapes covering weekdays, Saturdays, and Sundays for a year are prepared for use in production simulation.

A generation expansion program measures the adequacy of the installed generation to serve the forecasted loads through probability methods. When too high a risk of capacity shortages is encountered for a year, new generating capacity or interconnection capacity will be added, following the strategy supplied by the planner. Various ex-

SYSTEM PLANNING

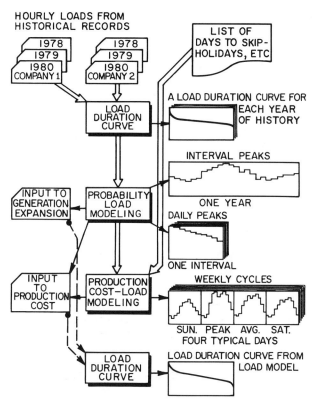

Figure 15.3. Building load-shape models from historical records.

pansion alternatives are formulated by using different strategies and the same probability measure of risk.

Transmission planning is used to examine the bulk power transmission network throughout the expansion period, and to determine the voltage level, location, and installation date for the future network additions through the application of linear programming methods. Alternate designs for a horizon-year condition are developed first to guide the scheduling of each new line addition.

The costs related to investments on generation and transmission for future years may be computed using an investment cost program once the expansion pattern and individual investments have been specified. The program uses fixed-charge rates and present-worth mathematics to combine the series of payments into an equivalent present worth of future revenue requirements.

Generation production costs and their present worths may be computed using a production simulation program. This program simulates the operation of future systems, computing fuel, start-up, operation, and maintenance costs. In addition to the annual cost summaries, the program calculates the present worth value for combination with the investment costing results.

A corporate or financial modeling program can then be used to predict the effects of investment decisions on the corporation's balance sheet, income statement and cash flows. The utility investments including those for generation and transmission and the production costs each year are input, along with a detailed description of the corporation's financial structure, accounting data, and management policy information. The program can simulate the company's activities on a monthly basis and provides forecasts of necessary capital acquisition to finance the system expansion. A corporate/financial modeling program is a useful addition to the normal economic evaluation methods involving the present worth of revenue requirements.

The generation and transmission planning methods will now be discussed in more detail.

Generation Planning

When planning future generation, the system planner can find himself confronted with a bewildering array of considerations. Should another nuclear plant be added, requiring extra capital to achieve lower fuel cost? Or is it time to take advantage of the capital savings available with simple-cycle combustion turbines and incur the additional annual operating expense of running the existing generation at a higher capacity factor? Perhaps the system is in reasonable balance from the standpoint of these two choices, and it is time to look at an intermediate type of generation that is more suitable for daily cycling operation. Essentially, the planner must ask how much generation is necessary, and what kind should be added next?

How Much Generation? The future load must be known to determine accurately when another generator must be added. Since future load cannot be precisely determined, it is forecast with the best information available. A forecast may be made with maximum load and another with minimum load growth. A major concern is how much of the installed generating capacity will be available to serve the load at any given time.

Generating capacity is taken out of service for scheduled maintenance or planned outage and is occasionally lost on an unplanned or forced outage. For this reason, there must be excess capacity available at all times. This is called reserve capacity. Not all types of generation require the same amount of reserves, which adds another evaluation factor between types.

To further complicate the issue, the generation outage characteristic of every system will change as new units are added. A helpful method is available to predict the system characteristics, including the effects of new types of units, different sizes of units, and different outage characteristics. This method uses probability mathematics and is called the loss-of-load probability (LOLP) method. The LOLP considers the forced outage of a generating unit. Essentially, this forced outage is a random occurrence, usually unrelated to forced outages of other units. The unit forced-outage probability is a statement of the odds that the unit will be forced out of service at any particular time. A forced outage probability of 10% for a unit scheduled to operate means that at this moment there are 10 chances out of 100 that the unit is on forced outage and 90 chances out of 100 that it is operating.

Consider the following example of two units of the same size, with the same outage characteristic, that are scheduled to operate with an 81 (.90 × .90 = .81) percent chance that they are both in service. At a time when the load was large enough that both units had to run to serve it, there would be a probability of 81% that the system would have sufficient generation to meet the load, or a 19% chance of insufficient capacity.

Now, suppose that the load was of such magnitude that one unit could supply it. Then the only time when generation would be insufficient to meet the load would be when both units were forced out of service. There is a one (.10 × .10 = .01) percent chance that both units will be forced out at the same time, or, there is a 99% chance that the system will have enough generation to meet the load.

On large systems, where many units are operating to serve load every day of the year, there are tens of thousands of possible combinations of events that are randomly distributed. Probability mathematics make it possible to compute these combinations for each day's operation and to calculate the odds of having sufficient generation to meet each day's load. Maintenance schedules are determined so that the probability of sufficient capacity tends to be constant.

A digital computer program has been created to model a power system. The program will cycle through each day of the year, changing the capacity models as called for by the maintenance schedule and calculating the LOLP. The sum of the daily LOLPs for a year gives the annual LOLP.

If the annual LOLP is excessive, capacity is increased by selecting some predetermined unit size. The annual LOLP is then recalculated using a new maintenance schedule and the correct capacity outage table for each interval. This procedure will be repeated until the system risk level for the year is satisfactory. The generally accepted design is a LOLP of 0.1 days per year.

An example of the LOLP measurement is illustrated in Figure 15.4. Here the probability of loss-of-load was calculated for an original 4600-MW generating system and a range of annual peak loads. A 4000-MW peak load for year one just about satisfies the LOLP risk index. A 300-MW load increase is predicted for year two. With no unit additions, the 4600-MW system would have a LOLP of about 1 day per year, far in excess of the design criteria. A 400-MW unit is added and the probabilities recalcu-

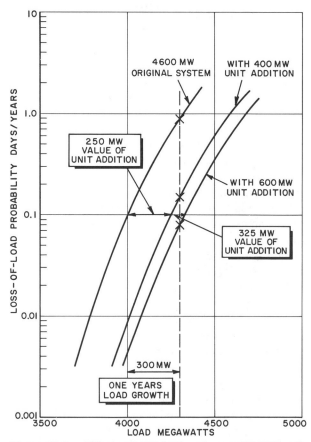

Figure 15.4. Effects of 400-MW unit or a 600-MW unit addition on a 4600-MW system.

lated to obtain the "with 400-MW unit" curve. The LOLP for the 4300-MW peak load is still greater than 0.1 days per year and so more generating capacity must be supplied. In the example, the new 400-MW unit was removed and a 600-MW unit substituted. The "with 600-MW unit" generating system satisfies the design criteria. Note that the 400-MW addition was only "worth" 250 MW and the 600-MW addition "worth" 325 MW of load growth. This is the effective capability for those unit additions for the particular system parameters in this example.

What Kind Of Generation?   The second step of the planning process is the determination of what kind of generation should be installed to minimize total system costs.

The future installed plant costs of all of the different types of units under consideration must be estimated. This requires some skill since material and field labor are present in different proportions in the various plant types, and the rate at which these two components escalate is generally considered to be different. Furthermore, the technology is not equally mature among competing types of plants, so that improvements in material content must be estimated separately for each type. The allowance for funds used during construction, the capital expenditures associated with meeting environmental considerations, and other charges directly attributable to the project are also different for each type of plant and should be included. The transmission needed for the station should be added to the cost of the plant.

Another component is the operating and maintenance costs associated with the unit. These costs should include the operational and maintenance labor required by the unit as well as the material needed to keep the unit in an operating mode. For base-load types of units, this is usually represented as a fixed cost for the year, independent of the energy produced by the unit during the year. For some peaking-type units, however, the operating and maintenance costs are a function of the number of hours that the unit is run and, for this reason, should be handled as a variable cost.

Fuel costs are a function of the type of unit and the type of fuel being used, as well as its source. Nuclear fuel costs are a special category since they consist of two components, the first being the variable cost of burn-up, expressed as ¢/MBtu, and the second being the fixed carrying charge on the inventory, expressed as $/yr/kW. In addition, both of these components change over the life of a nuclear unit. This is because of the physical changes that take place in progressing from the initial core, with all new fuel, to the equilibrium core in which there is a mixture of new and old fuel.

TABLE 15.1. Full-Load Heat Rates in Btu/kWh

Unit Type	Gas	Oil	Coal	Nuclear
Steam	9,800	9,400	9,800[a]	10,400
Combined cycle	8,400	8,400	—	—
Simple cycle Combustion turbine	11,500	11,500	—	—

[a] Heat rates for coal-fired steam plant do include an allowance for scrubbers.

Fossil fuel types and sources, as discussed in Chapter 3, have only a variable fuel component. The variable component of operating and maintenance (if there is one) can be handled by adding it to the fuel costs.

Table 15.1 gives approximate full-load heat rates for the various types of thermal units. Pumped-storage hydro units normally have overall conversion efficiency of about 70%. The heat rates of Table 15.1, multiplied by fuel prices in ¢/MBtu, give the basic fuel economy of the units. To the delivered price of each fuel type, it is also necessary to add those components needed for maintenance, any cost of fuel treatment, and variable costs associated with stack gas clean-up (scrubbers) or waste product handling.

Plant costs are quoted and discussed widely in the utility industry, usually in terms of dollars per unit rating, $/kW. Unfortunately, there is no consistency as to what plant cost components are included in the numerator, nor is there complete agreement as to what rating should be used for the kilowatts in the denominator. Together with these economic variables, assumptions about permissible unit sizes and earliest service year, planned and forced-outage rates, and unit retirements are required.

System simulation computer programs have been designed to aid in doing a complete job of answering the difficult questions of what type and when? A normal computer package consists of three different programs. The first of these is reliability, which determines when the capacity additions necessary to meet the risk criteria are desired, thus producing an

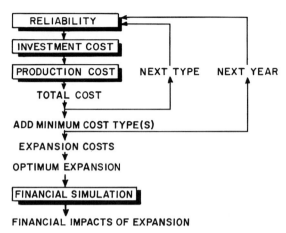

Figure 15.5. Optimum generation planning.

expansion pattern of acceptable reliability (Figure 15.4). The expansion is then inserted into the production simulation program which synthesizes the operation of the total system for each year of the expansion, calculating fuel costs and operation and maintenance expenses associated with the production of power during the expansion, and present worths these costs. The third required program determines the present-worth dollars associated with the investment costs of the expansion. A summation of these costs gives the total system 20-year cost for the one plan which was initially inserted into the reliability program. The system planner then repeats the procedure, inserting a different series of unit sizes and types and obtaining the cost for that plan. This procedure is repeated several times, and the costs of the alternate expansion plans are compared.

For system planning purposes, all expansion plans should have the same reliability so that differences due to unit size, forced-outage rate, etc., are taken into account. Given these equally reliable plans, a search is made for the one expansion plan that produces minimum total system cost.

Since the total system costs are a function of both the production and investment costs, and since the utility's load does not require that all units run the same amount of time, there is a balance and timing between types of units that produce a minimum cost expansion, the goal of system planning. The disadvantage in doing this with the computer programs shown previously is that many runs must be made before the optimum mix for just one set of input assumptions can be found.

Figure 15.5 illustrates the principles of an optimized generation planning program. The program goes to the first year of the study, checks the reliability for that year, and determines if the system capacity is sufficient to meet the load within the risk specified. If the study shows that no additional units are required that year, an investment cost and a production cost are made for that year with no new units installed. If the reliability is not met for the year, individual units are temporarily added until the desired risk level has been attained. The investment and production costs are then determined. Next, these units are removed from the system and enough units of another type are installed to meet the criteria for that year. Investment and production costs are again calculated with these units on the system, producing another cost. Similar calculations are done for all available thermal and energy storage technologies as well as various combinations. From all the costs produced for the year, the minimum cost is selected and the units associated with that cost are inserted permanently on the system. The next year's calculations are performed in a similar manner. Once again the minimum cost types are inserted on the system, with this process repeated until the 20 years of the expansion are completed.

Since the optimum type of generation based on 1 year's costs has been selected, it is apparent that cost inflation and the existence of high forced-outage rates in the first few years of a plant's operation can result in a decision which is not necessarily optimum over the life of the unit. It is for this reason that "look ahead" options base each year's decision on levelized future fuel costs and mature forced-outage rates.

The procedure using these options is the same as the procedure described earlier except that uniform future fuel costs are used for all units (existing and proposed new units), and mature forced- and planned-outage rates are used for all proposed new units in all calculations of annual costs (one for each type and combination of types). When the lowest cost type(s) is selected, calculations for the year are repeated, using immature outage rates, and actual fuel, operation, and maintenance costs. Future conditions are, therefore, recognized for decision making while still using current costs and outage rates for accumulating the annual costs of the expansion. By this year-by-year selection of a minimum cost type, an optimum generation plan is determined.

SYSTEM PLANNING

It is possible to determine other measures for ranking alternative expansion plans beyond the engineering economics revenue requirement results. This can be accomplished by simulating financial statements year-by-year through the expansion period. An integral financial simulation module representing a simplification of the corporate model enables planners to obtain output which describes the implications of plant additions in terms of new financing requirements, required rate increases, interest coverage, and earnings per share.

Transmission Planning

Transmission expansion planning is the study of long-range load and generation forecasts to determine a preferred network growth pattern.

On completion of the generation expansion study, an annual study of transmission expansion will include the necessary circuits; the transformers, both transmission-tie and generator step up; and the circuit breakers. These results may be for one or several different voltage levels.

The traditional method for transmission expansion planning is heavily dependent upon the system design engineer. A typical study is begun by sketching out some preliminary designs for load levels several times above today's loads. This look ahead may be termed a horizon year study and it serves to identify the long-range needs of the network. To check the designs, ac load flow calculations are made. The planning engineer modifies the network after reviewing the results of the computer calculations. As modifications are made, the planner has in mind many different requirements for the network that he has learned through his years of experience. For example, the network will be required to meet conditions of peak load, off-peak load, generation outages, and transmission circuit outages. Because of the experience of the planner, the additions selected for the network will be based on the results of many checks, not just the particular load flow being considered at the time. Also, several different designs are produced to meet one horizon-year load forecast by recognizing that different combinations of circuits are technically acceptable and deserve further study.

In the traditional approach to expansion planning, the digital computer is useful to the engineer only for checking his designs. The results of these checking calculations indicate whether the network

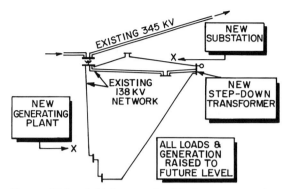

Figure 15.6. Initial network with future generation and load sites.

modifications should be made when the network is unacceptable.

A new approach to transmission expansion planning has been developed to implement the same strategy that the system engineer has traditionally used. This approach also begins with quick preliminary designs and there may or may not be circuits connecting all the buses. Steps in this approach include studying horizon-year conditions, designing a network, and then building toward that network; many normal and emergency checks are considered before making a circuit selection; and, finally, several designs are produced for one set of input.

The input data required for a typical transmission network expansion program is illustrated in Figure 15.6. Horizon-year designs are based on the following assumptions:

1. The existing network will be used as a starting point.
2. The location and magnitude of all future loads and generation are known.
3. The permissible voltage levels are specified.

One possible transmission plan for the horizon year, determined by the transmission network expansion program, is illustrated in Figure 15.7.

The transmission expansion method begins by stepping from the initial system to the horizon-year conditions, say 20 years ahead, developing one or several network designs for these conditions, and then backing up and moving through time, year-by-year, to study the network growth toward the horizon year system. The early years of the study are made in more detail, with the number of network

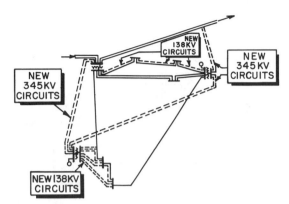

Figure 15.7. Horizon-year network planned by analytical transmission estimation network expansion program.

checks decreasing in later years. In some later years, however, there may be major changes in the system requiring detailed study; for example, the years just before and after a new generation site is opened up or a new voltage level is introduced.

As system planners have learned from the traditional approach to expansion planning, a long-range study is useful in examining the major influences in plant sizes and locations, the introduction of a new voltage, and/or the cost of increasing the contingency protection. It also aids the system planner in fitting today's construction decisions into tomorrow's network.

Investment costs are computed to guide the system planner in selecting the network expansion pattern with the lowest cost. Investment costing uses an equipment list and appropriate cost data such as dollars per circuit mile, dollars per breaker position, and dollars per transformer based on MVA rating. The results of the investment-cost calculation include the year-by-year listing of the annual costs and the present worth total of the yearly costs. This summary is quite similar to those supplied for generation expansion studies and can be combined directly with such expansion results.

Load-Flow and Stability Studies

Once a basic generation-transmission expansion plan is established, a more complete transmission system representation is used to define in more detail the nature of the circuits to be added. A large-scale load flow program is used which may accommodate up to 2000 buses, 4000 line sections, and 400 off-nominal, turn-ratio transformers. Cases are run as specified by the system planner to determine the adequacy of the proposed detailed transmission system design from the standpoint of voltages and overloads. As the system data are stored on magnetic tape, the only data that need be specified for cases following the first case are those necessary to describe the desired change. Rigorous representation of the stability problem, pursued before equipment characteristics are specified, can show a wide array of alternatives to achieve stability, and to allow the most economical choice while there is still time to adopt it.

System Dynamic Studies

In addition to stability, other performance considerations are studied using a variety of special digital and analog computer simulation tools which model specific components of a power system in considerable detail. Normally, these studies represent only modest size networks, i.e., less than 100 buses; however, very detailed, design-level models may be used for generators, turbines, excitation systems, power system stabilizers, transformers, series capacitor banks, etc.

Questions of the effect of generator and excitation system behavior following sudden load rejections are typically answered by these special tools. Other applications include the analysis of the interactions between transmission systems and turbine generator torsional dynamics caused by subsynchronous resonance, high-speed reclosing and HVDC controls. Other computer programs are used to study the performance improvements possible with static var controls, synchronous condensers, power system stabilizers and other devices. On a longer time scale, the dynamic behavior of power systems under emergency conditions can be studied using long-term dynamic simulation techniques.

Switching Overvoltage Studies

There are many switching overvoltage problems that must be explored in the design stages of an EHV system to assure intelligent specification and reliable operation of transmission line and equipment insulation requirements. A transient network analyzer is ideally suited to such studies. The transmission system, represented in miniature, can be subjected to all the switching operations anticipated on the future system. The surge levels can then be measured directly. For some design problems, e.g.,

SYSTEM PLANNING

transmission line insulation, it is usually appropriate to consider the surge levels as a statistical population, while on questions of equipment design, the maximum level is often more appropriate. The nature of the surge is often as important as its magnitude. Energizing a line and an integrally connected transformer, for example, may produce an overvoltage which will endure for many cycles and, therefore, bears directly on the surge-arrester rating at the terminal stations and indirectly on the basic insulation level of station equipment. Such phenomenon have in many cases imposed a more severe limit to surge-arrester rating than the traditional line-to-ground faults criterion.

Lightning Performance Studies

As voltage levels increase, the relative importance of lightning, as it affects transmission line outages, decreases. However, lightning remains a significant consideration for all voltage levels in use in the United States, especially in areas where lightning storms are frequent. There are two ways lightning can cause outages on transmission lines:

1. If it strikes a tower or a shield wire, a "backflash-over" can occur.
2. If the lightning strikes a phase conductor directly (shielding failure), the resulting traveling voltage wave on the conductor will usually flash-over the first insulator it passes.

In the past, backflash-over studies were performed using actual scale models of transmission towers. The model towers were subjected to simulated lightning strokes having different characteristics and points of incidence. The resulting voltages per ampere of stroke current were measured across the model's insulator string for various footing resistances. By comparing the voltages, resulting from many lightning strokes, to the critical flash-over voltage of the insulator string, an outage rate was calculated. Since then, digital computer programs have been developed which model the physical phenomena and simulate lightning strokes to a transmission tower. Outages, due to backflash-overs, can be reduced by improved grounding which reduces the tower footing resistance.

Shielding failures can be analyzed by means of a digital computer program or a hand calculation. The analysis relates the proper positioning of shield wires to the corresponding outage rate for that positioning.

METIFOR Optimization Studies

Lightning outage rate is only part of the line insulation problem. Insulation for switching surges and power frequency voltages also plays a part in determining a balanced design. Insulation of transmission lines is basically a problem in statistical methods and, therefore, very amenable to digital computer solution. The problem can most simply be visualized as an interplay between a probability distribution of electrical stresses of varying magnitudes imposed on an insulation system whose strength also varies statistically due to changes in wind, precipitation, air density, and other meteorological variables. The degree of overlap between these stresses and strengths defines the probability of flash-over for switching surges and the basis of comparing alternative designs. The greatest difficulty in this approach lies not so much in defining the electrical strength of an insulation system for prescribed conditions, but rather in converting appraisals of this type to a probability distribution useful for statistical solutions.

A program for performing this calculation was developed at General Electric's Project EHV in 1962. Given the name METIFOR (Meteorologically Integrated Forecasting), it uses sequential hourly weather observations from local weather bureau records to calculate the minimum switching surge strength at any point on a given transmission line for each hour during a 15–20 year period. This type of solution, repeated for a variety of design alternatives of increasing costs, provides the basis for the determination of cost versus performance curves. The METIFOR program, aside from leading to more economical, higher-reliability EHV lines, has provided new insights into insulation and clearance requirements and has proved extremely useful in the uprating of intermediate voltage lines to the next higher voltage level.

Voltage Levels

Introducing a new voltage level is a major step in the expansion of a utility system. The amount of power to be delivered is the primary factor in deciding when to change to a higher voltage level. The voltage level should be kept in step with load and generation growth. When the peak demand has grown to four times the value it was when the present highest voltage was introduced, a new higher voltage will likely be needed for continued economic system expansion.

Tradition has been to preserve a voltage level until it becomes appropriate to superimpose a new level of twice the old voltage rating. Doubling the voltage should give about four times the former capacity. All system expansion has not always followed this guideline. Superimposed voltage have ranged from about 1.4 to 2.5 times the previous level.

Interconnections between adjoining utilities have proven to be advantageous to both utilities. Interconnections can give reserve capacity or energy transfer or both. When a new voltage level is contemplated, consideration should be given to the existing neighbors and to planned voltage levels. Interconnection through a transformer would be more expensive and the transformer impedance would be added to the interconnection.

The ideal might be to install the entire higher-voltage network at the same time. While this approach would not be economical or practical, it is highly desirable to visualize the completely developed network before the first line is added. To achieve this, all transmission planning should be accomplished by keeping in view a horizon year that is far enough in the future so that the new voltage level is fully developed before the next higher voltage becomes necessary.

The use of high-voltage circuits is justified by the lower cost per unit of energy transferred. In the early years of its life, the load growth may require only a relatively light loading. This results in a higher cost per unit of energy transferred. Later in its life, when the line loading has increased, the unit cost is more favorable. Cost may not be a consideration when only very limited right-of-way is available.

Network strength, after an outage of the new higher-voltage circuit must be sufficient to prevent any cascading of the outage. The higher-voltage circuit with its high charging current, will probably require reactance compensation to keep the voltage at the terminal substations from rising beyond tolerable limits during light load conditions, such as early morning.

Reactive Power Supply

Fundamental to any discussion of the reactive power and reactive power supply is the full appreciation of the fact that they are separate and distinct from active power. The two commodities have radically different monetary values. They are generated differently. They have different effects on the system through which they are transmitted and distributed.

An important aspect of reactive power is that it cannot be transmitted the distances that active power can without incurring an objectionable voltage drop. The result is that reactive power and the related voltage problems must be solved on a local basis.

To correct load-power factor, shunt capacitors are located in the load area. If the load exhibits frequent active and reactive power variations, as is the case with arc furnaces and rolling mills, then compensation in the form of static var control or a synchronous condenser may be desirable. On subtransmission circuits, shunt capacitor banks are often employed to provide voltage support and allow higher line loading. On long transmission lines, shunt reactors are widely used to reduce high voltages under light load or open line conditions. The reactors may be on the line or connected through adjacent transformers. This line compensation is completed in spite of the fact that capacitors in the same system are applied in the load areas. Unfortunately, it is often impractical to transmit the reactive power from line to the load. If the transmission system is heavily loaded, shunt capacitors, static var control, and synchronous condensers are used to increase power transfer and power system stability. The type of compensation and its rating are often investigated using load flow and stability computer programs.

Corona Performance and Conductor Economics

Corona phenomena, such as audible noise, radio noise, television interference, and corona power loss, are important design factors at EHV transmission levels. The primary design method used to reduce these is bundled conductors. To make accurate decisions about the number and size of subconductors to put in the bundle, the line designer must make a careful study of corona phenomena. Computer programs are the major tool used to calculate these parameters. Programs have been developed that calculate and predict the corona performance of transmission lines before they are built. Because of governmental regulations on the amount of interference and noise a transmission line can produce, utilities will often measure these to check the calculated values after the line is built.

Short-Circuit and Relay Studies

Short-circuit studies are used to select relays that will be adequate for fault current magnitudes available at proposed relay locations. The information gained from short-circuit studies is then used to determine appropriate relay settings.

Many short-circuit studies must be run for all the types of faults that the relay must recognize; and also for faults where proper coordination of relays is important. Any possible change in system configuration, such as lines out of service, must be considered for each of these faults. Minimum and maximum system generation levels must be considered to determine the limiting fault currents at each relay location. Each short-circuit study must be done for single line-to-ground faults and for three-phase faults. The single line-to-ground fault information is used to select and set ground relays, while the three-phase fault information is used for phase relays. Considering all of the different combinations of system conditions and fault types which must be studied, these fault studies will result in a great deal of information being generated at each relay location.

The data generated by these short-circuit studies must be sorted to find minimum and maximum short-circuit currents at each relay location. This will enable ground and phase relay selection and settings to be made. While, this task is formidable if done by hand, computer programs exist which automatically do this sorting function. Some of these programs even incorporate a relay-setting function which can determine proper setting based on fault data and relay characteristics stored within the program.

For some types of relays, short-circuit studies do not provide enough information to set the relay. Distance relays, for instance, use system impedance to determine location of faults. This relay may react to apparent impedances which result when generators swing against each other. In this case, no fault exists, but the apparent impedance seen by the relay may indicate a fault. Stability computer programs can be run to determine the proper relay time delay necessary to prevent a misoperation for this system swing case. Other relays, called out-of-step blocking relays, may be set from this information to prevent misoperation.

Any of the above system studies should be delayed until the final system design and configuration is assured. Since all of these studies depend on system parameters, premature evaluation of the system may result in incorrect relay settings.

SUMMARY

In anticipation of continued growth in the loads served by the electric utilities, systems must be continually expanded in capability. Long-range planning is essential to assure that necessary additions are technically adequate, reasonable in cost, and fit into a growth pattern. The difficulties encountered by the long-range planner include:

1. Uncertainty of load growth, with respect to both geography and time. There is no certainty as to where in the system, or when, new loads will appear, or what their nature will be.
2. The probability that some new invention, discovery, or technological development will abruptly impose new conditions and limitations, or possibly bring freedom from existing limitations. One past example was the development of circuit breakers of higher interrupting ratings that postponed that necessity of designing systems around short-circuit magnitudes.

When it becomes necessary to superimpose a higher voltage on the present system, the voltage of the new system should be at least twice the present voltage.

System design should provide for the economical supply of the reactive current required by the customers and by the system itself. Preferably, this component of the current should be supplied near the point where it is needed, thus avoiding the extra I^2R losses and voltage drop caused by transmitting reactive current through the system.

Good system planning strives for optimum design on a system-wide basis, not necessarily for minimum cost in one part of the system without regard to the effect on the other parts.

In recent years, there has been an emphasis on economy in planning and operation. Now there is increased emphasis on reliability and environmental factors (physical appearance, air and water pollution.)

Modern analytical methods have permitted vastly increased sophistication in the planning and design of power systems, providing a means of quickly capitalizing on new test information and new conceptual approaches to engineering problems. It has proved beneficial to attack the problem of system expansion on a project basis where a full range of economic trade-offs between technical areas can be exploited.

Final stability and load-flow studies must often be pursued in the awareness of the transient or harmonic overvoltages which result from the system configuration and equipment ratings specified. The numerous cost influences of any system argue for the broad project approach to electric utility system expansion studies.

REFERENCES

1. Desell, A. L., Garver, L. L., and Adamson, A. M., "Generation Reserve Value of Interconnections," *IEEE Transactions,* **PAS-96,** (2), 336–346 (1977).
2. EBASCO Services, Inc., "Long Range Transmission Expansion Models," EPRI EL-1569, Electric Power Research Institute, Palo Alto, CA, 1980.
3. Edgell, R. M., and Bayless, R. S., "AC and DC Transmission Comparison for Kaiparowits Coal-Fired Generation Plant," *IEEE Transactions,* **PAS-95,** 1123–1135 (1976).
4. Endrenyi, J., *Reliability Modeling Electric Power Systems,* Wiley-Interscience, New York, 1978.
5. Felak, R. P., Marsh, W. D., Moisan, R. W., and Sigley, R. M., "Adding Financial Simulation to Long Range Generation Planning," *Proceedings of the American Power Conference,* **39,** 999–1009.
6. Garver, L. L., "Effective Load Carrying Capability Of Generating Units," *IEEE Transactions,* **PAS-85,** 910–919 (1966).
7. Garver, L. L., "Factors Affecting the Planning of Interconnected Electric Utility Bulk Power Systems," Vol. II of Factors Influencing Electric Utility Expansion, U.S. Department of Energy, Division of Electric Energy Systems, CONF-770869 Vol. II, National Technical Information Service, Springfield, Virginia, 1977.
8. Garver, L. L., "Transmission Network Estimation Using Linear Programming," *IEEE Transactions,* **PAS-89,** (7), 1688–1697 (1970).
9. Gens, R. S., Gehrig, E. H., and Eastvedt, R. B., "BPA 1100 kV Transmission System Development—Planning Program and Objectives," *IEEE Transactions,* **PAS-98,** (6), 1916–1923 (1979).
10. Landgren, G. L., and Anderson, S. W., "Simultaneous Power Interchange Capability Analysis," *IEEE Transactions,* **PAS-92,** 1973–1986 (1973).
11. LaForest, J. J., editor, *Transmission Line Reference Book—345 kv and Above,* Second Edition, Electric Power Research Institute, Palo Alto, California, 1982.
12. Marsh, W. D., *Economics of Electric Utility Power Generation,* Oxford, New York, 1980.
13. Moisan, R. W., and Kenney, J. F., "Expected Need for Emergency Operating Procedures: A Supplemental Measure of System Reliability," *IEEE Transactions,* **PAS-95,** (6), 1760 (1976).
14. NERC, "8th Annual Review of Overall Reliability and Adequacy of the North American Bulk Power Systems," National Electric Reliability Council, Princeton, New Jersey, 1978.
15. Sullivan, R. L., *Power System Planning,* McGraw-Hill, New York, 1977.
16. Whitehead, E. R., "Final Report of EEI Mechanism of Lightning Flashover Research Project," EEI Project RP50.

INDEX

Air quality, 40-41
Alternating current transmission, 2, 13, 17, 122-136
American Engineering Standards Committee, 11
American National Standards Institute, 12, 159, 199
American Standards Association, 11-12
Amortisseur windings, 119
Analog control systems, 264
Annunciating devices, 95
Association of Edison Illuminating Companies, 11
Asynchronous transmission system ties, 141
Automated distribution systems, 274-275
Automatic circuit reclosers, 213
Automatic generation control, 263, 273
Autotransformers, 147, 181
Auxiliaries, power plant, 53-60

Blackout, Northeast (1965), 277
Boiler controls, 92
Boiler feed pumps, 56
Boiling water reactor, 49-51, 98, 101, 105-106, 110
Bonneville Power Administration, 9
Brayton cycle, 37-39
Breeder reactor, 27
Bulk data transmission, 273-274
Bulk power system, 21-22, 274
Bus arrangements, 177-180
  breaker-and-a-half, 179
  double bus-double breaker, 179
  EHV bus design, 179-180
  main and transfer bus, 177-178
  ring bus, 178
  single bus, 177
Bus protection, 233-234

CRT's, 275
Capacitors, 136, 142, 182-183, 184-185, 209, 239, 261
Capital turnover, 6
Carnot cycle, 32-33
Circuit breakers, 18, 158-176, 180
  air blast, 165, 166
  capacity, 278
  EHV, 171-172
  magne-blast, 164-165
  oil, 161, 166-168
  SF_6, 169-170
  standards, 172

  types of, 172-174
Clean Air Act, 40
Clean Water Act, 42
Coal, 28
Coal burning stations, 45-47
Combustion control, 92-93
Combustion turbines, 37-39, 60-63
  combined-cycle, 39, 62-63
  simple-cycle, 61-62
Commitment to serve, 7-8
Communications, power system, 270-275
Competition, 8
Condenser, turbine, 52-53
Conductor economics, 286-287
Conductors, transmission line, 124, 131-135, 143
Conservator, 150
Controls and instrumentation:
  fossil-fuel power plant, 90-97
    boiler controls, 92
    combustion control, 92-93
    feedwater control, 93
    indicating and annunciating devices, 95
    man-machine interface, 95-96
    monitoring, alarming and tripping, 94-95
    recording devices, 95
    start-up and shutdown control, 94
    steam temperature control, 93
    turbine controls, 93-94
  nuclear plants:
    boiling water reactor plant control, 96
    normal plant shutdown, 96-97
    plant start-up, 96
    power operation, 96
    reactor protection system, 97
Converter stations, 140-141
Converter terminal, direct-current, 136, 143, 273
Converter transformers, 140-141
Converters, solid-state, 140-141
Cooling towers, 41-42, 53
Cooperatives, 9
Corona, 134, 142, 155-156
Corona performance, 286-287
Corporate model, 279
Current-limiting devices, 170-171
Cut-outs, 213

Damping, 255
Demand, 6-7, 23
Deuterium, fusion of, 27
Dielectric tests, 155
Differential replaying, 219-220, 232-233
Direct-current terminal, 136
Direct-current transmission, high voltage:
  conversion equipment, 140-141
  description, 136-144
  economics, 143-144
  installations, 138-139
  system control, 136-137, 140
  unique characteristics, 141-143
Directional relay, 219
Disconnect switch, high voltage, 163
Dispatch center, 270-272
Distance relaying, 220-221, 236-237
Distribution:
  continuity, 204, 210-214
  load type, 200-214
    commercial, 200-204
    industrial, 200
    residential, suburban, and local, 204-214
  metering, 214
  network protectors, 203-204
  network transformers, 202-203
  primary, 20
  primary circuits, 205
  secondary, 20
  secondary network, 20, 200-202
    grid network, 201-202
    spot network, 202
  substations, 20
  subtransmission system, 199
  system scope, 22, 199
  transformers, 213
  underground, 206-208
  voltage control, 204, 208-210
Doppler effect, 105, 107, 108
Dry-type transformers, 153
Dynamic stability, 249-250

EHV substations, 179-180, 181-182
Economic dispatch hardware, 266
Edison Electric Institute, 11, 123
Edison, Thomas A., 1-2, 10, 13, 276
Electric utility industry:
  business characteristics, 6-7
  genesis of, 1-2
  growth of, 2-3
  growth rate, 3
  power companies, 3-5, 270-272
  standards, 10-12
  technical characteristics, 5-6
Electricity:
  annual consumption, 5
  capacity, 5, 24, 28
Energy, definition of, 12, 24
Energy sources:
  coal, 28
  geothermal, 27, 30
  hydroelectric, 29
  magnetohydrodynamics, 27
  natural gas, 29
  nuclear, 27, 29
  petroleum, 28-29
  prime movers in U.S., 45
  solar, 27, 30
  utilization of, 27-28
  wind, 30-31
Energy systems, *see* Power systems
Entropy, 32-35
Excitation, 247-248
  loss of, 230-231
Excitation system, *see* Generator excitation system

Fans, forced- and induced-draft, 57-58
Fault, external, 232
Fault current, 159-162
Fault pressure relay, 232
Faults, 241-262. *See also* Short circuits; Stability
Feedwater control, 93
Feedwater heating, regenerative, 35-36
Feedwater system, turbine, 56-57
Field ground, 230
Firm power, 16
Fission, 99-103
Fossil units, 52
Fossil-fired plants, *see* Combustion turbines; Steam turbines
Franchise, 5-6
Francis-type turbine, 113, 115
Frequency:
  abnormal, 231
  standardization of, 2-3
Fuel handling, 45-47
Full voltage starting, 119
Fuses, 160, 170
Fusion, 27-28

Gas burning plants, 47
Gas detection, 232
Generating stations, thermal electric, 44
Generation, power:
  combined cycle power plants, 39
  combustion turbine power plants, 37-39
  costs of, 39-40
  definition of, 44
  economic allocation of, 267
  examples of stations, 13-16
  history of, 3
  planning, 278-283
  plant siting, 40-42
  reliability, 40
  sources of energy, 27-31
  steam-electric generating plants, 31-37
  unit size, 45
Generator, hydroelectric, 118-120
Generator configuration:
  electrical characteristics, 77-80
    reactive capability curves, 79-80
    saturation curve, 77-78
    synchronous impedance, 78
    "V" curves, 78-79
  four-pole, 76-77
  two-pole, 75-76

INDEX

Generator design:
  four-pole, 66
  rotor, 63-66
    forging, 63-64
    windings, 64-66
  stator, 66-75
    core, 66-68
    windings, 69-75
Generator excitation system:
  components, 81
    cooling systems, 81
    power source, 81
    rectification, 81
  control functions, 81-83
  performance requirements, 83-85
    prolonged disturbances, 85
    steady-state conditions, 83-84
    transient disturbances, 84-85
  power requirements, 80-81
  protective features, 83
  types of, 85-90
    alternator controlled rectifier, 86-87
    alternator rectifier, 85-86
    compound controlled rectifier, 89
    compound rectifier, 88-89
    dc commutator, 85
    potential source controlled rectifier, 87-88
    rotating rectifier, 90
Generator performance, 80
Generator protection, 229-232
Geothermal energy, 27, 30
Grounding, 188-189

HVDC transmission, *see* Direct-current transmission, high voltage
Harmonics, 143
Head, 116
Heat recovery steam generators, 63
Heat rejection, 53
Heavy water reactor, 29
Hewlett insulators, 124
High Voltage Transmission Research Facility, 17-18, 123, 130
High-head hydro plant, 114, 116, 118
High-temperature gas-cooled reactor, 101, 102
Hoover Dam, 8-9
Hydraulic turbines, types of, 114-116
Hydroelectric power:
  amortisseur windings, 119
  automatic hydro stations, 120-121
  auxiliary power, 120
  control of stream flow, 113-114
    pondage, 113
    pumped storage, 24, 113
    seasonal storage, 113
  cost per kilowatt, 112-113
  energy source, 29
  generators, 118-120
  hydraulic turbines, types of, 114-116
  hydroelectric development, 13-16, 114
  load factor, 111-112
  plant size, 3, 111-112
  power and energy equations, 114
  runaway, 118
  specific speed, 116
  speed regulation, 117-118

Impulse voltage, 154
Impulse wheel, 114-115, 118
Indicating devices, 95
Inertia, 252-253, 257-258
Insulation coordination, 186-187
Insulators, 124-125
  cap and pin, 124
  fog, 124
  Hewlett, 124
  high-leakage distance, 124, 126
Insull, Samuel, 10
International Electrotechnical Commission, 12
Interrupting devices, 159, 163, 165, 166, 168
Inverter, 136-137, 142
Investment costing, 279
Investor-owned utilities, 8, 22
Isotopic transmutation, 98-99

Kaplan wheel, 115-116, 118

La Que Panel, 11
Light water reactor, 3, 29, 47-49, 98, 101-102, 105-110
Lightning:
  characteristics of, 128
  performance studies, 285
  protection, 187-189
  stroke to shield wire, 128-129
  stroke to tower, 128-129
  traveling wave, 129
Lightning-proofing, 129-130
Line protection, 234-240
Load, 23-25
Load break switch, 163
Load curves, 24-25
Load shape model, 278
Load-flow and stability studies, 284
Load-frequency control, 265-266
Loss-of-load probability, 280-281
Low-head hydro plant, 116-117

METIFOR program, 285
Magnetohydrodynamics, 27
Medium-head hydro plant, 117
Metering, 214
Mollier diagrams, 37
Monopoly, regulated, 7
Motoring protection, 231
Municipal plants, 8

National Ambient Air Quality Standards, 40-41
National Electrical Manufacturing Association, 11
Natural gas, 29
Network protector, 203-204
Neutron interactions, 100-101
Nuclear plant auxiliaries, 60
Nuclear power:
  energy source, 27, 29
  fission reactor environment, 101-103
    fuel, 101-102
    neutron life cycle, 102-103
    reactor materials, 102

Nuclear power: *(Continued)*
  fuel costs, 281
  growth of industry, 3
  nuclear core design, 106-107
  nuclear fuel cycle, 110
  physics of processes, 98-103
    fission, 99-103
    isotopic transmutation, 98-99
    neutron interactions, 100-101
    nuclear reactions, 98-101
    radioactive decay, 98-99
  plant auxiliaries, 60
  plant control and instrumentation, 96-97
  radioactivity, 42
  reactor control, 107-110
  reactor core behavior, 103-110
  reactor core design, 106-107
  reactor kinetics, 108-110
  reactor safety, 42
  reactor types, 3, 29, 47, 49-51, 98, 101-102, 105-110
  steam supply, 47-51
  turbines, functioning of, 51-52

Oil, 28-29
Oil Embargo of '73-74, 7
Oscillations, damping of, 141-142
Overcurrent relay, 218-219, 234-236
Overexcitation, 232
Overvoltage protection, 185-187

Pearl Street Station, 1, 6
Pelton wheel, 114-115
Petroleum, 28-29
Phase currents, 231
Phase-angle transformers, 147
Pilot relaying, 221-226, 237-239
Pollutants, 41
Pondage, 113
Power:
  from combustion turbines, 60-63
  definition of, 12, 24
  from steam, 45-60
Power companies, 3-5, 270-272
Power generation, *see* Generation, power
Power plant controls and instrumentation, 90-97
Power plant siting, 40
Power pools, 272
Power system design:
  changing conditions, 276-278
  early systems, 276
  system planning, 278-287
    generation planning, 279-283
    transmission planning, 283-287
Power system operating states:
  alert, 269-270
  emergency, 270
  in extremis, 270
  normal, 269
  restorative, 270
Power system operation tasks:
  automation, 264
  communications, 270-275
  control functions, 264-268

  data display, 275
  functional service requirements, 273-274
  hierarchy of action centers, 270-272
    power companies, 270-272
    power pools, 272
    regional coordinating centers, 272-273
  operating states, 269-270
  system security, 268-270
Power systems:
  bulk, 21-22, 274
  definition of, 1
  design, 25
  distribution, 20, 22
  elements of, 13-14
  evolution of, 23
  generation, 13-16
  inspection and maintenance, 25
  interconnected, 277-278, 286
  investment, 22-23
  load, 23-25
  planning, 25
  switching stations, 17-19
  substations, 19
  subtransmission, 19-20
  transmission, 16-17
Power transmission, *see* Transmission, power
Pressurized water reactor, 49, 98, 101-102, 105-106, 110
Primary distribution circuits, 205
Production costing, 279
Protective relaying, *see* Relaying, protective
Pumped-storage hydro, 113, 119

Radioactivity, 42, 98-99
Rankine cycle, 33-35
Rates, electric service:
  block meter, 6-7
  flat demand, 6
  Hopkinson demand, 7
  regulation of, 7
  step meter, 6
  straight-line meter, 6
  time-of-day, 7
  Wright demand, 7
Reaction wheel, 114-115, 117
Reactive power, 142, 267-268, 286
Reactive power compensation, 181-182
Reactors, nuclear:
  breeder, 27
  control, 96-97, 107-110
  core, 103-110
  fusion, 27-28
  heavy water, 29
  high-temperature gas-ccoled, 101, 102
  kinetics, 108-110
  light water, 3, 29, 47, 49, 98, 101-102, 105-110
    boiling water, 49-51, 98, 101, 105-106, 110
    pressurized water, 49, 98, 101-102, 105-106, 110
  materials, 102
  safety, 42
  see also Nuclear power
Recapture clause, 8
Reclamation projects, 8

INDEX

Reclosing:
  high-speed, 258-260
  single-pole, 260
  transmission line, 240
Reclosure, immediate, 171
Recording devices, 95
Rectification, 81
Rectifier, 136-137, 142
Reduced voltage starting, 119
Regional coordinating centers, 272-273
Reheat, 34-35
Relaying, protective:
  application practices, 229-240
    bus protection, 233-234
    generator protection, 229-232
    line protection, 234-240
    power transformer protection, 232-233
  application principles, 226-229
    backup protection, 227-229
    primary protection, 227
  basic types, 218-226
    differential, 219-220, 232-233
    directional, 219
    distance, 220-221, 236-237
    overcurrent, 218-219, 234-236
    pilot, 221-225, 237-239
    static, 226
  definitions, 215-218
    electromechanical, 216
    static, 217-218
  single-phase, 239-240
  studies, 287
  substation, 185
Relaying protection:
  backup, 227-229
  generator, 229-232
  primary, 227
Reserve capacity, 1
Rotor, generator, 63-66
Rural Electrification Administration, 9
Rural lines, 21

Scroll case, 114
Secondary network, 20, 200-202
Separable connectors, 213
Series capacitors, 184-185, 239, 261
Series compensation, 239
Short circuits, 17-19, 158-161, 230, 278, 287
Shunt capacitors, 136, 142, 182-183
Shunt reactors, 181-182
Shut-down control, 94
Silicone-filled transformers, 152
Single-phase transformers, 147
Solar energy, 27, 30
Speed governor, 264-265
Stability:
  definition, 241-242
  dynamic, 249-250
  higher-transmission voltage, 260-261
  transfer of power, 242-243
  transient, 250-260
    damping, 255
    equal-area criterion, 253-255
    factors in, 256-258
    inertia, 252-253, 257-258
    methods of calculation, 255-256
    reclosing, high-speed, 258-260
    single-pole switching and reclosing, 260
    switching, 251-252
    transfer impedance, 250-251
    voltages, 252
  series capacitors, 261
  steady-state, 243-250
    application to complex systems, 248-249
    effect of saturation, 244-247
    excitation, 247-248
  voltages and reactances, 243
Start-up control, 94
Static var control, 183
Stator, generator, 66-75
Steady-state stability, 243-250
Steam field:
  dry, 30
  wet, 30
Steam production, 45-51
Steam station, 16
Steam supply:
  fossil-fired boilers, 47
  nuclear system, 47
Steam temperature control, 93
Steam turbine:
  auxiliaries, 53-60
  blades, 36, 51
  buckets, 36
  compound, 52
  condenser, 52-53
  controls, 93-94
  exhaust loss, 51
  functioning of, 51-52
  growth in size, 3, 45
  heat rejection, 53
  nozzles, 36, 51
  stage designs, 36-37, 51
  steam production, 47
Step-voltage regulators, 209-210
Substations:
  bus arrangements, 177-180
  circuit breakers, 180
  definition, 177
  design considerations, 187-189
  distribution, 20
  EHV, 179-182
  load center unit, 190
  mobile, 189-190
  overvoltage protection, 185-187
  primary, 19, 189
  protective relaying, 185
  reactive power compensation, 181-185
  SF_6, 191
  transformers, 180-181
  unattended control, 191-197
Subtransmission, 19-20, 199
Sulfur hexafluoride, 168-170
Superheat, 34-35

Supervisory control and data acquisition, 193-198, 271-273
Surge arresters, 185-186
Switches, 163, 172
Switchgear:
   arc interruption, 163-170
   breaker and switch requirements, 172-174
   components, 158
   control point, 175-176
   current-limiting devices, 170-171
   definition, 158
   EHV circuit breakers, 171-172
   fault current, 159-162
   leading current, interruption of, 162-163
   outdoor gear, 174
   POWER VAC®, 20
   short circuits, 158-161
Switching:
   fault prevention, 251-252
   single-pole, 260
   stations, 17-19
Switching overvoltage studies, 284-285
Switching surges, 129-130
Synchronism protection, 231
Synchronous condensers, 183-184
Synchronous machines, 241
System dynamic studies, 284

Tap changers, 146-147
Taxes, 8
Telemetering, 273
Telephony, 273
Teletype, 273-274
Tennessee Valley Authority, 9
Thermal dispersion, 41-42
Thermodynamics, steam-electric:
   Carnot cycle, 32-33
   entropy, 32-35
   feedwater heating, regenerative, 35-36
   generating plants, 31
   Rankine cycle, 33-35
   reheat, 34-35
   superheat, 34-35
Three-phase transformers, 147
Thyristor valves, 142
Tie line control, 265-266
Transformers:
   autotransformers, 147, 181
   cooling, 148-149
   cores, 147-148
   description, 146
   dielectric tests, 155-156
   distribution, 213
   impulse voltage, 154-155
   load-center unit, 190
   losses, 148
   network, 202-203
   neutral grounding, 186
   oil preservation, 150-152
   overheating, 233
   phase-angle, 147
   protection of, 232-233
   reduction of fire hazard, 152-153
   single-phase, 147
   substation, 180-181
   taps, 146-147
   temperature effect on organic insulation, 153-154
   three-phase, 147
   turn ratio, 146
   use, 146
   VaporTran®, 152-153, 192
Transmission, power:
   all-weather, 127-128
   alternating current, 2, 13, 17, 122-136
   bundled conductors, 134-135
   corona, 134
   definition of, 16-17, 122-123
   direct-current transmission, 3, 13, 17, 136-144
   economics, 135-136
   history of, 2
   lightning, 128-130
   line components, 124-127
      conductors, 124, 131-133
      insulators, 124-126, 131
      supports, 125-126
   line design, 130-133
   planning, 279, 283-287
   right-of-way, 6, 127, 142
   underground, 127
   voltages, 122-123
Transmission towers, 125-126, 131, 143
Turbine, *see specific types*
Turn ratio, 146

Underground distribution, 206-208
Underground transmission, 17, 127
United States National Committee, 12
United States of America Standards Institute, 12
Uranium, 29

Vacuum interrupter, 168
Valve hall, 141
VaporTran® transformers, 152-153, 192
Voltage, 122-123, 252
   control, 267-268
   extra high voltage, 16, 123, 130, 229
   higher transmission, 260-261
   levels, 285-286
   recovery, 161
   regulation, 146
   regulator, 247-249
   ultra high voltage, 16-17, 123, 130

Water discharges, 42
Water Quality Act, 41
Water wheel, 116-117
Wind (as energy source), 30-31
Wind turbines, 30-31
Windings, transformer, 146, 148